Ecosystems

Springer

New York
Berlin
Heidelberg
Barcelona
Budapest
Hong Kong
London
Milan
Paris
Santa Clara
Singapore
Tokyo

Kristiina A. Vogt
John C. Gordon
John P. Wargo
Daniel J. Vogt
Heidi Asbjornsen
Peter A. Palmiotto
Heidi J. Clark
Jennifer L. O'Hara
William S. Keeton
Toral Patel-Weynand
Evie Witten

Eco-systems

Balancing Science with Management

with contributions by
Bruce Larson,
Denise Tortoriello,
Javier Perez, Anne Marsh,
Miel Corbett, Kerry Kaneda,
Fred Meyerson,
Daniel Smith

Springer

Kristiina A. Vogt, John C. Gordon, John P. Wargo,
Daniel J. Vogt, Heidi Asbjornsen, Peter A. Palmiotto,
Heidi J. Clark, Jennifer L. O'Hara, Toral Patel-Weynand,
Bruce Larson, Denise Tortoriello, Javier Perez,
Anne Marsh, Miel Corbett, Kerry Kaneda,
Fred Meyerson, Daniel Smith
School of Forestry and Environmental Studies
Yale University, Greeley Memorial Laboratory
370 Prospect Street, New Haven, CT 06511, USA

William S. Keeton
University of Washington
College of Forest Resources
Seattle, WA, USA

Evie Witten
Executive Director
The Great Land Trust
Anchorage, AK, USA

Cover illustration: Views of Mount Hood National Forest in Oregon.
Images are contemporary (top) and simulated 20 years
in the future (bottom). The computer simulation depicts the landscape
as it may look if the Northwest Forest Plan is implemented.
Images courtesy of Robert Ribe, University of Oregon Institute for
a Sustainable Environment.

Library of Congress-in-Publication Data
Ecosystems: balancing science with management / by Kristiina A. Vogt
    . . . [et al.] ; with contributions by Bruce Larson . . . [et al.].
        p.   cm.
    Includes bibliographical references and index.
    ISBN 0-387-94752-3   (softcover : alk. paper)
    ISBN 0-387-94813-9   (hardcover : alk. paper)
    1. Ecosystem management.   I. Vogt, Kristiina A.
QH75.E3234   1996   96-19134
333.95--dc20

Printed on acid-free paper.

Acquiring Editor: Robert Garber.
Production managed by Frank Ganz; manufacturing supervised by Joe Quatela.
Typeset from the authors' original Word files.
Printed and bound by R.R. Donnelley and Sons, Harrisonburg, VA.
Printed in the United States of America.

9 8 7 6 5 4 3 2 1

ISBN 0-387-94752-3 Springer-Verlag New York Berlin Heidelberg   SPIN 10536045   (softcover)
ISBN 0-387-94813-9 Springer-Verlag New York Berlin Heidelberg   SPIN 10540549   (hardcover)

Acknowledgments

We would like to acknowledge several people who have contributed to the thoughts presented in this book, even though they were not active participants in the writing of this book. The following individuals have contributed enthusiastic discussions on several of the topics: Thomas Spies, Alan Covich, David Publicover, the Yale FES "Patterns and Processes in Terrestrial Ecosystems" class of 1995, Murray Rutherford, Thomas Siccama, Pat Peacock, Lisa Diekmann, Jeffrey Andrews, Janine Bloomfield, Joseph A. Miller, and Karen Beard. Special acknowledgment also has to go to Kipen Kolesinskas (Connecticut State Soil Scientist, Natural Resources Conservation Service, USDA), who contributed significantly to the development of the soil section. Philip Wargo (US Forest Service Northeast Center for Forest Health Research) also contributed exciting new perspectives on the role of disease in controlling ecosystem structure and function and we appreciated his input to the development of our ideas. The ideas developed in this book occurred while conducting research supported by the National Science Foundation on the long-term Ecological Research Program in the Luquillo Experimental Forest, Puerto Rico and as part of research support to Yale University and the University of Washington. Support from the USDA Forest Service, Northeast Global Change Program and USDA Forest Service Insect and Disease Lab were also instrumental in developing many of our ideas on ecosystems.

Contents

1 | Introduction

The management of natural resources has entered a period of unprecedented change and uncertainty. Our former paradigm, which stressed land allocation measures, maximum sustained yield principles, and multiple-use objectives, is being rapidly replaced by a new paradigm, which emphasizes *sustainable ecosystems* rather than *sustainable yield*. This new paradigm, termed *ecosystem management*, focuses on management of "whole systems" for a variety of purposes, rather than simply focusing on commodity production for a single resource.

Unfortunately old paradigms are quicker to die than new ones are born, and as a result a great deal of controversy exists regarding the definition and management implications of ecosystem management. For example, one of the most common misconceptions is that ecosystem management will be impossible to implement because it will require far too much data collection. The concern is that because this approach emphasizes management of an ecosystem, which by definition is "the whole system [in the sense of physics], including not only the organism-complex, but also the whole complex of physical factors forming what we call the environment of the biome" (Tansley 1935), in order to do ecosystem management managers will need to collect data on all of the biological, physical, and chemical attributes of an ecosystem, as well as data on energy flow and foodweb feeding relationships and interactions (Odum 1953). Although it is important to bear these ecosystem definitions in mind, it is not only impossible to collect data on all of these different components, but also not recommended. Instead, ecosystem management should focus on assessing those forces that drive or control an ecosystem of interest. Examples of driving forces, how they can be monitored, and the types of information that can be gained from their assessment are given in Chapters 2 and 3.

Another misconception is that ecosystem management must have a single specific definition in order to facilitate its implementation. Many individuals have attempted to define ecosystem management in concrete terms. These attempts have been met with skepticism because no single definition is applicable to the myriad of management scenarios and objectives. Within this book, a specific definition of ecosystem management will not be proposed; instead, a set of core principles integral to implementing this approach are discussed. The basic tenets of the ecosystem concept (see Chapter 2) should be retained in a way that allows for flexibility that can be applied on a case-by-case basis. These principles will not tell us what to sustain or produce in a particular ecosystem, but they will inform us about ecosystem potentials and consequences.

Ecosystem management is primarily about human values. An integral part of ecosystem management is the realization that the social system ultimately puts boundaries and constraints on our ability to manage the biological system (Figure 1-1; see Chapter 2). It has been common to pictorially show ecosystem management as the point where two circles overlap, with the commonly shared space being where ecosystem management occurs (see Figure 1-1). This type of diagram leads one to assume that most of the natural system and most of the social system not contained in the overlapping zones function independently from one another. This should be redrawn into a conceptual figure called the "egg" diagram (see Figure 1-1). There is no part of the natural system that is isolated from the

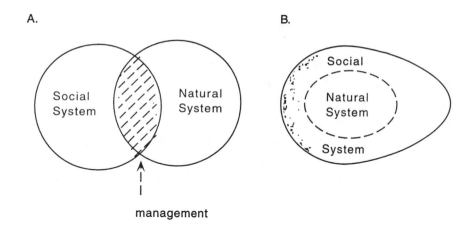

A.   B.

FIGURE 1-1 (A) The typical diagram showing where the natural and social syste ns interface (Bormann et al. 1993) and (B) our version of the "egg analogy."

social system because it is difficult to identify a natural system that has not been affected in the past or currently by humans. There are many legacies of human activities in systems that appear to be "virgin" or unaffected by humans. The yolk of the egg is the natural system, which is entirely embedded within the white of the egg, which is the social system. The overlapping circles diagram gives a false perception that there exists a part of the social system that is isolated from the natural system and functions independently from it, and also that there are natural systems that are isolated from the impacts of the social system.

At times this may mean that societal values for particular species or attributes of ecosystems will have to be incorporated into management, but with a clear understanding that other essential attributes may not be sustained in an expected or desired manner. Because of the dynamic nature of ecosystems and the degree to which humans have modified ecosystem structure and function, it may be difficult or the appropriate tools may not exist to maintain particular attributes of an ecosystem considered essential for the survival of some plant or animal species (see Chapter 3). Ecosystem management by definition means that all desired products, ecosystem integrity, and/or species cannot be maintained concurrently at a maximum level—a perception that multiple use instilled in our vocabulary (see Chapter 2).

Ecosystem management means having to accept the trade-offs that exist when society identifies that something is valuable and must be maintained in an ecosystem. This does not mean that the ecosystem management approach can be used to select species or ecosystem attributes (i.e., large, coarse woody debris in the old-growth forests of the Pacific Northwest United States) to be eliminated because they have been determined to be nonsustainable or undesirable. Ecosystem management needs to go beyond using our human values to identify what is or is not desirable in an ecosystem (Stanley 1995). This approach needs to be based on biological and ecological information that takes the discussions away from being primarily arguments centered around different values or peoples' perceptions of the system. This will require ecosystem ecologists to develop tools that begin to index ecosystem states and to allow for the detection of when different ecosystems move beyond the natural multiple states occurring during stand development or succession (Figure 1-2; see Chapter 4).

Our current tools for studying ecosystems (see Chapter 3) have not been effective in placing our decisions in a framework that can incorporate scientific information in combination with our values. This combination is critical to better understand the implications of our values and the trade-

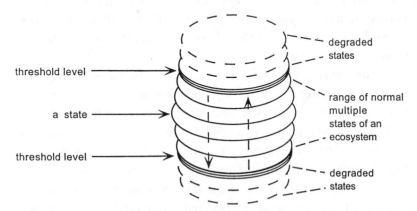

FIGURE 1-2 A diagrammatic presentation of potential multiple states through which an ecosystem may temporally fluctuate. The degraded states are conditions in which the system does not readily return to the normal multiple states without significant human input of energy.

offs of our decisions. This lack of understanding of our ecosystems has been frequently quoted as a reason to be cautious in our management of ecosystems (Ludwig et al. 1993). It is important to incorporate uncertainty into our management (this is explicitly dealt with as part of ecological risk assessment; see Chapter 4). However, it is insufficient to only deal with uncertainty without the development of tools that better identify (1) the current state of a system, (2) how far a current system has moved beyond its natural state, and (3) the ability to predict when the ecosystem state is shifting to a different equilibrium state prior to the appearance of several visual symptoms of ecosystem stress.

The ecosystem approach should also explicitly define the temporal and spatial scales of analysis. Ecosystem management defines the spatial scale of analysis to be the management unit, and the temporal scale is important because it incorporates disturbances that occur at longer time scales than the perceived management time frame (see Chapters 2 and 3). It is also important to understand that the use of the ecosystem management concept does not entail significant data collection—the sheer pursuit of data on all ecosystem structural and functional characteristics does not mean that you will have a better understanding of the factors impinging on that ecosystem or what will force parts of that ecosystem to shift to a different threshold state. In Chapter 3, past and current tools being used to study ecosystems are presented as well as the type of information that has been obtained from these studies. This is followed by Chapter 4, in which several different approaches and modifications of past approaches are presented to stimulate the development of methods that are better able to detect the risk of a system to a change and to index sites with regard to their location along a gradient of different human modifications of ecosystem state.

The ecosystem approach has also been perceived by some to ignore species and not to be an effective approach for conserving biodiversity (Franklin 1993). The nebulous nature of what ecosystem management means has created great discomfort in accepting this approach to protect species. This aspect, in combination with the current re-evaluation of the Endangered Species Act, has made people uncomfortable in accepting an approach that could be perceived to dilute the strengths of existing laws for protecting species. Current laws are structured to protect species but not ecosystems (see Appendix A). If there existed a strong link between biodiversity and ecosystem function, it would be easier to accept that the ecosystem approach would also function well in protecting species. However, few empirical tests of the biodiversity and ecosystem functional relationships have been conducted. A perusal of past literature shows most studies to be theoretical examinations of these relationships (see Chapter 4).

Scientists have attempted to show that there is a strong link between the number of species and healthy ecosystems, with the main idea being that the greater the number of species, the more resistant and stable these communities are to disturbances and the higher their productivities (see discussion in Chapter 4). If this was a generality across all ecosystems, ecosystem management would be greatly facilitated. However, it appears that positive species and ecosystem functional relationships may occur in some grassland or herbaceous dominated ecosystems but cannot be generalized to all ecosystems, especially those dominated by tree species (see Section 4.2.3). Even the grassland studies suggested the existence of a threshold in the number of species (> 10 species in this case) beyond which the addition of more species is not reflected in changes in ecosystem function or productivity (Tilman and Downing 1994). Our understanding of where species and ecosystem functional relationships exist has been hampered by the difficulty of designing experiments to conclusively show that sustaining ecosystem functions will also maintain biodiversity in ecosystems (Baskin 1994, 1995, Naeem et al. 1994).

Ultimately the importance of preserving habitats and their ecosystem integrities has highlighted the need for understanding and identifying tools that can be used to study species within an ecosystem context. There exist many examples in which a focus on a species would not have been sufficient to highlight their importance within the ecosystem or the ecosystem feedbacks that affect the species at the population level. The ecosystem approach forces one to understand how an organism fits into the ecosystem and attempts to decipher, using our historical backlog of information, whether a species serves a primary ecosystem function, is functionally redundant, or is a keystone species (Walker 1992, Grumbine 1990, 1994, Mills et al. 1993). Several examples follow that demonstrate the importance of understanding the links between species and ecosystems.

Frequently the role that an individual or a component of the system performs does not become apparent until a later date. A prime example of this is the case of the dodo (giant flightless pigeon) reported by Temple (1977) in Mauritia. In this example, 300 years after the extinction of the dodo, its role in the regeneration of a tree species (i.e., *Calvaria major*) was discovered. Large stands of this tree species were starting to disappear as natural tree mortality events occurred and no new regeneration was recorded. It was discovered that seeds of this tree species could only germinate after they were exposed to the grinding action of the dodo gizzard. It was found that no other animal species were present that could function in the same capacity (Temple 1977). This is a classic example of the lack of understanding of the functions of different plant or animal species in ecosystems. This is especially true for those species that are long lived such as trees.

If there were no positive and negative feedbacks in ecosystems, individual parts of a system could be studied and used as the only information needed to sensibly manage a system—no unexpected spatially or temporally disjunct repercussions would be expected from anywhere in the system. However, positive and negative feedbacks are an integral part of ecosystems, making it difficult to study individual parts to understand the whole system. There are several studies that highlight the importance of not focusing only on the organism itself or on one ecosystem component in isolation from the rest of the ecosystem. One of the most important reasons for using the ecosystem approach is because of the high interdependencies and strong feedback loops that exist in ecosystems. Ignoring these interdependencies or feedback loops results in the system becoming dramatically susceptible to other stresses that were not as significant prior to this change or having an increased susceptibility to insects or pests. There are many examples in the literature highlighting cases in which the interconnectiveness in ecosystems and the feedback loops resulted in ecosystem effects that might not have been predicted. Four cases are presented below that document well why an ecosystem perspective is crucial to use in managing our natural resources.

Links between ecosystem components are not always obvious. For example, Pastor et al. (1988) showed the existence of feedback loops for one ecosystem and how the activities occurring in one part of the system may control the response expressed at the whole system level. Pastor et al. (1988) documented how the activity of a consumer (i.e., a moose) selectively browsing on certain plant species dramatically changed plant community composition and soil processes. If moose were studied by focusing on the plant browse of choice only, the effect of moose on decreasing future soil nutrient availability would not have been identified. However, a study of vegetation dynamics of both browsed and unbrowsed species revealed that decreased soil nutrient availability occurred: The selective preferential browsing by moose of the hardwood species resulted in an increased dominance of the less palatable conifers (Pastor et al. 1988). Because conifer needles decay at slower rates than hardwood foliage, decomposition rates decreased, resulting in decreased nutrient availability to the plants. In turn, this decreased nutrient availability fed back to preferentially maintain the site dominated by conifers over the hardwoods because conifers are better able to occupy low nutrient sites.

Another excellent example of the interdependencies of ecosystem components is that reported by Graveland et al. (1994). In this study, the decline in forest passerines was indirectly linked to acid rain because of their impact on snail populations. In the Netherlands, decreased forest passerine populations were recorded, with 40% of the birds producing thin and porous egg shells that broke readily and exhibited a high incidence of clutch desertion (Graveland et al. 1994). In the past, the forest passerines had predominantly obtained their calcium (Ca) for producing nondefective eggshells from eating snails. However, snail population levels had decreased in response to insufficient soil Ca levels, mainly due to acid rain causing leaching of Ca from the soil (Graveland et al. 1994). These lower snail populations cycled back to reduce the reproductive success of passerines. This study highlights the problem of not being able to immediately identify all the components of the ecosystem that will be negatively affected by a disturbance. For example, acid rain effects on forests and lakes in Scandinavia and Europe had already been identified as being a critical problem in the early 1970s (Tamm 1976, Abrahamsen 1980, Roberts 1983), but the links between acid rain and populations of forest passerines was only recently reported. These time lags in identifying cause and effect relationships are normal. This occurs because once a problem is identified, a significant time allocation has to be made to specifically understand the direct, and not the indirect, impacts of the problem itself so that mitigation or restoration efforts can be quickly implemented.

The third example highlights the problem of not understanding the implications of several activities occurring at larger spatial scales because of inadequate information at that time and the focus on managing one parameter or resource in the ecosystem. In this example, management practices were implemented that were based on one or two desired commodities (i.e., wildlife, timber) that did not consider their long-term ramifications within the landscape. In some eastside forests of Oregon and Washington (especially the mixed conifer areas), two management activities were especially critical in changing the structure of the ecosystem and how it responded to disturbances: fire control and selective tree logging (Agee 1993). Fire suppression was part of management practices in these forests (going back 130 years, and especially within the last 60 years) and have changed tree species composition from being dominated by fire-adapted to fire-susceptible species (Agee 1993). In these forests, fires traditionally played an integral role in maintaining what has been classified as "healthy" and "productive" forests (Agee 1993). The selective timber harvesting of ponderosa pine also contributed to converting these forests to young stands of true fir and young Douglas fir, which are not fire tolerant and are more sensitive to pests.

Both activities have converted these forests from having an open, parklike appearance to dense stands (sometimes dog-hair stands) that are more susceptible to severe fires (Agee 1991, 1993). This shift in the dominance of many eastside forests to shade-tolerant and non–fire-resistant stands also coincided with their increased incidence of disease outbreaks and susceptibility to fire (Agee 1993). In the past, short-term fire cycles had been an integral part of these forests and had maintained them in open, parklike conditions that were less conducive to the spread of fire (Agee 1991) and helped to keep insect and disease outbreaks at lower levels historically (Harvey 1994). This means that the management activities have resulted in changing how the ecosystem responds to disturbance and have increased their susceptibility to insect pests and diseases compared with historical levels (Hessburg et al. 1993). The results of these management activities on the ecosystem were not apparent when fire control policies were first implemented, nor were the impacts of selective harvesting of tree species obvious early on. If the ecosystem approach had been the paradigm of choice, other impacts of these activities on the ecosystem and their landscape might have prevented the development of the current management problems.

In the final example, the importance of understanding that changes or feedback loops can exist in ecosystems that are not always obvious or visually apparent, is emphasized, as a result these relationships are not included as part of the suite of tools used to study ecosystems. For example, some disturbances will elicit an early ecosystem response at the population or process level that can only be detected by monitoring changes at the below-ground level (Vogt et al. 1993). The research conducted by Perry et al. (1990) is an excellent example of this. In that study, clear-cutting of a higher elevation forested area in Oregon resulted in repeated regeneration failures until ectomycorrhizal fungi were reintroduced to the site. Plants in these types of environments are highly dependent on having mycorrhizal fungi on their root systems to be able to grow (Vogt et al. 1991). As a result, regeneration was not successful until the appropriate symbionts absent from the site were reintroduced. This case illustrates the importance of understanding what legacies are eliminated from the site following a human activity and ensuring through management that these structures are reintroduced in order to re-establish critical functions within the ecosystem.

Why is the ecosystem approach important to use in understanding the implications of a management activity or human disturbance in ecosystems? Past approaches have emphasized examining parts of a system to try to isolate specific effects of a human activity, but they have been extremely site specific (i.e., they have not transferred well to other systems), were difficult to use to predict the response of a system to any disturbance, and frequently created ecosystems that were more susceptible to disturbances because other factors on the site were modified. Researchers and managers have accepted the need to get away from the piecemeal approach and to adopt a more holistic ecosystem approach. The holistic approach is perceived not to degrade ecosystems nor to change them sufficiently such that they are unable to maintain species or functions. There are several strong reasons to use an ecosystem approach to study systems, and the rationale for many of these is documented throughout this book. Several of these reasons are as follows:

1. The low level of understanding of processes occurring at an ecosystem scale and therefore inadequately developed tools to deal with

    (a) Uncertainty or insufficient data when assessing the ecological risk of a system

    (b) Poor ability to identify when an ecosystem has traversed from one biological threshold to another

    (c) How and what parts of an ecosystem are relevant to study to produce information adequate to test the trajectory that a system is following

2. The long temporal scales in which a system is responding, resulting in great difficulty in designing experiments to effectively mimic these temporal scales

3. The inability to effectively address or predict spatial scaling issues in which a biotic or abiotic structure and/or function at one scale can feed back to control processes occurring at another scale

4. The inability to measure ecosystem resistance and resilience, and to be able to identify what the system state being studied is in relationship to similar ecosystems

5. The poor ability to identify how past land-use activities or natural disturbances have changed the manner in which the present system is responding

Past management of natural resources has been unable to deal with the statements made earlier because management dealt with the specific part of the system that had been identified to be degraded or where a specific ecosystem function was no longer viable. This has resulted in a piecemeal approach that in many cases ended up being *crisis management*, which did not consider any of the long-term implications of a current focused activity on the maintenance of the entire ecosystem. Use of the ecosystem approach would encourage researchers and managers to examine a system holistically. Accordingly, the role of an individual species or the part of the system being degraded would be examined within a framework of identified key driving variables, forcing incorporation of the uncertainty existing in the knowledge base. Such an approach would facilitate the identification of feedbacks changing one activity in an ecosystem that can potentially reverberate through the system, causing unexpected changes in ecosystem function.

Throughout the following chapters in this book, why and how ecosystem management is an attainable goal are documented. Chapter 2 presents the historical timelines for the development of the *ecosystem concept* in ecology and how this concept is still evolving to better incorporate temporal and spatial scales. This is combined with the management and legal historical time lines to show how there has been little compatibility between the development of policy and management guidelines using the ecosystem approach. The fact that the ecosystem concept is still evolving and new approaches need to be developed to analyze ecosystems has made it very difficult to produce policy or management guidelines that are not outdated by the time natural resource managers feel comfortable with these approaches. Chapter 3 synthesizes the current approaches and tools being used to analyze ecosystems and presents their pros and cons.

Different approaches are presented in Chapter 4 that need to be tested for their applicability to effectively and adequately address the ecological risk of a system to disturbance and to the practice of ecosystem management. New approaches to studying ecosystems must allow researchers to identify the "present state" of an ecosystem and to determine when that ecosystem in moving in a trajectory away from its present state or towards another threshold state. This is followed by Chapter 5, which uses a case-study approach to illustrate how several private and federal organizations perceive and define ecosystem management, and how ecosystem management is being or could be implemented by these different organizations. The ability of these organizations to implement ecosystem management is examined by considering what factors have driven their management practices; this specifically focuses on the changing role of the public in modifying how management actually occurs. The ultimate goal of this book is to identify what is the minimum, reasonably obtainable ecological knowledge required to implement ecosystem management. This includes a suggested conceptual framework for conducting ecological research directed toward the needed implementation of ecosystem management.

2

Ecosystem Concept: Historical and Present Review of Definitions and Development of Ecosystem Ecology, Ecosystem Management, and Its Legal Framework

Managers and researchers are currently defining and setting objectives for ecosystem management; however, relatively few workers (NRC 1990) have attempted to determine what ecological information is needed to implement this new management strategy. While many workers agree that consideration of the scale of both natural processes and management practices is critically important under the ecosystem paradigm (Franklin and Forman 1987, Clark et al. 1991, Gordon 1993), and while many others have recently developed and refined theories about scale and natural systems (e.g., Allen and Starr 1982, O'Neill et al. 1986, Allen and Hoekstra 1992, 1994), few have directly applied these theories to guide management.

Our inability to implement sound ecosystem management in the natural ecosystems is partially due to the fact that it is only quite recently that the ecosystem concept developed in ecology (Golley 1993; Appendix A).

This means that the legal mandates and management guidelines are being developed within a framework that lacks a solid foundation upon which these activities can evolve. An examination of the historical development of the ecosystem concept in ecology, its utilization in management, and the legal mandates controlling management practices all highlight the difficulty of obtaining a consensus on what ecosystem management should entail and what the product outputs of using such an approach are (see later). This uncertainty is reflected in the strong desire by researchers and federal agencies to obtain a clear and very focused definition of ecosystem management and what this type of management entails. This has hampered the utilization of the ecosystem approach in management because considerable time is lost in producing a very explicit definition of ecosystem management that is supposed to be transferable to all ecosystems. This idea is further pursued in this chapter to demonstrate the inappropriateness of generating a sentence definition and a single model approach for ecosystem management if it is to be successful. There are costs as well as benefits from using the ecosystem approach. Indeed, one reason that multiple-use management did not work was that proponents sold it as offering only benefits for all parties concerned.

To begin any discussion of ecosystem management, it is essential to understand the historical development of ecosystems as an ecological concept and also the historical development of management. This type of historical analysis is very useful to help explain our present understanding of ecosystem management and why there is a lot of controversy over its utility as a management tool (Ludwig et al. 1993, Soulé 1993, Stanley 1995). The historical development of the ecosystem approach to conducting ecological research was also limited by the past emphasis on funding individually based research projects. Ecosystem-based research did not really become a reality until the 1960s, when several group research projects were established by federal government laboratories in the United States and as groups of university scientists linked up with U.S. Forest Service scientists and their long-term field research laboratories. In addition, the formation of the International Biological Programme in the 1960s helped to train a whole new generation of scientists who would contribute to ecosystem level research in the 1970s and 1980s, even though that program itself is not ranked highly for producing quality science (Golley 1993).

Part of this historical analysis also needs to examine the development of the theories and scientific understanding of species in systems because many legal mandates are species based. As part of this historical overview on species in systems, a discussion of microbial symbionts is included because they need to be incorporated as part of the information inputs for management. Most conservation and management efforts have focused heavily on animal and plant species that have been easier to identify as declining at the expense of those species (i.e., microfauna, microflora, liverworts, etc.) less studied, less charismatic, or less readily visible or understood by the public. These historical previews lay a foundation for the direction that people have pursued to implement ecosystem management.

The historical overviews are followed by a review of the operational definitions that have been proposed for ecosystem management. This considers what the benefits of such an approach are and the role that humans play in defining what should be managed in ecosystems. Because human values or perceptions help drive what is identified to be important, it is crucial to accept this as part of the formula for any implementation approach. For example, the public may identify a particular species as being important to maintain in an ecosystem. Even if there exists strong scientific evidence showing that a species has a clear link to maintaining a particular function in an ecosystem, managers have to fully understand what the implications are for the *whole ecosystem* if priority is given to trying to maintain that species and that ecosystem in a state that supports that species. The role of ecosystem management is to identify the trade-offs that are involved in accepting the socially driven option of maintaining the system for a species or for a particular ecosystem attribute because ecosystems are quite dynamic and fluctuate between several multiple states (see Figure 1-2; Chapter 3). Management should outline all the options and remove the decision from being predominantly based on resolving different values. Scientific information has to contribute to identifying thresholds and when a chosen action will result in the loss of other ecosystem attributes that are important in maintaining ecosystem integrity.

This chapter concludes with an analysis of how the public mood and need for forest products have been driving variables in determining how management has been conducted. Management has to contribute to this public dialogue by contributing scientific information that allows for the identification of the repercussions of implementing our values (see

Chapter 4). Management decisions cannot be made entirely on values. Ecosystem management by definition incorporates social and natural science values and information, and does not emphasize one at the expense of the other.

2.1   Roots of the Ecosystem Ecology Concept

The acceptance, utilization, and development of the ecological approach to analyzing ecosystems has been an ongoing process for the past 60 years. An examination of the historical time line for ecosystem ecology demonstrates our shifting ideas about the relative importance of species, plant communities, and ecosystem functions in maintaining systems and what parameters should be measured to understand natural systems (see Appendix A). Concurrent with the evolution of a framework of what and how much of the ecosystem needs to be studied has been the development of the research infrastructures to implement an approach requiring information from several disciplines. The role of different types of research organizations in conducting ecosystem level research will be examined, especially because many of the environmental issues have to incorporate information from several disciplines. A large stumbling block to utilizing the ecosystem approach in ecology has occurred because of (1) the difficulty of studying the complexity of systems without having tools to help explain why there are widely disparate results from systems previously identified as being similar to one another, (2) the difficulty of identifying what parts of the ecosystem to study to determine when the ecosystem is changing or when a species approach will be useful, and (3) the difficulty of having researchers shift from the individual scientist work mode to one in which scientists function as one component of an interacting team of experts in which an individual's core knowledge is only part of the solution.

Even though the roots of the ecosystem ecology concept can be traced back to 1935, when Tansley first used the word ecosystem, utilization of the *ecosystem* approach was not attempted as a way of examining systems until after the 1960s. Despite the fact that the ecosystem ecology concept is now accepted as an integral part of research and is incorporated into major programs such as the National Science Foundation's Ecological

Long-Term Ecological Research Program, established in 1980, researchers still have difficulty in conducting and being part of interdisciplinary research because of the strong emphasis placed on disciplinary training in educational institutions and because the success of an academician is typically based on evaluations of their individual accomplishments and not group activities. It is useful to examine the historical role of individual scientists and research groups or centers in studying ecological systems and how this has affected the framework chosen to assess the risk of a system to change.

Another factor that has contributed to some of the difficulties in building consensus on how ecosystems should be studied has occurred because of the strong bias held by researchers on how ecological systems should be viewed. For example, a distinction can easily be made between two major views of ecosystems: the *population-community* (or biotic) approach, in which the research can be easily conducted by individual researchers, and the *process-functional* approach, which is best implemented by several researchers working together (Table 2-1). The distinction between these two views has historically been reinforced and accepted in ecological thought and helps to explain some of the discordancies existing between the legal mandates (mainly a species approach) and ecosystem management. The population-community approach emphasized studying populations and interactions among them, especially focusing on studies of predation and competition. At the extreme end, this approach focused entirely on the biotic community and saw the physical environment as being an externality to the system. This contrasted with the process-functional approach, which emphasized measuring the flow of energy and matter through the environment and considered the physical environment to be an integral part of the system (Odum 1959). This means that the functional approach frequently perceived the biotic component or species to be an externality to the system; it didn't really matter who was performing a particular function as long as the individuals were present.

Despite the obvious differences between both approaches (see Table 2-1), both focused on studying interactions within their system, either at the biotic or energy and matter flux levels. These approaches may be considered as perspectives that emphasized differentially the role of the biotic and physical environments within a defined temporal and spatial context. In other words, these perspectives are a way to distinguish be-

TABLE 2-1   Comparison and contrast between the biotic and functional approaches to studying systems that prevailed until quite recently[a].

Biotic (species) emphasis	Functional emphasis
1. Assumes distinctive grouping of ecosystem components into the following so that interactions among components not examined: • *Producers* (energy fixing by autotrophs) • *Consumers* (could be mammals or insects) • *Decomposers* (fungi, bacteria, and soil animals)	1. Assumes that systems can be studied by focusing on: • *Energy fixation* and *storage* by the *producers* • *Energy* and *nutrient Transfers* from the *producers* to the rest of the ecosystem • *Nutrient acquisition* by the *Producers* from either the *soil* or from *nonliving organic material* through the mediation of the *decomposers* • *Energy* (carbon as a surrogate) and *nutrient cycling* due to *consumers* and *decomposers* • Assigned *consumers* and *decomposers* the role of *regulating* the *rate* at which processes occurred
2. Parts of the ecosystem not considered important to study (e.g., soil physical and chemical characteristics, nutrient cycling, etc.) they merely existed as the stage upon which biotic activities were played out	2. Unique biogeochemical behavior not considered (e.g., calcium and nitrogen affect ecosystems very differently)
3. Ignores redundancies and functional groups in the biotic component	3. Species changes in systems not examined and impacts of individual species on ecosystem function not considered.
4. Unable to deal with portions of the system, such as the phyllosphere or rhizosphere which act as an interface for many abiotic and biotic activities	
5. Ignores temporal and spatial scale controls on biotic activity	

[a] Adapted from Odum 1959, O'Neill 1976, O'Neill et al. 1986, with permission.

tween the background (typically identified as being unimportant or less important) and foreground information (identified as important or more important) for a particular study (Allen and Hoekstra 1990, King 1993). In actuality, three approaches are used to study the biotic components of systems: (1) the *ecosystem* perspective, emphasizing energy and nutrient fluxes; (2) the *community* perspective, emphasizing biotic interactions among plants, among animals, or both; and (3) variable combinations of the first two approaches.

A holistic view of an ecosystem requires a combination of the ecosystem and community perspectives. All functions of an ecosystem cannot be understood without considering community interactions. However, due to traditional scientific thought, combining the two perspectives is often, at best, difficult. Traditionally, biologists and ecologists have conceptualized a hierarchy of *organizational levels* in ecology, each with their own *emergent properties that don't necessarily transfer across the different organizational levels*. In this organization scheme, the biological spectrum is recorded to move from the smallest scale to the largest, starting with *protoplasm—cells—tissues—organs—organ systems— organisms—populations—communities—ecosystems—biosphere* (Odum 1959). This organizational scheme suggested that ecosystems and their processes are at a higher scale than communities and their processes, reinforcing the concept that they have to be dealt with separately. However, many community processes, such as Clementsian succession, and even population processes such as migration, often occur at larger spatial or temporal scales than many ecosystem processes such as nutrient cycling. The concept of ecosystem in this traditional hierarchy scheme is counter to the concept of ecosystem as a perspective that can be studied holistically and incorporates activities from different scales.

The next sections examine some of the key concepts that have been instrumental in the development of the field of ecosystem ecology. This begins with an examination of the historical roots of ecology that contributed to developing the idea of ecosystems as being dynamic. Much of this early literature focused on understanding how plant communities changed with time (i.e., succession)—explicit examination of temporal but not spatial scales of systems. This is extended to include the historical development of our understanding of the role of species (i.e., microbial symbionts, plants, and animals) within an ecological system. As part of this discussion, the relevance of redundancy, keystone species, and functional groups is examined within an ecosystem framework. Finally, much of the terminology used to describe ecosystem resistance and resilience characteristics is examined. Many of the terms reappear in Chapter 4 when different approaches are presented to detect changes in ecosystem states.

2.1.1   Scaling Ecology to Ecosystems

Community Ecology

The roots for examining systems with clear considerations of temporal scales can be traced back to early research conducted by plant ecologists on succession (see Appendix A). In the early 1900s, Clements (1916) was instrumental in presenting the idea that plant communities are dynamic (not static) based on research conducted on vegetation communities growing on sand dunes along Lake Michigan in the United States. This early contribution to the historical development of ideas on plant succession was important because it emphasized that communities are dynamic, and it stimulated significant research on this topic. However, this earlier research also created some major problems because it dominated how ecologists perceived plant community development for almost 40 years and suggested that communities changed in very identifiable and predictable ways, giving us an impression that it was easy to predict the direction of change in all plant communities. For example, Clements assumed that succession was the orderly and unidirectional replacement of species (or groups of species) by other species, as each in turn made the conditions less favorable for itself and more favorable for the succeeding species. The end result of Clements' ideas was the existence of a "climax" system—a stable and self-replacing vegetational association that could be determined over large regions using climatic variables. Today the climax ecosystems as defined by Clements are known not to exist—a system may have "arrested succession" in which a vegetative community may dominate a site longer than the short human time frame has decided is appropriate (ultimately, given the right conditions, the system would shift to another state). Clements (1916) also considered that disturbance was external to the process and succession was what occurred between disturbance events. Clements' scheme was a platonic ideal—the abstract essence of the climatic climax was the reality, while the complexity of the real world was "noise."

In the early to mid-1900s, Gleason (1939) and Whittaker (1953) introduced the population-based approach, in which individual plant species responded to environmental gradients in their own way. In contrast to Clements, they suggested that individual plants are not linked and do not respond as a "supraorganism" à la Clements (1916). Based on their ideas, recognizable plant associations were merely the coincidental result of several species having similar climate and soil preferences co-occurring on the same site. They also redefined "climax" communities as not always being the same plant community but as a complex mosaic of dominant species existing within the landscape. The climax, therefore, was no longer considered a stable, self-replacing entity, but more a dynamic pattern within the landscape. In addition, their ideas suggested that successional change in plant species could be more difficult to predict because the existence of a plant community varied based on the local conditions. Disturbance was also incorporated as an integral part of the landscape; the climax includes early successional plant communities, and chance creates a variety of successional pathways. These ideas were the forerunner of "patch dynamics" and landscape ecology. However, because their ideas were not mechanistically based, the type of plant community existing at any location could not be predicted with much confidence. This lack of a mechanistic understanding of the dynamic nature of plant community change with time and space was a definite weakness of their approaches.

In 1954 Egler introduced ideas that began to address the mechanisms that controlled why species replaced one another with time in a community. Egler's ideas contrasted with Clements' successive replacement of species (which Egler termed *relay floristics*) because he suggested that both early- and late-succesional species can become established during the initial stages of succession. According to Egler (1954), plant species have physiological attributes that determine when they will be present at different successional stages. For example, early successional species are fast growing but shade intolerant and therefore do not reproduce in a closed canopy forest. Thus early-successional species are replaced by late-successional species when they become shaded. This contrasts with late-successional species, which grow slower but can survive in the shade of the early successional species, and eventually replace them after small-scale disturbances. This scenario identified by Egler is a more realistic representation of most successional processes (especially secondary successions) and began to identify predictive tools for explaining successional changes in species.

Once clear stages of stand development were identified, a significant amount of research was conducted on this topic starting in the 1980s up to the present. Oliver's (1981) four-stage stand development model (e.g., stand initiation, stem exclusion, understory reinitiation, old growth) is an extension of Egler's model, which specifically focused on identifying the different stages that one plant community experiences. These stand developmental stages have also been identified to have specific carbon and nutrient cycling patterns that change in response to the growth of that system (Vogt 1987). These different stages are characterized by many factors changing, especially during the stem exclusion phase, where the lowest mass of understory plants exist, seedling establishment ceases, litterfall inputs plateau, highest tree mortality rates are recorded, and the lowest abiotic resource availabilities occur (see Section 2.1.1). Much of this information is valuable in understanding the dynamic nature and the multiple states that a particular system will filter between. This information, combined with the changes that occur in the carbon and nutrient cycles, allows for the identification of the natural levels of change for ecosystems within these multiple states—critical information needed to assess the ecological risk of a system to change.

Starting in the late 1970s up to the present, the role of disturbances in changing both abiotic and biotic resource availabilities, and therefore in controlling plant community dynamics, has also been extensively examined. The differential roles of small- to large-scale disturbances in maintaining species composition has been shown in many forests. For example, at the small-scale disturbance level, gap dynamics has been studied at length as being the major mechanism by which regeneration occurs in mature forests subject to infrequent major disturbance (such as many tropical forests or the coniferous forests of the Pacific Northwest) (Denslow 1980, Pickett and White 1985, Brokaw 1985). Much of this research primarily focused on identifying plant species-level succession into these gaps and the relative importance of above-ground abiotic variables (especially light) on controlling species establishment. These studies were not holistic and ignored the below-ground parts of the ecosystem. Not until the mid-to late 1980s was there a realization that below-ground parts of plants are able to affect species establishment and resource availabilities in the soil (Vogt et al. 1995a). Similar to most stand level studies conducted during the same time period, gap dynamic studies emphasized treating the gap as a distinctive environment to be compared with a distinctive forest and ignored the spatial heterogeneity existing within the gaps (Spies personal communication, Vogt et al. 1995a). It was not until the 1980s that an understanding was being developed by the research community of the importance of legacies (Franklin et al. 1981) within gaps or forest areas in mitigating and modifying the response of a system to disturbances.

The role of large-scale disturbances such as fire has received considerable study since the 1970s (Grier 1975, Oliver and Larson 1990, Agee 1991, 1993). Most of these studies have mainly occurred at the community ecology level, where a solid understanding of the role of fire in controlling the vegetative composition at the landscape level has been developed (Agee 1993). Other studies emphasized examining the impact of fire on specific ecosystem attributes, such as nutrient losses after a fire (Grier 1975). Because fire is a management tool, considerable research has been conducted on fire fuel loads and how to most effectively use fire as a management tool (Oliver and Larson 1990). However, few holistic studies of fire in ecosystems have been conducted. Even the public has a poor understanding of the role of fire in ecosystems, as evidenced by their negative reaction to the 1988 Yellowstone fire (Christensen et al. 1989, Schullery 1989, Knight 1991).

An examination of the development of our ideas on succession shows our increased understanding of plant community changes with time, but information is also needed on more ecosystems to be able to predict what is controlling these changes for particular systems. It is obvious that the spatial scale of systems has been inadequately addressed by researchers in the past; however, this is currently a very active research area in which new tools are being developed (see Chapter 3). Successional changes in species make us realize the dynamic nature of natural ecosystems and accept the fact that it may be extremely difficult to select one state and try to maintain that system state. It also makes us realize that much of plant composition that comprises an ecosystem resulted from small- to large-scale disturbances. It is not until quite recently that we have begun to develop our mechanistic understanding of why succession of plant species is occurring.

This type of information is valuable because it allows us to begin predicting how plant communities might change with time or in response to a changing environment. Several successional models (JABOWA, Gap Dynamics) have been developed and are being used by academicians, natural resource agencies, and policymakers to predict how species composition might change in response to a changing climate (Kiester et al. 1990).

Ecophysiology

Having a mechanistic understanding of how plants responded to their environment based on physiological measurements has been a core research area since the 1960s (Billings 1985; see Appendix A). There are several good reviews covering the theory and research in plant physiology (Mooney 1972, Larcher 1975, Kramer and Kozlowski 1979, Waring and Schlesinger 1985, Chabot and Mooney 1985). Physiological research in forests has mainly focused on studying seedlings under controlled conditions because of the difficulty of studying mature trees (see Chapter 3). Forests contrast with other systems (i.e., arctic tundra, grasslands, marshes, etc.), in which good ecophysiological research has been conducted on intact systems in the field. In these nonforested systems, or where trees are naturally quite small (i.e., tundra), researchers have been able to completely encase part of the system in a measurement chamber (Mooney et al. 1991, Chapin et al. 1995). Recent developments in techniques, such as stable isotopes, will allow better temporal integration of physiological processes at the scales of interest (see Section 4.2.4) and hold great promise for linking physiological and ecosystem processes for mature trees in the field.

Billings (1985) identified physiological ecology as addressing the following three core problem areas:

1. Attempting to explain the reasons for the present geographical distributions of populations of wild plant taxa and crop plant cultivars by measuring their physiological, morphological, and reproductive adaptations of environmental factors acting in concert;

2. Definition of the potential environmental tolerance ranges of plant taxa (native, adventive, and agricultural) and the evolution of the responsible character traits; and

3. Determination of how ecosystems operate in a changing biosphere.

These three core problem areas identified by Billings (1985) are still relevant today because the same questions are being asked but using different tools to try to address them. According to Billings (1985), some of the reasons why physiological ecology really developed as a field since the 1960s are because people had better training in several disciplines (i.e., physical sciences, mathematics, and physiology) and due to the development of good physiological (e.g., the infrared gas analyzer to measure photosynthesis) and meteorological instruments, combined with access to personal computers. He emphasized the importance of having individuals with training in more than one discipline being critical for the development of physiological ecology—the exact same problem identified today with implementing the ecosystem approach in management. It is important to train researchers to comprehend the complexity of the natural systems without being overwhelmed by it, to understand how different disciplines contribute to understanding the environment, and to be able to decide what tools from which disciplines will facilitate our ability to detect ecosystem change.

Even though plant physiological ecology has long historical roots in the field (Billings 1985), much of the earlier physiological information was collected under very controlled conditions because of the difficulty of utilizing some of the equipment to take physiological measurements in the field. According to Lassoie and Hinckley (1991), forest ecophysiologists became too preoccupied with techniques, especially once a piece of physiological equipment became portable. They further stated, "There have been many descriptive studies where physiological parameters were related to environmental variables for one species after another without real consideration for the role physiology was to play in forest ecology—specifically, in helping to understand the distribution, abundance, and productivity of forest trees and stands." They go on to further state that starting in the 1980s, there was a shift in emphasis away from the descriptive studies of the past; several good examples exist in which ecophysiology was integrated into modeling plant productivity, plus examining the effects of pollution, nitrogen fertilization, and irrigation on plant growth (Lassoie and Hinckley 1991). The number of studies utilizing physiological techniques to monitor mature plant responses to a changing environment are limited and difficult to generalize. Few studies have concurrently exposed seedlings and mature field plants of the same species to similar treatments and monitored their physiological responses. The typical approach has been to study pollution impacts on seedlings or saplings grown under very controlled conditions, and then to use models to transfer the physiological responses of seedlings to predict impacts on mature forests (this is the approach used in some of the earlier studies such as NAPAP; see Section 4.2.9).

Recently there have been several publications that have synthesized the pros and cons of the major physiological techniques for studying forest ecosystems (Lassoie and Hinckley 1991, Smith and Hinckley 1995). These books were written to identify what and when specific physiological techniques can be used to study forests. With some exceptions (i.e., Brix 1971, Grier and Running 1976, Waring and Schlesinger 1985, Smith and Hinckley 1995), physiological techniques have been mainly used under controlled conditions and on small seedlings. Physiological techniques are increasingly being utilized in ecosystem studies. A large stumbling block to this has been the difficulty of determining how to transfer physiological information to the ecosystem or landscape scales (see Section 3.4). The tools for scaling physiological measurements to the ecosystem level is an ongoing activity that has to deal with the fact that physiologists typically have to collect more information because of diurnal fluctuations in physiological processes (Smith and Hinckley 1995). A discussion of the physiological differences existing between seedlings and mature plants, and how readily information collected at one physiological state can be transferred to another state, is presented in Chapter 3. Chapter 3 also includes a discussion of which physiological measurements have been utilized to effectively explain plant responses to their environment.

Soils

Knowledge related to how soils control the growth of plants growing on their surface, and therefore ecosystem processes, has mainly derived from two sources: farmers' (and later foresters') experiences based on centuries of trial and error, and scientific investigations of soil characteristics and how they changed in response to management (Brady 1990). Most attempts to understand soil biological, physical, and chemical characteristics and how these related to plant growth are quite recent, even though history has documented the role of the degradation of agricultural lands in contributing to the collapse of civilizations. The demise of many past civilizations has been partly attributed to their misuse of soils because they did not understand the impacts of their land-use activities on the maintenance of soil integrity and productivity.

Because sufficient food production is a requirement for the development and maintenance of large populations, ancient civilizations were commonly centered around fertile river valleys (e.g., the Tigris and Euphrates in Mesopotamia, the Nile in Egypt, the Indus in India and Pakistan, the Yangtze and Hwang Ho rivers in China, etc.) (Hillel 1991). The high fertility of the soils in these valleys was the reason that large populations were sustained for long periods in these areas; however, the fertility of these soils was sustained only as long as the river banks were flooded with the nutrient-rich, silty soil. As the population size of these communities increased, many people had to move further away from these river banks, requiring the construction of irrigation and drainage ditches. This movement of people away from areas annually enriched with nutrient-rich flood waters meant that they became dependent on using and managing a *finite soil resource base*. The ignorance of the proper management tools for crop production that would not degrade the soil resource base or result in the soil becoming salt affected due to irrigation (Szabolcs 1989) contributed to the rapid decline in the productivity of these fields and an inability of these ancient civilizations to produce enough food.

Despite the poor understanding of human land-use activities on soil productivity, several civilizations were quite aware of the relationships between plants and soil. For example, the Chinese (4200 years BP) used a schematic soil quality map as a basis for taxation. In both the *Odyssey* by Homer (about 1000 years BC) and in the bible, references are made to the use of manure on the land and "dunging" around plants. The Greeks and Romans used leguminous plants for replenishment of soil nitrogen as well as ash and sulfur for soil amendments (Brady 1990). But it was not until the 17th and 18th centuries that scientific inquiry became an integral component of studies relating plant growth to soils. The relationship between crop productivity and soils (edaphology) led to the discovery of the importance of soil, and ultimately to the development of soil science and the study of soil as a natural body (pedology). With this knowledge it was found that the sustainability of crop production could be approached or at least extended.

The history of forest soils theory is not as old as its agricultural counter-part and only covers a period of about 100 years. As early as 1840, Grebe, a German forester, introduced the role of ecology in silviculture, which was strongly based on an understanding of forest soils. His doctoral dissertation (published in 1844) has been considered to be the cornerstone of forest soils (Wilde 1941). Wilde (1941) quoted Grebe as saying, "As silvicultural horizons widen, the importance of environmental conditions becomes more sharply pronounced. It appears clearly to the foresters that the form of forest management is determined by a number of physical influences related to topography, geology, type of soil and climate." Grebe, 30 years later, wrote, "In short, almost all of the forest characteristics depend on the soil, and, hence, intelligent silviculture can be based only upon a careful study of the site conditions" (Wilde 1941). Even the dominant silvicultural text today supports Grebe's opinion concerning the importance of soils in silviculture and the need for its conservation (Smith 1986). Smith (1986) stated that the main objective of forestry is the maintenance of forest productivity, and thus conservation of the site (and, hence, its quality) is one of the principal tenets of good silviculture. Smith (1986) suggested that the soil is one of the site factors that is most subject to long-lasting harm and is the most nearly nonrenewable of the resources used in silviculture. Oliver and Larson (1990) went on to state that even the patterns of stand development are based not only on tree physiology and micrometeorology but also on soils.

Leonardo Da Vinci once said that more is known about the movement of celestial bodies than about the soil underfoot (Hillel 1991). Even though soil science is a relatively young science, our understanding of soils today is quite strong. In 1980, Jenny stated, "soil is more than farmer's dirt, or a pile of good topsoil, or engineering material; it is a body of nature that has its own internal organization and history of genesis." This concept of soil genesis, development or formation was initiated in the middle 19th century by an American geologist, E.W. Hilgard, and independently by a Russian geologist, V.V. Dokuchaev. They proposed that the soil and its development was influenced by the soil-forming factors: climate, parent material, organisms, relief (or topography), and time (Joffe 1936). However, it was Dokuchaev who first expressed this relationship in an equation or mathematical format: $S = f(cl, pm, o, r, t, \ldots)$, where S = rate of soil development, cl = climate, pm = parent material, o = organism, r = relief or topography and t = time (Glinka 1927). But it was not until the work of the American soil scientist Hans Jenny that the quantification and solving of this equation was even attempted (Jenny 1941). Since then many soil scientists have routinely (and in some cases, not so routinely) determined the effects of some environmental parameter on the soil or some soil property, and subsequently its effect on crop sustainability (Brady 1990). This soil-forming equation can be readily converted to an equation expressing the parts of ecosystems that should be considered as part of management: ecosystems = $f(s,cl,pm,o,r,t\ldots)$.

In the late 1930s Hambidge (1938) suggested that the major factor explaining the causes of waste of human resources and misuse of soil was lack of knowledge on the part of individuals of their relative importance or functions. However, he went on to say that even if the knowledge was available to these individuals, social and economic limitations could still prevent them from applying that knowledge effectively. Hambidge (1938) explained that one factor hindering the proper use of soils was the traditional attitude of Americans toward the land. They had not only an abundance of resources but also a strong tendency toward individualism, believing that every person had a right to the unrestricted ownership of a piece of earth. Our land policy in earlier times consisted of disposing of the public domain as speedily as possible. Regretfully, our early land policy was not comprehensive or well-thought-out in respect to land conservation. It was not uncommon then for corporations and speculators to acquire large acreages, much of it consisting of excellent forested lands, which were then usually exploited for their natural resources and devastated in the process (Hambidge 1938). However, all this does not mean that ecosystem management, facilitated by the use of soil indicators, could not or should not still be used and integrated with other disciplines for a more holistic management.

Today there are several scenarios that have been suggested to be relevant for monitoring soils as "indicators of: (a) ecological site integrity for comprehensive prediction of risk or susceptibility with respect to broad-scale forest management strategies; (b) natural and human-induced physical stresses from regional and localized site manipulations; (c) long-term forest sustainability, with specific value as a buffer for water quality effects such as acidic deposition and as a sink for airborne metal contaminants or deleterious organic compounds; (d) soil biological health with respect to long-term nutrient cycling; and (e) soil as a carbon sink" (Van Remortel 1994). Further discussions of soil parameters and soil nutrients in controlling ecosystem productivity are presented in Sections 3.6.2 and 3.6.3.

Stand Dynamics, Watersheds

Stands

Overview of Past Studies.   There are many excellent reviews of stand level studies that incorporated analyses that varied from the physiological to the ecosystem levels, too many to really discuss in this book (Harris et al. 1978, Grier et al. 1981, Waring and Pitman 1985, Fahey and Knight 1986, Gholz et al. 1986, Van Cleve et al. 1991, Vitousek and Walker 1989, Sprugel et al. 1994, Chapin et al. 1995, Lugo and Scatena 1995, and others). Many of these studies have focused on studying a few different ecosystem types in the world. As an example, readers should familiarize themselves with the research conducted on one ecosystem type (*Pinus contorta* Dougl.) in Oregon (Waring and Pitman 1985) and in Wyoming (Running 1980, Romme and Knight 1981, Fahey 1982, Knight et al. 1985, Fahey and Knight 1986).

Stand level studies typically emphasize examining an ecosystem during a particular stage of development. However, even though the dominance of the site by particular plant species does not change, the structural

TABLE 2-2   General characteristics of early and late successional plant species

Early successional
Shade intolerant
Fast growing
Short lived
Produce many small seeds dispersed by wind, birds
Need open areas or bare mineral soil to regenerate
Late successional
Shade tolerant
Slow growth, longlived
Large maximum sizes
Produce fewer, heavier seeds dispersed by gravity or animals
Germinate large seeds on organic layers
Advanced regeneration

and functional relationships are altered dramatically in the different stages (Oliver and Larson 1990) and plant successional characteristics change (Table 2-2); they must be understood as part of ecosystem management. These stages of stand development are designated as the *multiple states* (Figure 1-2) that a system may naturally experience. An ecosystem should move readily between the thresholds of the multiple states; in fact, this is a common occurrence because disturbances (to which the system is adapted) will frequently restart the system back to an earlier stage. However, once a system moves beyond the range of multiple states common for that system, it should be more difficult to recross the last threshold. It is this change in the ecosystem state that should be of great interest for managers to identify and to assess what has changed in the system that has allowed this to occur.

Several hypotheses have been published in the past on the expected changes in ecosystems with stand development and succession that were based on research in forests and abandoned fields. These hypothetical patterns of change with development can be generally grouped into several categories: plant growth (i.e., biomass, net primary production, height, and stratal development), diversity and stability of foodwebs, succession of plant and animal communities, and nutrient cycling.

Odum (1969) presented the following generalizations of how changes in plant biomass and net primary production should occur during plant succession. Odum (1969) suggested that Gross Primary Productivity (GPP)—the total amount of carbon fixed by a plant during an annual cycle—(see Section 3.6.1 for further clarification) would peak during the middle stages of succession (at the end of the closed canopy or stem exclosure phase) and then decline to lower levels ("steady state"). He further hypothesized that Net Primary Production (NPP) would also peak at the middle stages of succession but would eventually decline to zero, as respiration by plants would equal the total amount of carbon fixed by the plants (i.e., GPP). Total plant biomass would then increase asymptotically to a maximum—leveling off when "steady state" conditions were reached. Bormann and Likens (1979) agreed with Odum's hypothesis of how GPP should change with time; however, they disagreed with his predictions for how NPP would change with time. Bormann and Likens (1979) hypothesized that NPP should be negative at two successional stages—immediately after a disturbance and during the transition to the old growth stage. Their rationale for the negative NPP values was that these are the time periods in which respiration would exceed GPP because a large amount of dead organic matter would contribute to increasing the ecosystem respiration.

The hypothesized successional patterns of change in GPP and NPP suggested by Odum (1969) were upheld during the earlier but not later stages of stand development in the Pacific Northwest United States (Table 2-3). The predicted patterns suggested by both Odum (1969) and Bormann and Likens (1979) have not been found to generally occur in forest ecosystems. Comparison of data from the old-growth forests (Grier and Logan 1977) and younger stands shows that they have total productivities that are still a third to a half of the first two stages of stand development (see Table 2-3); they are not zero or negative. The hypothesized patterns with GPP are more difficult to evaluate because direct measurement of plant respiration of mature plants are sparse and it is quite recent that data have become available that have systematically documented how respiration rates vary within an individual plant growing in the field (Sprugel et al. 1994).

Odum (1969) also suggested that species richness and equitability would increase during succession. However, this prediction has proved to be much too simple. For example, California coastal sage scrub communities have the highest species richness and equitability in the first couple of years after a fire (Specht et al. 1958). Even Whittaker (1975) noted that richness and equitability were the highest during the middle stages of succession (6 to15 years) and declined as a temperate oak-pine forest matured. More recently, it has been suggested that size and frequency of disturbance may influence diversity patterns during succession. Denslow (1980), for example, noted that plant communities exposed to frequent large-scale disturbances had fluctuations in species richness that were related to the cycles of gap formation and loss as canopies closed. Species richness would increase as shade-intolerant pioneer species grew into light gaps created after a disturbance, but species richness declined as the canopies closed.

Odum (1969) also hypothesized how susceptible ecosystems would be to losing nutrients at different developmental stages. He felt that ecosystems would tend to be open and "leaky" to nutrients during the early stages of succession because plant biomass is low and so they are unable to function in retaining nutrients. These ideas were further used to sup-

TABLE 2-3   Species diversity and ecosystem attributes of Douglas-fir dominated forests in Oregon and Washington grouped by stand developmental stages (according to Oliver and Larson 1990), with approximate stand ages in parentheses (Vogt 1987, Schoonmaker and McKee 1988, Schowalter 1989, Grier et al. 1990, Spies 1991, Franklin and Spies 1991, Vogt 1991, Hansen et al. 1991, O'Dell et al. 1992).

Attribute	Stand initiation stage (< 40 years)	Stem exclusion stage (40–80 years)	Understory reinitiation stage (80–250 years)	Old-growth stage (200–1000 years)
Total living biomass (mg/ha)	Lowest	Low-intermediate	Highest	High
	206–231	200–446	904–1103	871
Total organic matter accumulation (mg/ha)[a]	Intermediate	Intermediate	Intermediate	Highest
	134,890	80–280	80–140,831	379
Coarse woody debris mass (mg/ha)	High	Lower	Lower	Highest
Net primary Production (Mg/ha/yr)[b]	Highest	Intermediate	Intermediate–high	Lowest
	Above = 32.2	Above = 5.1–15.8	Above = 9.6–22.8	Above = 8
	Total = 34.9	Total = 11.5–18.8	Total = 10.9–25.4	Total = 10.9
Plant diversity (No. species)[c]	Highest	Intermediate	Intermediate	High
	(40–60)	(40–47)	(43–52)	(37–56)
Mammal diversity (No. species)	Highest	Lowest	Intermediate	Intermediate–high
	(45)	(15)	(35)	(40)
Invertebrate diversity (No. species of arthropods)	Low	?	?	High
	(16)			(62)
Mycorrhizal diversity (No. species)	Low	Intermediate	?	Highest
	(1)	(7)[7][d]	[3][d]	(12)[8][d]
Structural diversity	High	Lowest	Low	High
Retention of nutrients from leaching	High	High	High	High

[a] Total dead biomass includes surface organic matter accumulation, soil organic matter, coarse woody debris.
[b] Above = above-ground productivity only; total = above- and below-ground productivity.
[c] Includes trees, shrub and herbaceous species.
[d] Hypogeous sporocarp producting ectomycorrhizal species only.

port the idea that mature stages would be strongly retentive of nutrients because the functional role of plants in retaining nutrients would be quite high (Odum 1969). The research conducted at the Hubbard Brook watershed (see Section 2.1.1) supported these hypotheses for the earlier stages because clear-cutting and and removal of vegetation from a site resulted in high nutrient losses from the watershed (Bormann and Likens 1979). However, these relationships are not as simple as originally hypothesized because other forests were also clear-cut without the resultant losses of nutrients (Sollins et al. 1980). Vitousek and Reiners (1975) disagreed with Odum's (1969) hypothesized pattern of low nutrient losses in the older stages of stand development or succession. They felt that the middle stages were the most retentive of nutrients and that the mature stages would be most leaky (Vitousek and Reiners 1975, Gorham et al. 1979). All these hypotheses have not been upheld as generalities for all ecosystems because they only considered the role of living plants in regulating nutrient cycles and totally ignored the fact that detritus may have a large role in determining how leaky an ecosystem is. The low loss of nutrients in the Pacific Northwest United States is probably related to the role of detritus in conserving nutrients in ecosystems during the early and old-growth stages (see Table 2-3).

Current Overview—PNW Coniferous Forests Case Study.   This dynamic nature of stand development also suggests the difficulty of identifying one particular stage and trying to manage to maintain a stand in that stage, an objective suggested for old-growth forests in the Pacific Northwest United States and for many restoration projects. It also makes it difficult to implement restoration activities to mimic particular stand stages because one may have to restore those functional and structural links and relationships that may have been lost and that typically evolve with time. Restoration activities also are being designed so that once stand structural and functional characteristics have been re-established, that stage is arrested in that one state—again, difficult to accomplish because of the dynamic nature of systems. Furthermore, it is important to understand that not only will the internal characteristics of a system state change, but how that system responds to or changes the expression of a disturbance will be quite different. Even the maintenance of a desired system state at the stand level will vary based on how the stand level characteristics mitigate disturbances in a landscape context.

As a way of demonstrating how species diversity and other ecosystem characteristics change with the developmental stage, data have been synthesized by stand developmental stages for Douglas fir (i.e., *Pseudotsuga menziesii*) dominated forests in the Pacific Northwest United States (see Table 2-3). This ecosystem type is an excellent place to explore how ecosystem attributes change with developmental states because data are available for most of these stages. It is important to realize that some of the data presented in Table 2-3 were obtained from only a few different studies so the patterns should be accepted as conditional and potentially changing as more data are collected. Despite the low number of studies, the changes in species diversity and in ecosystem characteristics clearly show that these relationships are not linear with time, and each attribute has a different pattern of change linked to the characteristics of each stage.

As should be expected, total living biomass and total organic matter accumulation are the highest at the later stages of development for Douglas fir forests (see Table 2-3). The changes in the amount of total living biomass increased significantly from the *stand initiation*, to the *stem exclusion*, to the *understory reinitiation* stages but had leveled off by the *old-growth* stage. The changes in organic matter accumulation did not follow any particular pattern, except for the apparently higher carbon sequestering capacity of the old-growth forests. Key attributes of old-growth forests have been identified to be "large live trees, large snags, and large logs" (Franklin and Spies 1991). Today, there has been a carryover of coarse, woody debris into the early stages of stand development, which is highlighted by the high amount of coarse wood at this stage, with the amount varying dramatically with the moisture status of the site (Spies and Franklin 1991). Despite this high presence of coarse wood in young stands, Spies and Franklin (1991) and Hansen et al. (1991) were unable to statistically separate the different developmental stages using coarse, woody debris attributes.

Ecosystem managers need to be aware of structures or functions that may have a greater significance at particular stages of development. This becomes especially important for coarse, woody debris, which becomes a legacy that is carried over into the young developmental stages and can serve as a function for retaining nutrients in ecosystems because they decompose so slowly (Harmon et al. 1986, Franklin and Spies 1991). The high ability of all developmental stages to retain nutrients from being leached from the ecosystems (see Table 2-3) may be due to these legacies that continue to perform nutrient conservation functions at developmental stages other than the stage at which they were added to the ecosystem. If this is the case, the role of coarse woody debris in helping to maintain ecosystem nutrient cycles has to be considered as part of forest management. The loss of these legacies could result in the system becoming more leaky to nutrients than had occurred in the past.

In Douglas fir forests, many of the changes in ecosystem functions are keyed into when the stand achieves optimal foliage biomass, just prior to the stem exclusion stage. Once canopy closure is reached, the photosynthetic surface area becomes relatively constant, resulting in a leveling off of the amount of carbon being fixed by plants at later stages. This explains why forests with closed canopies have to be thinned to obtain the full effect of increased growth with the application of fertilizers (Waring and Schlesinger 1985). This canopy closure period coincides with the stage at which optimal root biomasses are also recorded in these ecosystems (Vogt 1987). These changes in the ecosystem are reflected in the total NPP values recorded with stand development—the highest total productivities measured during the end of the first stage, followed by a decrease in the stem exclusion stage when trees compete with one another for growing space, an increase in the understory reinitiation stage as the trees respond to gap openings, and finally a decrease in productivity to the lowest levels in the old-growth stage. Part of the increase in total productivity during the understory reinitiation stage reflects the increased dominance of western hemlock in Douglas fir stands. When hemlock contributes more to the total stand biomass, NPP values in ecosystems are higher because hemlock is shade tolerant and is able to retain more foliage per given land area than Douglas fir. This results in higher NPP values being achieved after canopy closure in some forests (Vogt 1987).

These changes in NPP with stand development are an indication of the ability of a site to respond to disturbances (see Chapter 4). The approach presented in Chapter 4 for indexing site susceptibility to disturbances and their recovery after a disturbance will probably be more effective at detecting the impact of a disturbance when stands have reached the stem exclusion to old-growth stages. Prior to canopy closure, the trees growing immediately on a site after a disturbance are too adapted to disturbances occurring at short temporal scales so that it may be very difficult to detect their response to additional disturbances. For example, the early stages of stand development are typically dominated by plants species that grow quite rapidly (Oliver and Larson 1990; as seen in the high NPP values) and may not reflect changes that may be occurring at the ecosystem level (see Table 2-2).

Changes in the number of plant and animal species with stand development are difficult to generalize because these patterns are very much driven by which plant and animals species are being discussed and the availability of habitat. In hemlock/Douglas fir forests in Oregon, the highest diversity of plant species occurred at the earliest stage of development (15 to 20 years of age) and reflected the invasion of the stand by many annuals, herbaceous, and shrub plant species after a disturbance (Schoonmaker and McKee 1988). However, the number of tree species (six) did not change with stand development; most of the changes in number of species occurred with shrubs and herbaceous species (Spies 1991). "Thus the early successional herb/shrub seral stage has a structure and composition that differs substantially from all older forest stages, a fact that should be considered in forest management" (Hansen et al. 1991).

These changes in plant species diversity somewhat track the changes recorded for the number of mammal species and also the changes in ecosystem structural diversity recorded with stand development (Franklin and Spies 1991). The highest mammal diversity was observed during the stand initiation stage, followed by the lowest numbers in the closed-canopy stage, and increased again when the canopy opened and the amount of structural habitat increased in the old-growth stage. There are certain animal communities that are mainly found in old-growth forests so that the loss of old-growth forests have direct impacts on their survival (Hansen et al. 1991). For example, bird species that overwinter in the Cascade Mountains in Washington and several herpetofauna are mainly found in the old-growth forests (Hansen et al. 1991). Even though there is thought to be a general relationship between structure and animal community composition, the coniferous forests in the Pacific Northwest only have a few significant differences between these different developmental stages (Hansen et al. 1991). Hansen et al. (1991) suggested that the lack of strong differences in animal community composition with developmental stage may reflect the fact that natural stages of forest development may contain the diversity of habitats and structures needed by the animals. In this case, management should then be concerned with maintaining the degree of spatial heterogeneity of habitats in different developmental stages so that animal community composition can be sustained in all stages. These types of analyses can be obtained using several of the tools presented in Chapter 4 for documenting the variation in spatial scales.

The changes in mycorrhizal diversity reported for one study in Oregon are difficult to generalize at this point (see Table 2-3). That study suggested the highest diversity of mycorrhizal species based on total sporocarp counts occurred in the old-growth stage and the lowest during the stand initiation stage. Because the mycorrhizal diversity recorded in the earliest stage will vary depending on the type of disturbance that resets the developmental stage, the dominant tree species, and the abiotic characteristics of the site (Vogt et al. 1992), one cannot generalize that early stages of stand development will have a low diversity of mycorrhizal fungi. Another study, which only assessed the mycorrhizal community that produced hypogeous sporocarps, suggested that there were no differences in the number of species between the stem exclusion and old-growth stages. Even though one cannot presently state how the diversity of mycorrhizal fungi changes with stand development, most studies conducted in other ecosystem types suggest that mycorrhizal species diversity appeared to be the highest just prior to canopy closure. How mycorrhizal diversity changes after canopy closure is still unclear because results have shown both a decrease (Dighton and Mason 1985) and an increase in diversity (O'Dell et al. 1992).

The number of mycorrhizal species present on a site may not be as relevant as which species are present. Several studies have documented that not only does plant succession occur but even mycorrhizal fungi existing on the root systems of individual trees change with time (i.e., fungal succession; Last et al. 1984). Dighton and Mason (1985) showed results in which the succession of mycorrhizal fungi varied from non–host-specific symbionts prior to canopy closure to host-specific mycorrhizal fungi at later stages of development. These were suggested to track the decreased nutrient availability that occurred as litter layers accumulated and were of poorer nutrient quality (Frankland 1992). As Frankland (1992) wrote, these changes in the type of mycorrhizal associations have great relevance to forest management because older stands may be the reservoirs from which the inoculum becomes available to younger stands. If the dominant mycorrhizal inoculum for regenerating stands is only available from mature stands, this inoculum may not be as effective because these fungi appear to be host specific (see later section on symbionts).

These studies also highlight the fact that there is not a strong relationship between the diversity of species and the productivity of a site (see Table 2-3). At some level, animal species diversity is related to the structural heterogeneity of the site and plant species diversity also tracks the changing heterogeneity of the ecosystem. If one only examines the diversity of the tree species, there is no change in species diversity across a long time scale of hundreds of years. Plant diversity changes are mainly occurring at the shrub and herbaceous layer levels. These contrast the diversity of mycorrhizal species, which appear to more closely track the productivity of a site. This is not unexpected considering that mycorrhizas are dependent on tree photosynthate for their growth, therefore, whatever affects the ability of plants to photosynthesize will directly feedback to affect the mycorrhizal fungi. None of the attributes presented for these Douglas fir ecosystems appear to have any relationship to how effectively the ecosystem retains nutrients against leaching losses. This would make it very difficult to utilize these variables to predict what controls nutrient leaching or conservation in these ecosystems. Other variables would have to be pursued if this subject is of interest. Substrate heterogeneity appears to be quite useful to document in these ecosystems to track some of the changes in animal diversity. It is critical to identify the type and the quality of substrate to obtain a mechanistic understanding of these relationships.

These changes in species and structural and functional characteristics with time highlight the importance of knowing the current stage of stand development for a system. These different stages are the natural changes that an ecosystem readily shifts between and define the multiple dynamic states for that system (see Figure 1-2). The changes in the ecosystem characteristics of each of the multiple states can be quite dramatic as highlighted by the Pacific Northwest data (see Table 2-3). The state in which a system is functioning may have very strong implications for how that system should be managed. It is important to realize the *dynamic nature of an ecosystem* and to acknowledge the fact that it is difficult to arrest the system in one of the multiple states. It is also important to recognize that when a system shifts among these multiple states, it does not mean that an ecosystem is degrading. Only when a system moves beyond the range of these normal multiple states do managers need to worry about the system potentially degrading.

Watersheds

Early studies (1950s) in watersheds mainly focused on particular issues, such as (1) how the utilization of forest lands is affected by the magnitude of water outflow out of different watershed and (2) how water quality was affected, especially by sedimentation (Hornbeck and Swank 1992). The research conducted in the 1960s was expanded to begin addressing questions related to:

1. Understanding the role of the vegetation and the ecosystem on the cycling of nutrients;

2. The importance of this approach to identifying how clear-cutting in forests was impacting the sustainability of ecosystem functions (this was very relevant during this time because clear-cutting issues were at the forefront of public attention); and

3. The utility of watershed studies to examine atmospheric deposition issues (Hornbeck and Swank 1992).

One of the first watershed ecosystem studies was initially established in 1934 by the Forest Service as a forest hydrology laboratory at the Coweeta Experimental Forest, North Carolina, United States (see Appendix A). These watershed studies were specifically established to address two problems considered important at that time for forests in the southern Appalachian mountains: "(1) how is streamflow regulated by vegetative cover, and (2) how can erosion be controlled by vegetative cover?" (Swift et al. 1988). Many early experiments were established at the watershed level that examined the influence of different vegetative covers on the amount of water outflow from each watershed—white pine compared with hardwoods, or a mixture of yellow poplar and white pine mix, or unregulated agriculture (Swank and Crossley Jr. 1988). In 1959, experiments were established to determine if water outflow into streams could be increased by converting forested areas into grassland. The surprising result was that the water outflow did not initially vary between the watershed dominated by forests and grasses. This study also began to show the importance of fertilization on controlling plant growth and the link between fertilization and the maintenance of grass productivity for determining how much water outflow occurred into the stream (Swank and Crossley Jr. 1988). Other experiments tested different forest management practices (i.e., clear-cuts, selective logging, herbicide treatments, burning) on streamflow and the impact of insect infestations on regulating nutrient cycles (Swank and Crossley Jr. 1988). Research at Coweeta used the paired watershed approach, in which one watershed is manipulated and another is used as a control. Currently there are 16 gauged watersheds in Coweeta. Coweeta became a Long-Term Ecological Research site with funding from the National Science Foundation in 1980. Joint research efforts were established among Forest Service scientists and many university scientists (Franklin et al. 1990). Long-term climatic data collection has been continuously collected since 1934 in The Coweeta Experimental Forest (Swift et al. 1988). These long-term data sets are extremely valuable to determine the fluctuation of climatic variables over time and if there are general temporal patterns that occur over longer time scales in controlling functions in these ecosystems. These watershed level studies were also combined with stand level studies to examine the mechanisms of change at smaller scales.

Watershed level studies at Coweeta were instrumental in showing that in the southern Appalachian Mountains many of the different forest management activities had minimal impact on changing nutrient outflow into the streams and little effect on water quality (Swank 1988). These studies showed the long-term impact of forest management activities (especially clear-cutting) on nutrient cycles in these watershed. Even though the levels of nitrates in stream water were quite low, they did persist for at least 20 years after the cut (Swank 1988). The ability to study watersheds dominated by either hardwood or conifer ecosystems was also important in showing the effect of plant species on the water budgets and nutrient cycles in these forests, and also the role of biological processes in nutrient conservation and leaching losses from these forests. The role of insects in nutrient cycles in these watersheds was also demonstrated when nutrient losses into the streams increased during insect infestations.

Watershed studies were also established at the Hubbard Brook Experiment Forest in New Hampshire, United States by the U.S. Forest Service in the early 1960s (see Appendix A). These studies were specifically established to examine the impact of disturbance on watershed hydrologic cycles. As part of this, watersheds were clear-cut or strip-cut, followed by herbicide applications to minimize all plant growth (Bormann and Likens 1979). In addition, no logging roads were constructed into the watershed and all downed trees were left on the site. This research was established to examine the effect of vegetation on nutrient retention and the extent to which hydrologic input–output balances in undisturbed watershed are controlled by plant growth. However, this research occurred at the same time as the clear-cutting controversy became a major public issue in the national forests. The experimental results from Hubbard Brook were used to demonstrate the negative impacts of clear-cutting because these experimental manipulations resulted in significantly higher amounts of nitrate being detected in the streams, at levels higher than are allowable in drinking water. Even though these experiments were not specifically designed to examine the clear-cutting issues, they were used to support the idea that all clear-cutting was bad and should not be allowed in forests. This was an inappropriate use of data to examine forest management issues because the experiments were not designed for that purpose. For example, under most clear-cutting situations, there is rapid regrowth of vegetation, which can be relatively effective in preventing the leaching losses of nutrients from the terrestrial area.

Despite the inappropriateness of using the Hubbard Brook study to examine clear-cutting issues, concerns were immediately raised as to how applicable the results obtained at Hubbard Brook were to other forest types around the country. In contrast to the Coweeta study, the Hubbard Brook study showed a higher level of nutrient losses to streams in response to removal of the vegetation. The losses of nitrate into stream water for several years after treatment showed the potential long-term losses of limiting nutrients in these ecosystems. The nitrate losses in stream water exceeded the U.S. Public Health Service standards for drinking water for several years after the treatment was established (Bormann and Likens 1979). However, the clear-cutting conducted at the Coweeta Experimental forest did not mimic the patterns observed at Hubbard Brook, suggesting other factors needed to be pursued to identify why leaching losses occurred in some watersheds but not others. In response to the public attention given to the studies conducted at the Hubbard Brook, many clear-cutting experiments were carried out in several places in the country and showed that high leaching losses of nutrients was not a universal phenomenon after clear-cutting. Negligible nutrient leaching losses were recorded in coniferous forests in the Pacific Northwest United States and in hardwood forests in Virginia and Michigan (Brown et al. 1973, Aubertin and Patric 1974, Richardson and Lund 1975, Sollins et al. 1980, Freedman 1981). The general conclusion that was drawn from all these watershed studies was that forests growing on more nutrient-rich or higher quality sites (e.g., red alder sites with nitrogen-fixing nodules) not only had more leaching losses of nitrate into the streams following a disturbance, but also had generally higher cycling rates of nitrogen within the ecosystem.

The Coweeta and Hubbard Brook studies were instrumental in identifying that there was a regularity in the hydrologic and nutrient input–output budgets that is predictable and therefore can be used to study ecosystems (Bormann and Likens 1979, Swank 1988). The role of vegetation in controlling water and nutrient fluxes was also clearly identified using the watershed approach. Watershed studies are very useful for examining what controls leaching losses from larger spatial scales, such as drainage basins. Though with this approach it is difficult to examine the impact of smaller scale changes within the watershed mosaic or manipulate small parts of the watershed and detect the impact of that manipulation disturbance. The watershed studies have great value for detecting changes in the atmospheric inputs of nutrients or pollutants over longer time scales because these studies include long-term (50 years) continuous collection of water and nutrient data. The recent publication by Hedin et al. (1994) demonstrates the great utility of detecting measurable changes in base cation inputs into parts of Europe and North America using watersheds.

2.1.2 Species as Part of Ecological Systems

Public awareness and understanding of the connection between land-use activities and overhunting on decreasing game animal populations has existed since the late 1600s in North America (see Appendix A). This emphasis on animal species contrasted the lack of knowledge of human impacts on ecosystems and on particular components of ecosystems (i.e., microbial symbionts; see Section 2.1.2), which did not develop until after the mid-1900s (see Appendix A). For example, protection of game animals was already being implemented in the late 1600s (i.e., in 1694 Massachusetts established closed seasons for deer hunting; in 1776 the first Federal Game Law was passed which required closed seasons to control overhunting of deer). By the 1800s it had become clear that wildlife resources were not infinitely abundant in the United States and that human activities were directly causing the extinction of species. A classic example of the extinction of a game species was the passenger pigeon, which was once the most common bird species in North America. This species was commercially hunted because it was tasty and easy to hunt (the last pigeon died in 1914). Back in 1844, the New York Association for the Protection of Game was created, and this highlights how game protection was already an organized effort. In 1871 the U.S. Fish Commission was established in response to a national concern about the relationship between agriculture and decreased wildlife populations. This was first time that the U.S. federal government became involved in protecting fish and wildlife. This was followed by the formation of the American Ornithologists Union in 1883. They estimated in 1886 that 5 million birds were being killed annually for the purpose of supplying feathers for women's hats. There are many other examples of public, business leaders, and government involvement to protect wildlife and their habitats during the late 1800s and early 1900s. Again, much of this was geared to protecting game animals (see Appendix A).

In 1956 the U.S. Fish and Wildlife Service was formed with more expanded responsibilities including: (1) managing wildlife refuges; (2) protecting migratory birds, endangered species, and marine mammals; (3) conducting research to prevent animal diseases; (4) enforcing federal fish and wildlife laws; (5) advising other federal agencies on ways to avoid harming fish and wildlife while carrying out responsibilities; and (6) assisting state and foreign governments. This was followed by the Endangered Species Preservation Act of 1966, which authorized the Secretary of the Interior to acquire land and to fund studies to protect endangered species. In 1969, the Endangered Species Conservation Act authorized the Secretary of the Interior to generate an endangered species list and to ban imports of endangered species. This was eventually followed by the Endangered Species Act of 1973, which banned the taking of species on the endangered list and allowed the preservation of critical habitats.

The historical synthesis of species as part of ecological systems shows that preferred game species were the major emphasis when maintaining wildlife populations. Much of the management of animal species occurred under a scenario in which management was responding to the loss or decrease in game animal populations. This management was not driven by any ecological understanding of these species within the ecosystem. The concepts of how species interacted with their environment only began to develop after the mid-1900s (discussed later in this section), while the relationships between species and ecosystems is very recent (see later). Our understanding of the ecosystem roles of microbes and other less obvious or charismatic species and their diversity is even less developed. A brief discussion of symbiotic microbes with trees and their role in ecosystems follows.

Microbial Symbionts

The role of symbiotic microorganisms on plants roots in controlling nutrient uptake, and therefore the amount of carbon fixed by a plant, is a well-established concept (Harley 1969, Marks and Kozlowski 1973, Silvester 1977, Harley and Smith 1983, Dawson 1983, Allen 1991). Despite this, "Mycorrhizal associations represent an enigma to most ecologists. For decades, theoretical ecologists have treated mutualisms, including mycorrhizae, primarily as interesting oddities. Moreover, one of the partners is a fungus, microorganism, and therefore too small to be seen and bothered with (Allen 1991)." In the 1970s most ecologists considered symbiotic associations not relevant to ecology because the perception existed that few examples of their importance existed in nature and it was wrongly assumed that symbionts were restricted to a few natural ecosystems (Allen 1991). Today it is known that it is the exception to find plants that do not have symbiotic associations on their roots, and it has been hypothesized that plant establishment on land required these symbiotic microorganisms (Pirozynski and Malloch 1975).

The symbiotic association between a fungus and plant was first described back in 1840 by Hartig, but it was not until 1885 that a study by Frank (1885) coined the term *mycorrhizas* and identified this to be a mutualistic relationship (see Appendix A). Since first being identified, a significant number of studies were conducted in an attempt to understand the physiological relationships between a fungus and plant, and to understand what benefits each symbiotic partner accrued from this relationship (Harley 1969). The development of research on mycorrhizas was limited by the inability to originally cultivate the fungus in isolation from the plant. Once techniques were developed for culturing some of the symbionts, the most common approach to their study was to grow the fungus in isolation from the plant in petri plates and to use seedlings inoculated with the fungus grown under very controlled conditions. This has limited our understanding of the ecological roles of symbionts because few studies were conducted under natural conditions.

The symbiotic association between plants and actinomycetes or bacteria (called *nitrogen-fixing plants*) was first recognized as early as 1695 (Binkley 1992). It was not until 1895 and 1904 that it was identified that elementary atmospheric nitrogen (N) was being fixed by nodules of *Alnus glutinosa* grown in N-free solution (Hiltner 1896, Nobbe and Hiltner 1904). Nitrogen-fixing bacteria can exist as free-living forms (asymbiotic) in the soils, or they can form symbiotic associations with plants. Nitrogen-fixing plants are less common in established forests and are generally found in disturbed forests or in those forests where soil N reserves are low. Land managers have used this information for a long time by planting N-fixing trees or leguminous agricultural crops on sites with low N levels and where the availability of fertilizers is a concern (Tarrant and Trappe 1971, Burns and Hardy 1975, Dobereiner and Campello 1977, Franco 1978, Gordon and Dawson 1979, Binkley 1992). Nitrogen-fixing plants have also been identified to be important in early successional stages of unmanaged forests and play an important role in ameliorating site conditions after a disturbance.

There are three main types of symbionts that are most relevant to forest ecosystems: the ectomycorrhizas, the vesicular-arbuscular mycorrhizal associations (Harley and Smith 1983), and the N-fixers (Binkley 1992). Detailed descriptions of these relationships and the attributes of each are available in Harley and Smith (1983), Allen (1991), and Binkley (1992). A brief summary follows to illustrate the different attributes of these symbionts.

Ectomycorrhizas are mostly restricted to woody plants and are an important component of temperate and boreal forests. The diversity of mycorrhizal species forming the ectomycorrhizal associations is quite high—over 2000 species of fungi are capable of forming a symbiotic association with one tree species (*Pseudotsuga menziesii*) (Trappe 1962).

Vesicular-arbuscular (VA) mycorrhizas have a very broad geographic distribution and are the most ubiquitous association in the plant kingdom. They are found in almost all the herbaceous families, ferns, and many woody plants, including tropical and temperate hardwood forests. However, the diversity of VA fungal species forming the relationship with a plant is low, while the plants species diversity is high. There does not appear to be a host preference for these fungi.

Nitrogen-fixers symbionts are found on tropical and temperate rainforest species, in temperate and boreal forests, subalpine pioneer species in the temperate and tropical zones, arctic-alpine pioneer species, temperate wetland trees, xerophyte shrubs, many diverse herbaceous species, coastal shrubs, grasslands, chaparral, and subalpine and desert shrubs. Similar to the VA mycorrhizas, the species diversity is not high for the N-fixing symbionts.

Several benefits to plants from maintaining symbiotic associations with fungi and bacteria have been suggested by different researchers:

1. *Increased effective root surface*—increased absorption of nutrients (particularly P) and water.

2. *Longer rootlet function.*

3. *Increased heat and drought tolerance.*

4. *Increased availability of soil nutrients,* making plants more competitive on the sites because they grow faster. This is important in seasonal climates where nutrient availability is cyclic. Symbiotic relationships (both mycorrhizal and N-fixing) are maintained by low nutrient conditions and are eliminated from the roots if the nutrient status of the soil is significantly increased (Bowen 1984). Nitrogen fertilization depresses the fixation of N by both symbiotic and free-living bacteria (Tamm 1991).

5. *Increased availability of atmospheric nitrogen.* Legumes can fix from 22 to 178 kg N^2 ha/yr, actinorrhizas (e.g., *Casuarina*) may fix up to 218 kg N^2 ha/yr on poor soils. Asymbiotic N fixation has been documented to be relatively low (ranging from 0 to 3 kg/ha/yr; Norstedt 1982).

6. *Deterrence of Infection by disease organisms*

7. *Increased plant tolerance to heavy metals* (Vogt et al. 1987, Wilkins 1991) due to being colonized by mycorrhizas.

8. *Providing a food source for small mammals and humans.* Humans collect mushrooms as a nontimber forest product that is comprised mainly of the reproductive structure of mycorrhizal fungi (Vogt et al. 1992, Molina et al. 1993). The digestive tract composition of small mammals in coniferous forests in Oregon (Maser et al. 1978) is very high in mushrooms, which are the reproductive structures of ectomycorrhizal fungi.

9. *Ecosystem stability.* See article by Perry et al. (1989, 1990).

Even though we can list nine separate categories in which symbionts contribute to ecosystem processes, few field studies have examined mycorrhizas in the context of an intact forest ecosystem. For example, there is good laboratory evidence that mycorrhizas obtain their carbon for growth from plant photosynthate and that they increase the uptake of soil nutrients. However, except for a study by Fogel and Hunt (1979, 1983) and Vogt et al. (1982), others have not examined mycorrhizas as an integral part of the forest ecosystem. In the Vogt et al. (1982) study, it was estimated that 13.9% and 15% of total net primary production was allocated to just the fungal symbiont in 23- and a 180-year-old *Abies amabilis* stands in Washington, respectively. Both studies also examined the contribution and role of the turnover of mycorrhizal structures on the nutrient cycling occurring in a forest ecosystem, which was more than twice as high as their contribution to carbon cycling. There is a large body of literature examining mycorrhizas in forested ecosystems, but most of this has examined mycorrhizas under very controlled conditions in order to determine their response to fertilizers and pesticides (Vogt et al. 1996a).

The importance of mycorrhizas for plant growth was identified as part of forest management in the 1930s to 1960s when many exotic tree plantations failed because of the lack of the appropriate mycorrhizal fungus in the soil where the trees were being planted (Mikola 1973). This resulted in a large research effort to try to identify the appropriate fungal species to inoculate seedlings as part of reforestation efforts. This research produced interesting results because it clearly showed the high diversity of mycorrhizal species capable of forming a symbiotic association with individual plants (Trappe 1962). It was not until the 1980s that field trials with mycorrhizal fungi were initiated. However, many managers were discouraged from expending financial resources to produce mycorrhizal plants for outplanting because it was shown that most plants had the inoculated fungus replaced by other species existing in the soil (Richter and Bruhn 1993). It is now accepted that the presence of the mycorrhizas on outplanted seedlings does act as a buffer for the seedlings during the short time in which they exist on the roots before being replaced by other species in the field. Other types of land use, that reduce the survivability of mycorrhizal inoculum in the soil and where translocation of inoculum to the site is not occurring, have been shown to cause regeneration failures of seedlings (Amaranthus and Perry 1987, 1988, Perry et al. 1989).

There have been conflicting results on the effectiveness of different mycorrhizal species on plants. One fungal symbiont can cause growth reductions and growth increases on different plants (Mikola 1973). This highlights the importance of a diversity of fungal species existing in natural ecosystems because the effectiveness and presence of fungal species varies by tree species and abiotic site conditions (Harley 1969, Harley and Smith 1983, Allen 1991). Because the soil environment is highly heterogeneous, one can hypothesize that a diversity of fungal species is needed for plants to access abiotic resources from the soil and to be able to respond to a changing environment that may modify the resource availability in the soil (i.e., acid rain). Even in a subalpine environment dominated by one tree species, half of the root tips were occupied by one mycorrhizal fungus, but on average seven different species were found on the root systems (Vogt et al. 1981). In these systems, the diversity in the ecosystem is not at the plant species level but at the level of the mycorrhizal fungal species. The kind of species diversity that exists for mycorrhizal fungi cannot be stated with much confidence today because only a minuscule proportion of the vascular plants species have been examined for their mycorrhizal status. Only recently have techniques been developed that enable the study and identification of mycorrhizal fungi to the species level in the field using molecular techniques (Gardes et al. 1991) that do not base species counts on physical and color characteristics of fungal mantles.

It also has become recently clear that human land use changes the predominance of mycorrhizal fungi on plant roots and may result in their elimination from root systems. It has been suggested that pollution in Europe is causing the decline in fruiting bodies produced by mycorrhizal fungi (Arnolds 1991). That pollution caused decreases in the species of mycorrhizal fungi present on the root systems of trees has been well documented in Poland (Kowalski 1987). At one level, having mycorrhizal fungi on plant roots is important to allow plants to maintain their growth and to be resistant to the impacts of pollution. However, after some threshold level or duration of pollution additions to a forest, the mycorrhizal associations appear to lose their ability to mitigate the impacts of pollution and are eliminated from the root systems (Persson 1990).

These shifting stages of vulnerability to a disturbance or pollutant were also hypothesized for soils impacted by acid rain (Ulrich 1987, 1989). This type of response is seen with the addition of N as part of atmospheric deposition where initially N functions as a fertilizer, increasing plant growth rates, but then shifts to produce a negative effect when the ecosystem becomes N saturated. There is a large literature base on fertilizer studies that suggests pollution inputs of N will reduce fine root biomass and eliminate the mycorrhizal association from root tips (Persson 1990, Vogt et al. 1990, 1996bc, Gower et al. 1992). Because pollution ultimately decreases the presence of mycorrhizas in forests, monitoring roots for early detection of the system response may be the best early indicator that some threshold is being reached in the ability of the ecosystem to continue to respond to that disturbance (Vogt et al. 1993).

In 1991, Allen identified that one of the problems in dealing with mycorrhizas in an ecological framework "resides in our inability to perceive the scale at which the two symbionts interact with each other and their environment." This arises because it appears that the environment of the fungal associate may be a few millimeters to centimeters in size, while the associated tree may occupy a space that may be 30 or more meters in area. This has propagated the idea that something that small really cannot be controlling how the plants are responding to disturbances and mitigating the negative impacts of some disturbances. However, the size of an ecosystem component is a poor indicator of its role or function in the ecosystem. For example, Rygiewicz and Andersen (1994) showed that even though only 5% of the total dry weight of seedlings was in fungal tissues, the presence of mycorrhizas on the root systems increased the allocation of carbon to roots by 23%. Allen (1991) proposed the utilization of a hierarchical system developed by MacMahon et al. (1978) to examine mycorrhizas. This system would move away from the problem discussed earlier of changing information needs when crossing scales and the difficulty of transferring information across scales as one moves from the populations to the landscape level. In this system, each hierarchical level (i.e., population, community, ecosystem, landscape) would elicit a different set of interactions with the fungal associate (Allen 1991), and therefore a different set of attributes of the fungal associate should be examined at each level. In fact, what this approach emphasizes is the importance of identifying the scale at which the problem needs to be addressed and the variables that are relevant to that scale. This statement will be further expanded upon in Sections 3.2 and 4.2.2, where the importance of defining the appropriate scale while not worrying about scaling information from small to larger spatial scales is discussed. Hierarchical theory suggests that one should attempt to develop tools to scale information among different hierarchical levels (O'Neill et al. 1986), but this is not the approach to take. It is interesting that few concrete examples of studies using the hierarchical approach have been published.

When community ecologists began studying plant succession in the early 1900s, succession or change in species composition of mycorrhizal fungi on plants was also being examined and recorded (Stahl 1900). A well-designed study conducted in the field by Janos (1980) showed the importance of the presence or absence of mycorrhizal fungi on plants roots in controlling which plant species dominated at what successional stages in these wet tropical forests. Other studies have shown that there is a succession of mycorrhizal fungi on the same plant species with age (Lapeyrie and Chilvers 1985). The significance of succession of mycorrhizal fungi on plant roots can be hypothesized as a response to the changing abiotic environment in which the plant is growing. Our understanding is hampered by a lack of knowledge of the functional role of many of these mycorrhizal fungi so that we do not understand the rationale or the factors causing these species shifts and how they ultimately affect plant growth. Because different fungi have very different abilities to acquire nutrients, understanding when species shifts are occurring and which specific species become dominant could be valuable in understanding when the effectiveness of the symbiont might be changing. There is already some evidence that pollution has changed the type of relationship that existed between plants and fungal symbionts where the symbiont has begun to function more like a pathogen (Kowalski 1987).

Even though many of the studies with symbionts have emphasized utilizing seedlings to test the role of symbionts in affecting plant growth, these studies have strongly suggested that plant associations with microbial symbionts: (1) increase the ability of plants to access nutrients (i.e., N, P) that typically limit their growth, (2) protect plant roots against fungal root pathogens, and (3) diminish the toxicity effects of trace elements on plant growth by either complexing the element or reducing their uptake by plant roots because of the fungal sheath that surrounds roots (Vogt et al. 1991). These results strongly suggest the importance of incorporating them as part of the analysis framework for ecosystem management and suggest a role for them in helping to maintain healthy tree growth. The many documented cases of plants being unable to grow under highly altered environments (i.e., mine spoils), or when plants have been introduced into areas where they normally do not grow and the symbiotic association was not present (Allen 1991), also strongly suggest their importance in modulating plant growth. More studies need to be conducted on mature plants that determine how the complex environment of an ecosystem modifies or modulates the expression of the symbiont in the field. Our understanding is not at the stage where a prediction can be made as to how plant growth will be affected if the intensity of mycorrhizal development, species composition, and/or diversity decreases at the root level.

Plants and Animals

Species and Spatial Scales

In contrast to plant studies, much of the early history related to understanding the abundance of animals in a given habitat explicitly examined the importance of spatial scales on controlling animal abundance, with little attention given to temporal scales (see Appendix A). Zoologists have noted since the late 1950s that there appears to be a relationship between the abundance and distribution of species and the size of the spatial scale (Smith 1990). The reader is referred to Smith (1990) for a good, more detailed discussion of the development of the ideas related to species and spatial scale relationships. A brief summary of the development of the ideas on species–spatial scale relationships follows.

Preston (1962) first formalized the species and area relationship ideas into a mathematical equation when he suggested that there is a linear relationship between the size of an island and the number of bird species breeding on each island. This concept was further developed by MacArthur and Wilson in 1963, when they published the island biogeography theory. That theory stated that there is a dynamic equilibrium in the number of species existing on an island that is related to the rate at which new species migrate to the island and the rate of extinction of the pre-existing species. This relationship was also affected by the degree of isolation of the island from the mainland (MacArthur and Wilson 1963). Many studies were conducted in the late 1960s and 1970s to try to determine if the island biogeography theory would be upheld experimentally. Some of these studies were quite creative and used experimental manipulations to systematically eliminate all fauna or selective faunal groups from islands with adjacent control islands (Simberloff and Wilson 1969, Rey 1981). One study used a long-term data set (26 year period) to examine the immigration and extinction rates of bird species in an isolated forest (Williamson 1981).

Most of the studies specifically examining the species–area relationships did not support the patterns proposed by the island biogeography theory (Smith 1990a). Despite this lack of strong support for this theory, it was crucial in focusing research efforts to understand the different relationships that can exist between species and their spatial scales. According to Smith (1990a), the island biogeography theory failed to consider the following:

1. The model was deterministic. It did not incorporate chance immigrations and extinctions, or the effect of species turnover on the composition of the species. According to Roughgarden (1995), one of the main problems with this theory is that the turnover rate of species on islands is suggested to be continuous while examples in the literature do not verify this.

2. The models ignores the degree to which an island is isolated from the mainland. The degree of isolation is a relative term and varies with the species being examined. (Patterns were more dependent on life history requirements and population dynamics.)

3. Immigration and extinction rates may not be interdependent of one another. Each can feed back to affect the other and how the rate is expressed.

4. Habitat heterogeneity is ignored. Habitat heterogeneity is critically important in determining how many species can exist on an island and not just the area.

Lugo (1987) also suggested that the factors most important in controlling plant species richness on islands vary from small islands (which are more homogenous), to intermediate-size islands (which are more heterogeneous), and large islands (which encompass microsite to landscape level heterogeneity). But instead of the number of tree species being strongly related to the size of different-sized Caribbean islands, Lugo (1987) showed the best correlation existed between the number of tree species and elevation, which reflected the increased diversity of microsites that occurred with elevation.

On islands and continental areas, most of the earlier studies in conservation biology attempting to determine what controlled the abundance and distribution of animal species did not explicitly incorporate the ecology of vegetative communities as part of their studies. Concomitantly, those scientists who were interested mainly in the relationship between plant distribution and the abiotic environment considered animals to be peripheral to their studies, except in the cases where introduced animal species were noted to be overgrazing native plant species. This is not a negative statement on the type of research being conducted at the time but merely reflects a research climate funding small, individual-based proposals. Furthermore, our baseline understanding of both species and ecosystem interactions was in the early stages of development so that their importance in addressing conservation issues was not apparent.

A major publication on the plant ecology of islands was the book edited by Mueller-Dombois et al. (1981), which synthesized the ecology of the island of Hawaii and demonstrated the more individualistic nature of much of the research prior to the 1980s. Since the 1980s, the National Science Foundation has contributed significantly to supporting ecosystem level research by interdisciplinary groups of scientists (Franklin et al. 1990). This shift in emphasis to specifically supporting interdisciplinary research has resulted in sufficient financial resources being allocated to examine the mechanisms by which introduced plant species are changing species composition on some islands using more interdisciplinary approaches. For example, an interesting case is the introduction of a N-fixing plant

species to control soil erosion in Hawaii that has dramatically changed ecological processes and plant species composition on the islands (Vitousek
and Walker 1989). In that example, the main mechanism by which the
N-fixing species impacted these ecosystems was due to their ability to significantly increase soil N levels, thereby decreasing the competitive ability
of native plant species not adapted to the higher N regimes (Vitousek and
Walker 1989).

Not all island ecosystems respond in the same manner to disturbances,
introduction of plant or animal species, or other human activities (Mooney
and Drake 1986, Lugo 1988, Vitousek et al. 1995). Unlike Hawaii, which
had little resistance to the introduction of both plant and animal species,
many islands are not easily invaded by new plant or animals species (Lugo
1988, Roughgarden 1995). The perception that all islands would generally
respond in a similar fashion to the immigration of species is quite understandable. For example, islands do share a common feature in which not
all organisms are represented on most islands (i.e., the only native mammal is the bat) and particular niches may be empty because of the isolation
effect (i.e., reptiles, amphibians) (Howarth et al. 1988, Loope et al. 1988).
However, this lack of representation by certain organisms does not make
islands more susceptible to introduced species. There is no generalization that can be made for how resistant islands are to immigration by
non-native species (Roughgarden 1995).

Despite the fact that the island biogeography theory has not been
supported by research, the simplicity of the concept has resulted in people attempting to use that relationship (i.e., the simple area dimensions
to species abundance) to assess the impacts of landscape fragmentation
on species abundance and survival, and to design animal and nature reserves (Smith 1990a). According to Roughgarden (1995), "Conservation
biologists tend to accept all of island biogeography theory as an integral
whole, not realizing that some parts may be correct and others incorrect for various taxa." Some of the rationale for the continued use of the
species–area relationship is quite apparent. Managers are frequently faced
with insufficient information to identify the impacts of different land uses
or fragmentation schemes on species maintenance and survival. Therefore, this simple relationship was adapted as a tool to model potential
management scenarios and to identify the thresholds or safety nets for
managing species. A further confounding factor in identifying and design-

ing reserves for conservation purposes is the fact that species-rich areas for different taxa do not coincide and many species-rich areas do not contain a substantial number of rare species (Prendergast et al. 1993). Research by Prendergast et al. (1993) suggested that it may be difficult to identify a few reserve areas where one or two species are located and automatically assume that other taxa will also be included in that reserve.

A simple species–area relationship also fails to consider the wide variability in gradients or mosaic patterns of resource availabilities that can exist for two spaces of equivalent size or the changes in the amount and quality of edge environment that can exist for similar-sized areas with different shapes (Saunders et al. 1991, Gosz 1992). Plant growth, species composition, and other ecosystem processes also fluctuate along these gradients and edges (Chen et al. 1992). Similar to plants, animal distribution and diversity also vary in relationship to the heterogeneity of resource availabilities that exist on a given land area (some animals are specialists and are restricted to particular habitats, while others are generalists and can be found across a range of habitats; Hunter 1990).

A good review of the abiotic and biotic impacts of the edges created by fragmentation on ecosystem processes and plant growth is given by Saunders et al. (1991). Some of the physical changes occurring along fragment edges that were reported by Saunders et al. (1991) are as follows:

1. Alterations in the fluxes of radiation hitting the ground. These result in different plant species occupying a habitat, changes in the nutrient cycling rates, alterations in decomposition rates of litter, and changes in soil moisture.

2. Altered wind conditions. These increase the exposure of edge vegetation to wind damage, such as wind pruning, windthrow, or increased tree mortality.

3. Alterations in water flux. This changes the rate of rainfall interception and evapotranspiration.

The importance and varying impacts of edge environments are becoming increasingly apparent. For example, Franklin and Forman (1987) documented the importance of increased wind damage along old-growth forest edges compared with interior forest areas due to major windstorms. Forty-eight percent of the blowdowns from a 1973 storm and 81% of the blowdowns from a 1983 storm were adjacent to existing clear-cuts and roads. Chen et al. (1992) reported that the edge effect can go quite deep into a forest. For example, the edge effect of 10- to 15-year-old clear-cuts next to old-growth forests extended 137 m into the "interior" forest, suggesting a very strong influence of edges into these forests. The degree of edge influence into a forest will vary based on the structure of a forest and what is creating the edge. The greater edge effects measured in the old-growth forest were not reported in eastern United States hardwood forests that had a depth of influence of 10 to 20 m, while a tropical rain forest in Panama was recorded to have an edge influence that extended 15 to 25 m into the forest (Chen et al. 1992).

As our understanding of how species relate to land areas and what factors influence these relationships is further refined, it should be easier to predict the impact of human activities on species diversity and abundance. This will require a better definition and detection of spatial scales and the edges or ecotones that can occur simultaneously at multiple scales in a system. These tools are being presently developed. (See Chapter 3 for a discussion of tools available to analyze scale.) It is important to move away from utilizing the simple relationships identified by island biogeography theory and to not be content "with knowledge that is one to two decades out of date" (Roughgarden 1995).

Diversity, Redundancy, and Ecosystem Function

The loss of biological diversity is currently an issue of great concern and debate among ecologists and managers due to the rapidly accelerating rate of species extinctions (Wilson 1988, Myers 1988). While some argue for the need to be concerned about the possible loss of any and all species, others advocate focusing on the relationship between diversity and ecosystem function (Baskin 1994, 1995), and on species that play key functional roles (Kremen 1992, Walker 1992, Schluter and Ricklefs 1993). Noss (1990) pointed out that most definitions of biodiversity do not include critical ecological processes that require a diversity of both biotic and abiotic functional components. Solbrig (1993) suggested that diversity plays a dual role in an ecosystem, in maintaining ecosystem functional properties (energy and materials flow) during periods of stasis, and in the resilience of the system after a perturbation. In theory an ecosystem is more resilient to perturbations if it contains a diversity of species, abiotic components, and functional types. If one species, component, or func-

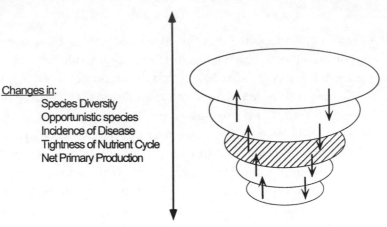

Changes in:
Species Diversity
Opportunistic species
Incidence of Disease
Tightness of Nutrient Cycle
Net Primary Production

FIGURE 2-1   Factors that can move a ecosystem to a different state. Each factor will not cause the system to change in a comparable direction (i.e., increasing net primary production can decrease species diversity, but increase the incidence of disease).

tional type is eliminated, then the existing diversity can substitute for or maintain the functioning of the system (Figure 2-1).

Initial studies suggested that the number of species in a foodweb was indicative of the stability of a system (see Appendix A). MacArthur (1955) first suggested a direct positive relationship between the number of species in a foodweb and the stability of the community in response to a disturbance, mainly because of the greater number of alternate pathways existing for energy flow. Thus if one species were to be decimated by stress, a predator had more alternative feeding options in a species-rich foodweb than in species-poor one. From a conceptual viewpoint, genetic diversity can be seen as the *information* in a biological system. To have much information may be to have a greater capacity to evolve and may lead to a greater chance that some species will be able to tolerate a given disturbance. This has led some authors to propose that biological diversity promotes ecosystem stability. Although debate on this topic began about 20 years ago, it has been renewed recently by growing concern over human-induced species extinction and heightened public awareness of biological diversity (Wilson 1988, Tilman and Downing 1994, Naeem et al. 1994).

Two competing ideas are central to the debate; they stem from theoretical considerations of the relationships between diversity and stability. The first idea is that if species differ in their traits, then systems that are more biologically diverse will be more "stable" or resistant to change, because they are more likely to contain some species that can survive during a given environmental perturbation. In this scenario, surviving species compensate for those that are reduced or lost by that disturbance (May 1973). The implication is that biological diversity should contribute to ecosystem resistance to disturbance or the ability to remain functionally unchanged in the face of a perturbation.

In contrast, the species-redundancy hypothesis (Walker 1992) states that many species are so functionally similar that ecosystem function is independent of diversity as long as there are a few species performing the major functions, such as carbon fixation. In this sense, species that do not play major roles in system structure or function are *redundant*. From this viewpoint, the important questions regarding diversity revolve around the important functions or functional groups and their component species. If an ecosystem contains more than one structure, process, or group of species that is capable of carrying out a given function, then the ecosystem is redundant with respect to that function. The point at which only one functional group remains to perform a specific function represents a threshold, below which the function cannot occur and the ecosystem state may change. Depending on the nature of the function and how many functions are affected by the loss of ecosystem components, the ecosystem may then be considered degraded, as it no longer provides desired goods or services.

The recent findings of Tilman and Downing (1994) support the theory that diversity imparts resistance and resilience to environmental perturbation. In grassland communities, they demonstrated that primary productivity is more resistant to and recovers more fully from a major drought when the community is diverse than when it is less diverse. Tilman and Downing also found that the stability associated with species diversity reached a point of saturation a t 10 species, and that the addition of more species did not significantly contribute to stability possibly because the essential functional niches were filled. Tilman and Downing (1994) did not consider their findings to support the species-redundancy hypothesis, however, their findings do support the functional group theory as presented in the above paragraph. According to Tilman and Downing (1994), "the species-redundancy hypothesis asserts that many species are so similar that ecosystem functioning is independent of diversity as long as all the major functional groups are present." Tilman and Downing (1994) stated that their findings support the diversity-stability hypothesis, which holds that species differ in their traits and that more diverse ecosystems are more likely to contain species that are able to thrive during periods of stress and thus compensate for competitors who do not do well. However, to fill the major functional groups requires a certain amount of diversity. Therefore, according to our definition of a species' functional role in an ecosystem (which states that a species may participate in many ecosystem functions) and because a functional group is defined by the intersection of functions in which the species participate, species comprising a functional group are not necessarily "so similar" so as to be insignificant in terms of diversity. No matter which competing idea is accepted, the diversity debate is important, and these divergent views indicate the need for further increasing the understanding of diversity as it relates to ecosystem function. Also, the notion of a threshold below which the stability of an ecosystem state is diminished is important to pursue in the process of further refining and identifying where both hypotheses may be relevant.

Although it is often assumed that biological diversity provides services crucial to energy flow and biogeochemical processes (Karieva 1994), there is little documented research on the subject. A fairly small amount of field research has been done to examine the role of biodiversity in ecosystem function (Ricklefs and Schluter 1993, Schulze and Mooney 1993, Johnson et al. 1996). However, studies originally designed for other reasons have provided insights on both the role of biodiversity and where research in this area might best be directed. Agriculture provides a model of the effects of extreme, deliberate reductions in biodiversity. In monocultures, plant diversity is nill, while below-ground and herbivore diversity (and biomass) is reduced via intensive management. Monocultures are often associated with increasingly intensive human inputs, suggesting a lack of self-sustaining ability. Moreover, less diverse (simpler) communities are thought to be more easily invaded than more diverse communities (Pimm 1991), suggesting decreased stability. This is a problem in agriculture, where herbivores are especially abundant and increasingly resistant to chemical control measures. Similarly, less diverse plant communities have been found to respond to fertilization through increased productivity more than do communities of higher diversity (McNaughton 1993).

Tilman and Downing's (1994) report on grasslands was originally designed to examine the effects of nitrogen fertilization on plant communities of varying diversity. A drought that occurred during the course of the study provided a perturbation, the effects of which were evaluated with respect to plant diversity and above-ground biomass. The study showed that above-ground primary productivity in more diverse plant communities was less severely reduced by the drought than that of the species-poor plots. The authors felt that this showed that the more diverse communities were more "resistant" to and recovered more quickly and fully from a severe drought. The authors measured resistance to drought by calculating the relative rate of plant community biomass change from the year before the drought to peak of the drought period. Drought resistance was found to be a significantly increasing but saturating function of predrought plant species richness.

Tilman and Downing's 1994 study suggested that species richness led to greater drought resistance because species-rich plots were more likely to contain some drought-resistant species. Herein lie two important points: The first is that it is the characteristics of the species present, rather than the sheer numbers of species, that determine system resistance or resilience to a particular disturbance. Tilman and Downing (1994) claimed to have evaluated the functional redundancy hypothesis in that they grouped plants by whether they used the C3 or C4 photosynthetic pathway. This grouping was not related to above-ground biomass. However, the authors did not consider that those species that have deep roots can better access water at greater depths in the soil and can therefore better maintain photosynthetic activity when a drought occurs. Perhaps a different functional grouping would have led to different conclusions. The second important lesson from this study is that the perturbations considered (drought as opposed to fire or grazing) will influence judgment as to system stability and "importance" of species or guilds. One other factor that must be considered in their study, that strongly confounds their results, is the fact that only above-ground biomass was examined. In grasslands, a high proportion (up to 80%) of biomass production occurs below-ground (Johnson et al. 1996), so the Tilman and Downing (1994) study may have missed a great deal of the ecosystem response to the drought.

The role of biodiversity in determining the primary production measured on a site has recently become a topic of interest in the literature (Baskin 1994, 1995, Naeem et al. 1994, Karieva 1994). Naeem et al. (1994)

established experimental systems to assess the role of species numbers on biogeochemical processes. Laboratory microcosms were set up so that lower diversity systems contained a subset of the species in the higher diversity systems. All units contained four trophic levels: primary producers, primary consumers, secondary consumers, and decomposers. Of five processes evaluated (e.g., community respiration, decomposition, nutrient retention, plant productivity, and water retention), only plant productivity was significantly positively associated with species diversity. The authors suggested that higher diversity ecosystems with fuller, space-filling plant architecture allowed greater light interception and thus greater photosynthesis. Karieva (1994) noted that, just as interesting as the association between productivity and biodiversity is the lack of relationships between diversity and any other ecosystem processes. Nutrient and water retention did not seem to be positively related to biodiversity. This study represents one of the few controlled experiments on biodiversity and ecosystem function. The experimental setup was such that the "loss of biodiversity" was evenly distributed across trophic levels. This, the authors claim, is analogous to what is currently occuring globally in natural ecosystems (Naeem et al. 1994). However, many species extinctions may not be random at all: Selective logging, changing the hydrologic regime, and pollutant discharge may cause nonrandom patterns of extinction. Further research might focus on assessing the effects of certain human activities and the effects on extinction, with special attention to the patterns of extinction and associated functional ecosystem changes.

Much of the debate on biodiversity/ecosystem function relationships has been in the realm of theory (Connell and Sousa 1983) and relatively little experimental field or laboratory microcosm work has been done to test these hypotheses. Despite this, the diversity and stability hypothesis has been accepted as a concept that is generally transferable across ecosystems and has been pushed as a useful tool for policymakers. However, today it has become apparent that the stability of webs may actually decrease as species are added (see the earlier island discussion); having more species present on a site does not automatically translate to a more "stable" ecosystem. It is important to accept the fact that the relationship between species and stability is much more complicated and may be related to (1) the numbers of species, the number of connections between species, and the intensity of interaction between members; and (2) which species are being examined. The strong or weak intensity of foodweb links is more important in the stability of a system than the mere presence or absence of a species.

It is also important to begin determining in which ecosystems strong relationships exist between species diversity and ecosystem functions. The ability to group ecosystems into categories of response or no response is valuable information for managers who need to stay away from managing systems using the same generic approach for all systems. Interestingly, the few research studies that examined the diversity–stability relationships in ecosystems have all been conducted in either grasslands or herbaceous dominated communities (see Johnson et al. 1996). Despite the fact that even McNaughton (1993) clearly stated in his paper that results from his research projects were not generalizable to all grasslands, these results have been used as evidence to support the generality of the positive linear relationship between the number of species present on the site and the stability of a system to disturbances. Even within the grassland communities, the successional stage of the community and the magnitude and duration of the interannual cycles of precipitation determine whether a positive, negative, or no relationship exists between ecosystem diversity and stability (Johnson et al. 1996).

Furthermore, it is important that results obtained in grasslands or herbaceous dominated communities are not automatically transferred to ecosystems dominated by woody plants. For example, the diversity of adaptive responses available to trees for increasing their access to site abiotic resources, changing their growth rates, or just how they respond to a disturbance is not linked specifically at the species level. Individuals within a species may have a broad range of reactions to a change in the environment (Knowles and Grant 1981, Sultan 1987, Johnson et al. 1996). In fact, trees have tremendous phenotypic and genotypic plasticity at the species level so that one species will demonstrate a multiple of responses to any given disturbance. One has to group the responses of trees at the level where the functional attributes are expressed. As stated by Sultan (1987), "Because it is at the level of individual phenotype that the genotype has ecological meaning, the adaptive response capacity of the individual is of critical significance in the study of ongoing and evolutionary relationships between plants and their environment." Sultan (1987) further argues that plants have an "adaptive phenotypic repertoire" that allows them to respond to the "unpredictable temporal and spatial heterogeneity of the plant environment." This means that a high diversity of perennial woody species will not be necessary for a tree to successfully occupy a site and to respond to disturbances. It is the high diversity of genetic responses at the individual species level that should be examined. An excellent example of this is seen in the National Atmospheric Precipitation Assessment Program (NAPAP) research (NAPAP 1993), in which seedlings were tested for their tolerance to acid rain and ozone. In the NAPAP studies, the variable responses to these pollutants occurred at the family level, with some families showing no response, while others had negative growth responses to the pollutants that they were exposed to. Because of this, trees should not be expected to respond to disturbances at the species level in a manner similar to some grasses.

In addition, it is important not only to examine the diversity at the plant species level but to document the diversity of mycorrhizal species colonizing the roots of tree and grasses. The mycorrhizal associations found on tree species (i.e., ectomycorrhizas) tend to have high diversities, while those found on grasses (i.e., vesicular-arbuscular mycorrhizas) tend to be of lower diversity (see earlier discussion on microbial symbionts). Trees, therefore, have a larger number of mycorrhizas with different capabilities of responding to the heterogeneous environment of the soil or to disturbances that are not uniformly expressed in the soil environment of the individual tree (Vogt et al. 1995a). Mycorrhizal associations on plant roots do appear to increase the number of adaptive responses that the individual plant species is able to express. Because human activities appear to be changing the presence and diversity of symbiotic associations recorded in natural ecosystems (Arnolds 1991), it is important to incorporate symbionts as one of the factors that are considered to be part of any management framework.

It is important to document the level at which diversity determines how plants and animals adapt to their ecosystem. This type of information is very useful for management purposes because it allows one to group plants or animals adaptive mechanisms that are functional. One can then focus on monitoring changes in the persistence of diversity at the different grouping levels in order to identify when the system is changing in response to a perturbation. It is obvious that if one does not produce these groupings, the result will be an impasse because researchers or policymakers, depending on their goals, will be able to create controversy by identifying similar systems where in one case there exists a strong relationship between species and ecosystem function, and other examples where there are no relationships. This leaves the controversy unresolved and results in decisionmaking occurring at the level of values, with the choice reflecting the personal preferences of the individuals for a particular desired outcome that has no relevance to the actual situation.

Keystone Concept

The term *keystone species* was first coined by Paine (1966) in reference to species whose activity and abundance determined "the integrity of the community and its unaltered persistence through time, that is, stability." A keystone species is a species that has a disproportionate effect on the persistence of other species and whose removal leads, often indirectly, to the loss of many other species in the community (i.e., decreased diversity). Identifying keystone species has been endorsed by some conservation biologists as a useful approach to determine where conservation efforts should be placed to have the greatest potential for protecting more species at any given time. This especially becomes important when financial resources are limited. However, Mills et al. (1993) argued that the term *keystone species* has been "broadly applied, poorly defined, and nonspecific in meaning" and "the type of community structure implied by the keystone-species concept is largely undemonstrated in nature." Also, they stated that the keystone species concept should not be accepted as a tool by managers due to the difficulty of identifying keystone species and our poor understanding of systems and their feedbacks (Mills et al. 1993).

As part of their rationale for why the *keystone species* term is too ambiguous and therefore is not useful in conservation, Mills et al. (1993) identified several broad categories of keystone species separated into groups according to the character of the crucial role the species plays in the community or ecosystem. Mills et al. (1993) identified five groups of keystones: predators, prey, plant, link, and habitat modifiers. Their broadening of the definition of keystone species helps to take the keystone species concept from being primarily focused at the level of examining interactions between different groups of animals to linking species with their impact on the habitat, and also linking plant–animal interactions. However, except for the modifier category, the other categories are still very species based and mainly link plants to animals at the level of loss of animals critical for pollinating plants. The five categories they identified do not explicitly integrate species to ecosystem level processes or identify a keystone organism by the degree to which it changes energy transformations and flows within the ecosystem.

Researchers have theorized that there are several other categories of keystone species that more explicitly relate them to their importance in modifying ecosystem level carbon and nutrient cycles. For example, the notion that ecosystem processes, such as litter decomposition and nutrient cycling, may be controlled by keystone species has been suggested (Perry et al. 1990, Read 1993) but has yet to be conclusively demonstrated (Bond 1993). Brown lemmings are a good example of animals that function as keystone species that indirectly affect nutrient cycling rates in tundra ecosystems. Lemmings control the activities and reproductive success of many other animal species in the tundra but at the same time function in increasing plant productivity and nutrient availability in pulses related to their population fluctuations (Remmert 1980). The population levels of brown lemmings fluctuate in 3- to 5-year cycles, during which time they reach a high level followed by a crash. When population levels are high, they can remove 25% of the above-ground net primary production and transfer this plant material earlier into the decomposition cycle. This plant material is much higher in nutrients because the plants have not been able to retranslocate nutrients out of the leaves, resulting in a pulse of nutrient availability and faster decomposition rate of this material (Bunnell et al. 1975).

There are several other examples of keystone species and how they provide a critical link in the structure of an ecosystem. These examples are listed in Mills et al. (1993) under the modifier category. All these examples show clearly how the removal of the structural link provided by a keystone species alters ecosystem function and changes species composition and/or density. For example, Brown and Heske (1990) identified three species of kangaroo rats (*Dipodomys* spp.) as keystone habitat modifiers in North American deserts. They demonstrated that experimental exclusion of the kangaroo rats from their shrubland habitat led to a marked conversion of the ecosystem from shrubland to grassland over a 12-year period. They identified food caching and burrowing by the kangaroo rats as critical activities in maintaining species composition and diversity in the shrubland ecosystem. In addition to the importance of kangaroo rats to the physical structure of North American desert communities, Thompson et al. (1991) demonstrated the importance of several different species of rodents, which act as keystone predators, to the trophic structure of the same desert communities. Rodents consume large seeds, thus reducing the competitive advantage large seeded plants have over small seeded plants. Small seeds support desert ant and bird populations, thus the foraging activity of rodents directly influences the relative abundance of small seeded plants and indirectly influences ant and bird populations.

As the above-mentioned examples demonstrate, identification of a species' keystone status contributes to our understanding of ecosystem structure and function. Keystone species are strong interactors and have a disproportionate effect on (1) the persistence, abundance, and distribution of other species; (2) carbon and nutrient cycling rates; and (3) ecosystem resiliency to perturbations. For these reasons managers need to be aware of the presence of keystone species to make informed management decisions. Removal of a keystone species generally results in decreased species diversity and a change in the ability of the ecosystem to respond to disturbances. The keystone concept may also serve an important role in identifying critical functions in an ecosystem; this has not received much attention by the research community. Such keystone functions can be imagined as those dealing with a limiting factor for production, such as nitrogen, phosphorous, light, or water.

The use of keystone groupings can be quite useful for management at the ecosystem level. Those species that have a more significant influence on controlling some structural or functional aspects in ecosystems need to be identified and the manner in which their activities reverberate throughout the ecosystem elucidated. This does not mean that all ecosystems have keystone species (most, in fact, do not), but it is important to identify when they do exist so that consideration can be given to explicitly incorporate them as driving variables in that system.

Functional Groups

A more complete view of an ecosystem may emerge when ecosystem attributes, such as structures, species, and processes, are grouped according to a specific functional role that they perform in that ecosystem. Such functional groups may be made up of any elements at a specified organizational level that "bear a certain set of common structural and/or process features" (Körner 1993). Functional groups may contain both structures and processes, as both may have dynamic influences on an ecosystem (Körner 1993), or may be composed solely of different species (Walker 1992, Dawson and Chapin 1993). Each species, structure, or process may participate in many different ecosystem functions, and each species, structure, or process plays a role in an ecosystem according to the set of functions in which it participates. Here, the concept of dual hierarchies (O'Neill et al. 1986) in an ecosystem, one of the population-community perspective and the other of the function-process perspective, can assist in meshing the role of structure and processes in functional groups. The functional groups are constructed to help understand ecosystem functions of interest as directed by management goals.

Solbrig (1993) reviewed simple functional classificatory schemes (such as Raunkier's life forms) and explained that because most are based on a single functional characteristic, they may yield information about the ecological adaptation of a species but are not useful in predicting the many facets of a species' functional role. The functional group approach differs from simple classification schemes in the degree of flexibility required in defining groups. Functional groups must be defined according to a specific ecosystem and scale. The challenge lies in the recognition of the many aspects of ecosystem function in which a species or process may be involved, and in the identification of all the key organisms and processes involved in the specific ecosystem function of interest.

One functional group approach focuses on species assemblages. In a discussion about the importance of biological diversity, Walker (1992) advocated grouping species according to their functional roles to determine which are most important for maintaining overall ecosystem function. Under this approach the appropriate basis for defining species functional types (or guilds) is the way biota regulates ecosystem process. Walker (1992) suggested a four-step process for defining functional guilds: (1) identify rate-limiting or otherwise important ecosystem processes, (2) develop corresponding functional classifications of biota through guild analysis, (3) further subdivide functional group on the basis of functional attributes related to critical ecosystem process, and (4) assess redundancy within functional groups (complete functional redundancy occurs when there is density compensation among group after removal of one member).

A second type of approach to grouping ecosystem components or attributes is described by Körner (1993), who stated that, at any scale, elements that have a set of structural or process features in common may be treated as a functional group. Körner suggested five (structural) grouping criteria: (1) the degree of integration within a hierarchy of ecosystem complexities, (2) quality criteria (e.g., life form, life history, size), (3) the spatial distribution of plants and plant organs within an ecosystem, (4) phenological sequences, and (5) large-scale distributional characteristics.

The functional group approach will contribute to better understanding of ecosystem function only to the extent that significance of functional group is understood in relation to the behavior of the whole system. At present full implementation of the functional group approach to management is not possible in most situations due to a lack of scientific information about ecosystem processes and species roles. Better understanding the multiple facets of a species' or process' functional role in an ecosystem will require further research at several scales of integration.

2.1.3 Ecosystem Concept Defined

Structural and Functional Attributes of Ecosystems

Since the coining of the term *ecosystem* by Tansley (1935) to include both organisms and the abiotic environment, ecosystems have been perceived and defined in numerous ways. According to Odum (1959), an ecosystem is "any area of nature that includes living organisms and nonliving substances interacting to produce an exchange of materials between the living and nonliving parts," or as Odum stated in 1971, "any unit that includes all of the organisms (i.e., the community) in a given area interacting with the physical environment so that a flow of energy leads to a clearly defined trophic structure, biotic diversity, and material cycles (i.e., exchange of materials between living and non-living parts within the system)." Holling (1986) stated that "ecosystems are communities of organisms in which internal interactions between the organisms determine behavior more than do external biological events." Whether the focus is on material and energy fluxes or on interactions within biotic communities, the term *ecosystem* is often used to refer to the largest scale in which an entity is self-sustaining. This can vary in size from a few centimeters to hundreds of kilometers. Whether the focus is on material and energy fluxes or on interactions within biotic communities, the term *ecosystem* is often used to refer to the largest scale entity on a hierarchical scale. This traditional hierarchical scale may range from organs to individuals, populations, communities, and perhaps to landscapes, as referred to earlier.

It has been stated that an ecosystem boundary can be defined to exist at the organism, individuals, populations, or communities levels as long as the biotic and abiotic processes are self-sustainable within the confines of that boundary. Therefore, even lichens growing on tombstones or rock walls can be classified as an ecosystem because all the biotic (i.e., primary producers, consumers, and decomposers) and abiotic processes and functions are self-contained on the surface of that rock. While referring to a system as being an ecosystem at any of these levels is not necessarily inaccurate, depending on how the ecosystem of interest is defined, doing so implies an ecosystem is more an entity than a construct. We believe this emphasis on the ecosystem as an entity has lead to some of the current confusion concerning the goals and objectives of ecosystem management. Therefore, it is important to stress the view that ecosystems should be seen as a construct, a way of looking at the biotic and abiotic components of the natural world and the interactions among them. As interactions are the focus of the ecosystem view, the use of the term *ecosystem* refers primarily to processes and functions. The spatial extent of the ecosystem is of secondary consideration and is imposed with a proper consideration of the function scale.

Odum (1959) identified four constituent components of all ecosystems that need to be considered when studying energy flow through systems: (1) abiotic substances, (2) producers (the autotrophs), (3) consumers, and (4) decomposers. Kimmins (1987) defined the term *ecosystem* as a concept that consisted of six major attributes: (1) structure, (2) function, (3) complexity, (4) interaction and interdependency, (5) "no inherent definition of spatial dimensions," and (6) temporal change. Kimmins' list not only consisted of the four constituents identified by Odum but went further by identifying that feedbacks exist in systems and that there are no defined scales. Odum (1971) also had a very homogenous perception of the ecosystem unit. Today it is recognized that an ecosystem as an unit is quite heterogeneous and it is necessary to document this heterogeneity in order to understand ecosystems (Turner et al. 1995, Vogt et al. 1995a).

The term *ecosystem structure(s)* is used to refer to physical components of the system that are integral to their function. The term *ecosystem function* refers to the dynamics of the holistic system that are associated with fluxes in energy and material transfers, while the term *process* refers to the mechanisms that contribute to overall ecosystem function. Ecosystem structure and function are very closely related so that a change in function may alter some structural components within that ecosystem or a change in structure will change some functional aspect of the system (King 1993). Knowledge of ecosystem structure can usually provide insight to ecosystem function. Realistically, the difficulty in defining and measuring ecosystem functional processes requires the search for measurable parameters to manage ecosystems. Because structural attributes can reflect functional processes, they were considered to be the logical place to collect data in many past studies. However, structural attributes can be misleading. For example, plant biomass reflects a measurement that has accumulated over longer time scales (i.e., 50 years or more), and all of this biomass does not respond as a unit to the disturbances being imposed on an ecosystem.

The following have been identified to be important ecological components of any ecosystem:

1. Integration of *all* biological (*biotic*) and nonbiological (*abiotic*) parts.

2. Monitoring the *movement of energy* and *materials* (including water, chemicals, nutrients, pollutants, etc.) into and out of its boundaries.

3. Utilization of a *common currency* called *energy* to measure ecosystem function and the strength of the links between different ecosystem components. In practice, changes in organic matter or carbon accumulation (i.e., net primary production) over a defined space and over a given time period is used as a surrogate for energy because photosynthesis, which fixes carbon, is an energy-assimilating process.

4. Boundary definitions—a site that can be *bounded* by identifying the *smallest unit that is self-sustaining*.

5. Explicit incorporation of *spatial and temporal scales*.

6. Encompassing system or species characteristics that are highly *interde-*

pendent and have *strong feedback loops*. Feedback loops can be expressed at the species or ecosystem level (i.e., keystone or functional groups) and can be *driven by microbes and/or consumers*.

7. Incorporating *disturbance cycles* at defined temporal and spatial scales, explicitly acknowledging that disturbances can occur at varying scales and are an integral part of ecosystems. Identifies the importance of *legacies* (imprint of past disturbances or structures) and how they have contributed to the development of the current ecosystem structure and function.

8. Characterization of all of the above for the *multiple states* that a system can fluctuate between as part of the *natural development* of that system.

The eight components just detailed identify tools and attributes of ecosystems that should be considered as part of any ecological analysis framework for studying ecosystems. Once the characteristics or attributes of each component have been recorded, they function as an effective checklist to ensure that important parts of the ecosystem have not been missed.

Many of the components listed cannot be examined in isolation from one another if the desire is to predict the functioning of the entire ecosystem. For example, predicting future productivity on a given site cannot be only determined from past records of site productivity. Despite the difficulty of using past records of growth on a site to predict future productivity, this approach has been used in forestry to predict the growth of managed stands using *site index curves*, which assumed that the abiotic resource base and disturbance cycles would remain constant and similar to the past. However, more site-specific knowledge of abiotic resource availabilities at a site combined with productivity estimates can be much more useful to predict future productivities because tree growth rates are highly sensitive to site nutrient availabilities (see Section 3.6.3). In turn, because decomposers regulate the rate and the magnitude of abiotic resource availabilities to plants from decaying litter material (Swift et al. 1979), site-specific information on decay rates allows one to estimate the nutrient availability to plants and the potential productivities achievable for the site. Many of the current production models have explicitly incorporated the nutrient availability that accrues to plants from the decomposition of litter material.

An examination of the historical changes in our knowledge of ecological systems highlights how recent our understanding of the complexity of systems, their dynamic nature, and the role of species in ecosystems is (see Appendix A). It is quite recent (within the last 10 years) that variables have been identified that are important as *constraint points* or *bottlenecks* in different ecosystems. Managers will have to be sensitive to the fact that the knowledge base of the important components in ecosystems to focus on evolve with time. As part of ecosystem management, managers will have to be ready to filter through new information as it becomes available and to incorporate those parts that are relevant. That fact that the development of the knowledge base will be ongoing with the development of an ecosystem management framework means that there is no "real" stable endpoint that can be aimed for by managers. This fact has resulted in the statement being made that the knowledge base will always be inadequate to use the ecosystem approach in management. This is an inappropriate statement to accept. It is important to use the best information that is available at the time and to realize that mistakes will be made, and sometimes these will not be apparent until much later. Ecosystem management, however, has to explicitly deal with the uncertainty in the information base that results in it being more conservative because many more variables have to be included as part of the framework. It is important to incorporate the basic tenets of ecosystems into the management system, even if the term *ecosystem management* falls into disfavor.

It is necessary to synthesize and analyze much of the data that already exists in the literature because it can be quite useful in helping to understand what variables are important to focus on in systems (see Section 4.2.7 on large data sets). There exists a diverse and extensive database in the different fields of ecology, varying from community ecology, species as part of communities or ecosystems, and ecosystem ecology. Many permanent study sites have been collecting ecosystem level information for over 10 years. In the United States, the following are a few examples of sites where a significant database exists at the ecosystem level: *Pinus elliotti* in Florida; mixed hardwood forests in North Carolina and Tennessee; mixed northern hardwood forests in New Hampshire; *Pinus contorta* ecosystems in Wyoming; *Abies amabilis* ecosystems in Washington; *Pseudotsuga menziesii* ecosystems in Oregon and Washington; and boreal and arctic ecosystems in Alaska (Grier and Logan 1977, Bormann and Likens 1979, Sollins et al. 1980, DeAngelis et al. 1981, Grier et al. 1981, Edmonds

1982, Vogt et al. 1982, Gholz et al. 1986, Fahey and Knight 1986, Van Cleve et al. 1991, Chapin et al. 1992). There are also many permanent research sites outside of the United States. The following few are given to demonstrate the diversity of sites that have been studied: tropical rain forests in Brazil, Venezuela, and the Ivory Coast; coniferous and hardwood forests in England, Germany, the Netherlands, Poland, and Karelia, Russia; coniferous forests in Japan, southern Sweden and Scotland; and tundra sites in northern England (Cole and Rapp 1981, DeAngelis et al. 1981). Many of these sites are representative of the major vegetative communities globally. However, these sites are frequently not represented by the several dominant stages of stand development or successional stages particular to that vegetative community, nor do they encompass the different soil environments within each vegetative zone. It is important to expand these studies to understand larger temporal scales in the development of these ecosystems but to also extend similar approaches to study less well-known sites.

All this information collected at these different and relatively permanent sites has contributed to building a baseline knowledge and mechanistic understanding of processes occurring at different scales for these ecosystems. They are extremely useful for identifying what variables are valuable to monitor in specific ecosystems. This type of data will not, however, identify when a system state is changing beyond the normal range of multiple states or to predict thresholds in ecosystems (see Chapter 4). These earlier studies were not designed to address these points so it is unfair to assume that this type of data can be manipulated to answer current questions concerning ecosystem resistance or resilience (see Sections 2.1.3 and 4.2.1). To answer questions related to resistance and resilience characteristics of ecosystems, it will be necessary to develop creative new approaches to study ecosystems or more creative modifications or combinations of techniques than what have been used in the past.

Ecosystem Terms: Resistance, Resilience, Retrogression, Buffering Capacities, and Legacies

The changing state of ecological systems has long been of interest to ecologists; the early plant successional studies are a prime example of this. Whether their interests lie in natural history, population dynamics, or energy/nutrient cycling, ecologists are often interested in questions as to why there is a particular distribution of species or energy flow in an ecosystem and how this distribution persists (or changes) over time. Recently environmental managers, regulators, and others have become interested in this same question as human activity is increasingly seen as the causal agents of large-scale environmental changes. These changes are also perceived to be occurring at a faster rate than considered normal, and there is concern whether the changes are irreversible. In addition, many of these changes are surrounded by a great deal of uncertainty as to their long-term effects. In ecosystem management, there is concern with estimating the probability of moving from a desired biological state (one in which desired system outputs, i.e., harvestable fish, clean water for swimming, are abundant) toward a less desired state (in which the desired amenities are no longer available from the system).

In ecology, many terms have been defined to explain the state of a system over time, how readily that system state changes, and the trajectory of change in response to a disturbance (Westman 1985). It sounds extremely simple to monitor state changes in ecosystems because the perception exists that it only requires the identification of a number of state variables that are descriptive of the system state and monitoring these over time. If these state variables can be tracked over time, a system trajectory is assumed to be something that can be determined. However, most of these terms used to define system state changes are still in the theoretical sphere due to our inability to measure small-scale and early changes in ecosystem functions. Large-scale structural changes in systems can typically be detected, but usually this is after major changes have already occurred in the ecosystem over longer time scales and the system is already highly stressed. The ability to measure when changes are occurring in ecosystems would be very useful for managers who need to be aware of when the system state may change due to a management activity. These tools are also essential for implementing ecological risk assessment in the field. Several new approaches for detecting ecosystem state changes are given in Chapter 4.

It is extremely useful to define the terms commonly used to discuss system states, how readily they change, and their direction of change, because these terms are used interchangeably in the literature. Several of the terms defined later are used rather loosely in the literature and there needs to be more consensus as to their definitions if they are to be effective. Often our ecosystems are classified in terms of their *stability* and *their movement away from a stable state*. In the case of ecological systems (which are highly dynamic) the term refers not to stasis of all state variables but to variations within some defined bounds. Although there has been some confusion stemming from the variable use of terms, a number of system descriptors serve as organizing concepts for the study of systems dynamics. Ecological systems can be described in terms of the degree to which they vary from some steady state, their capacity to absorb or otherwise dissipate perturbations (resistance), the speed with which they return to a preperturbation state following a perturbation (resilience), and the length of time a system remains in a defined state (persistence). These terms are often used to describe stability with respect to a defined equilibrium or condition under which system outputs, such as high-quality drinking water or sufficient harvestable fish, are maintained at desired levels.

A very useful concept, called *resilience* by Holling (1973), refers to the bounds in state space around which a system will vary but return to preperturbation state. If the system moves outside these bounds, it will move to another state. Ecosystem properties that control the "pace, manner, and extent of recovery following disturbance" (acute or chronic) are components of ecosystem resilience (Westman 1985). Resilient ecosystems are those systems that can be altered relatively easily (not very *resistant* to disturbances) but will return to the initial state more rapidly, while resistant ecosystems are those systems that exhibit relatively little response to disturbance and a severe disturbance is required to change the state of the system. Westman (1985) separated resilience into four different components, which can be individually measured as distinctive responses by the system as follows:

1. *Elasticity*—time required to restore the system to its predisturbance state.

2. *Amplitude*—degree to which the system state has changed from its predisturbance condition and from where it will still return to its initial state.

3. *Hysteresis*—extent to which the trajectory of "degradation" under chronic disturbance and the trajectory of recovery after the removal of the disturbance are mirror images of each other.

4. *Malleability*—degree to which the new system equilibrium state estab-

lished following the recovery of the system after a disturbance differs from the original state.

These terms are useful organizing terms to identify how a system state is changing; however, few if any irrefutable examples exist where they have been used to evaluate a system. Most examples only exist for one of the resilience components (i.e., amplitude), where repeated cycles of the same disturbance have been used to define when the system will not return to its original condition (Westman 1985). The following examples were presented by Westman (1985) for *amplitude*: (1) marsh grasses treated with crude oil at a frequency of 2, 4, 8, or 12 times over a 14-month period. Here, increased duration of oil stress (8–12 successive oilings) resulted in a threshold beyond which marsh grasses did not recover from being inundated with crude oil; (2) wet tropical forests exposed to repeat burning cycles as part of slash-and-burn agriculture; (3) aquatic communities subjected to pollutant discharge. These examples are not addressing the mechanisms by which the system state may or may not recover after a number of repeat cycles of the same disturbance. Unless a mechanistic understanding is developed of why and when a system reaches a threshold beyond which it will not recover, this information will be of little utility to predict how other similar systems will respond. The other three components identified by Westman (1985) do not have good examples, partially due to our past inability to identify when a system state is changing. It is for this reason that several different approaches to detect ecosystem resilience are introduced in Chapter 4 to stimulate further research in this area.

The resistance and resilience characteristics of ecosystems are nicely illustrated in Figures 2-2 to 2-4. These figures highlight the fact that just because an ecosystem has high resistance to stress, it does not automatically have a fast recovery rate from the disturbance. Figure 2-2 represents a generalized template of ecosystem resistance and resilience characteristics. In this figure, resilience is illustrated as the time period that lapses between a disturbance at point (a) and recovery at point (b). The ecosystem is represented as having a normal operating range in which the multiple states of an ecosystem are expressed. In this case, when a disturbance (i.e., natural or anthropogenic) is imposed on the ecosystem, the ecosystem may be shifted to the initial stages of the first developmental level. Those commu-

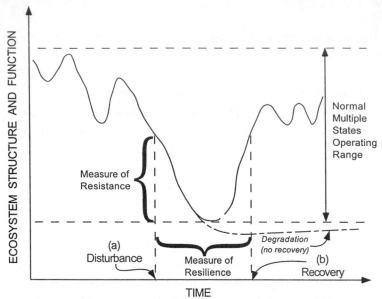

FIGURE 2-2   Conceptual framework representing the responses (resistance and resilience) of ecosystem structure and function to a disturbance. (Adapted from Leffer 1978, Odum 1985, and O'Laughlin 1993, with permission.)

nities that have wide fluctuations in ecosystem dynamics due to frequent disturbances over short time frames were classified by Holling (1973) and O'Laughlin (1993) as being most resilient. The next two figures represent two different ecosystems with different resistance and resilience characteristics: (1) California redwood forest ecosystems, which have a thick bark, making them very resistant to fires, but if they burn the recovery rate is slow and they may not recover at all (Figure 2-3); and (2) California chaparral vegetation, which is not very resistant to fire and burns easily but recovers very rapidly, thus being highly resilient (Figure 2-4). At the landscape level, managers should manage for resilient ecosystems by maintaining a diversity of successional or developmental stages and species across the landscape to reduce the homogeneity of vegetation over large areas. In this way, a catastrophic disturbance will be expressed across a variety of ecosystem stages, resulting in a mixture of earlier stages, which are more resilient to these disturbances, and later stages, which may not be as resilient but have desirable characteristics that minimize the impact of disturbances.

Two other terms that frequently appear in the literature are *buffering*

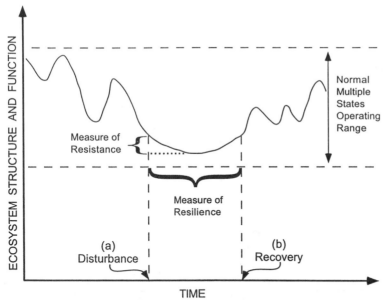

FIGURE 2-3 Ecosystems with a high resistance and lower resilience. (Adapted from Leffer 1978, Odum 1985, and O'Laughlin 1993, with permission.)

and *retrogression*. According to Westman (1985), a gradual breakdown of ecosystem integrity under the influence of chronic stress is called the process of *retrogression*. When a disturbance is continuous but at a low level (i.e., chronic), the system changes slowly and usually in manner difficult to detect as a significant chance in the system; frequently, the impact becomes a management problem when the system has become saturated with whatever is causing the disturbance. The term *retrogression* is used since the changes in the ecosystem (i.e., decline in biomass, forest floor accumulations, or plant cover) have been perceived to somewhat mimic the reverse changes that occur with succession (Westman 1985). Retrogression can be induced by several disturbances that are classified as chronic: air pollution, low-level ionizing radiation, toxic substances, grazing, soil compaction, and soil erosion. For aquatic ecosystems, chronic disturbances causing retrogression can consist of many different types of activities: increased frequency and strength of currents or waves, introduction of toxic compounds at low but continuous levels, introduction of nitrogen compounds from sewage or fertilizers, or increased sedimentation rates.

Most ecosystems have one or several internal mechanism that allow them to buffer or dampen the impact of a disturbance—its *buffering capac-*

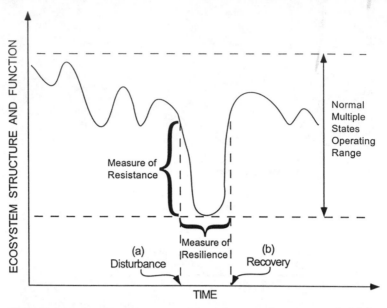

FIGURE 2-4   Ecosystems with a low resistance and high resilience. (Adapted from Leffler 1978, Odum 1985 and O'Laughlin 1993, with permission.)

ity. Buffering capacity can occur in parts of ecosystems where there exist a large storage capacity and a slow turnover rate within that pool; the slow turnover rate of a large pool can buffer a system against sudden or large changes in the magnitude of inputs to that pool due to a disturbance (Westman 1985). A prime example of an ecosystem buffer is soil organic matter because it is relatively stable over long temporal scales—changes in organic matter accumulation rates are low in the soil. In *Pseudotsuga menziesii* forests ranging in age from 10 years to over 400 years and of the same site quality index, neither the clear-cutting and harvesting of above-ground parts of trees nor the successional development of vegetative communities through time were detectable as changes in soil organic matter accumulations (Vogt et al. 1995c). This may reflect the large amount of coarse woody debris that has been traditionally left in forests converted to younger developmental stages. Except for a major disturbance, such as conversion of a forest to agriculture, most forest harvesting activities have a low impact on changing soil organic matter levels (Johnson 1994).

   This means that soil organic matter functions as a *legacy* because it carries over as a relatively *intact memory* between different vegetative community states and as part of succession. Other legacies that have been identified in ecosystems that can affect how the system responds to a disturbance are as follows:

1. Large-sized diameter coarse, woody debris in streams and terrestrial areas

2. Chemical or physical imprints left by some plants within their ecological space (i.e., those plants associated with N-fixing microorganisms and that increase the amount of nitrogen in the soil; those plants that accumulate heavy metals (i.e., aluminum), which can be toxic to other plants; those plants that accumulate nutrients, such as phosphorus and calcium, in higher concentrations than other plants)

3. Persistence of mycorrhizal inoculum (i.e., sclerotia) in the soil

4. Persistence of plant propagules in the system

5. Persistence of the soil chemical and physical characteristics (Moran 1987)

Discussions of these different legacies are given in Binkley et al. (1982), Perry et al. (1989), Halpern (1988), Spies et al. (1988), Vitousek and Walker (1989), Dahlgren et al. (1991), Swanson and Franklin (1992), Bormann et al. (1993), and Vogt et al. (1995a). Understanding the role of legacies in controlling ecosystem function is still in its early stages, but they appear to have important roles in contributing to ecosystem resilience and resistance.

2.1.4 Ecological Research Infrastructures

The historical time line given in Appendix A not only shows how our ideas in ecology have developed through time but also how important they are in highlighting the consequence of our shift from the individual researcher mode to teams of researchers working on the same question. A lot of the original ecological research that focused on the dynamic nature of systems and functioned as a foundation for the development of the field of ecosystem ecology was conducted by individual university scientists in association with their graduate students (Appendix A). Even though much of that research was not conducted with large research budgets, it was instrumental in training the next generation of scientists studying the community ecology of vegetative systems (Billings 1985). A more holistic and team approach to research was initiated in 1934 when the U.S. Forest Service opened the Coweeta Experimental Forest hydrology laboratory in North Carolina. This research was not conducted at the ecosystem level but began to examine the role of different plant species in controlling the hydrologic cycles using the watershed approach. This research explicitly integrated management and hydrology. Interestingly, many of the federal laboratories in the United States were the first bastions of interdisciplinary research and are still continuing to function in this capacity.

Once multidisciplinary research teams were formed, our ability to understand natural ecosystems and the impacts of particular management activities on ecosystem structure and function was greatly facilitated. The transition of individuals working by themselves to functioning under the umbrella of group research has been a shaky road. It has been very difficult for individuals to shift from being in charge of their own research to being part of a larger group where group dynamics control the direction of the research. It was not uncommon in the past for individual scientists to continue their own favored individual research projects, even though they were supported under the auspices of interdisciplinary group research to conduct research in specific ecosystems. A prime example of this was the International Biological Programme (IBP), which was launched in 1964 and has been criticized for not developing any new theories or revolutionary ideas as part of their activities (Golley 1993). Most of the published results that were a direct offshoot of this program were individual publications, even though there were some notable exceptions, such as the 1981 IBP book, which collated and synthesized information from sites scattered around the world (Cole and Rapp 1981, DeAngelis et al. 1981).

The IBP may have been criticized for not resulting in significant leaps in our understanding of ecosystems, but it did an excellent job of being the graduate training ground for future scientists around the globe who learned to work within interdisciplinary teams (Golley 1993). Most of the research being conducted by the IBP was supervised by university scientists. Even though most of the research conducted under the IBP was considered to be descriptive and not mechanistic, the graduate students who received their training under the auspices of the IBP are those who really have contributed significantly to strong interdisciplinary ecological research from the early 1970s to the present. As with any large program, scattered groups (i.e., interior Alaska and the Pacific Northwest United States, etc.) existed within this program which were conducting very mechanistically based research (Cole and Rapp 1981, Edmonds 1982). For example, the research in interior Alaska has continued since the IBP days and culminated in 1986 in a published book that is an excellent synthesis of these systems and that is ecosystem, based (Van Cleve et al. 1986).

The IBP programme also contributed to enhancing science research by focusing on the importance of developing common methods to study ecosystems and by increasing the interactions among scientists around the world. The IBP programme compelled people to utilize similar techniques across several different ecosystem types so that systems could be compared. The program also emphasized the collection of biological, ecological, physical, and chemical data on each site (this is an early attempt to measure some of the key components of ecosystems). In fact, the first phase (i.e., the first 3 years) of the IBP was dedicated entirely to the planning and development of methodology to be used at all sites around the world that were part of this program. This program also facilitated interactions among scientists from many different disciplines and from many different countries, and facilitated the synthesis of large data sets from around the world (DeAngelis et al. 1981, Cole and Rapp 1981, Cannell 1982).

About the same time that the IBP programme was established, more integrative research was also being conducted under the auspices of the United States Atomic Energy Commission (Golley 1993). This research was mainly located in the United States and Puerto Rico, with scientists from the United States Federal Laboratories performing the research. This research program studied several different ecosystem types (i.e., from tropical forests to coral reefs) using an interdisciplinary team approach that went beyond the generally more descriptive research of the IBP. In fact, the research conducted in an eastern deciduous forest in Tennessee was the first in which a total ecosystem budget was obtained for a forest system in the world (Harris et al. 1978). For example, instead of just including the above-ground parts of the ecosystem (an approach that permeated all other research at the time), researchers included fine-root biomass and productivity as an integral part of the ecosystem and their responses to stress (Harris et al. 1978). The research conducted under this program in the 1960s to 1970s in Puerto Rico (Odum 1970) still forms the core foundation and baseline information for much of the ongoing ecosystem level research presently in these wet tropical forests (Waide and Lugo 1992).

In the 1970s, several research groups formed that were comprised of scientists from more than one university and from the U.S. Forest Service. In most cases, the Forest Service contributed established research sites or watersheds with an infrastructure that was capable of maintaining the long-term collection of environmental data. The university scientists were instrumental in bringing in research funding, which allowed the expansion of the type of research questions that could be pursued in conjunction with the baseline monitoring already being conducted by the Forest Service. A classic example of one of these research groups is the interdisciplinary group of scientists (i.e., limnologist, ecologist, hydrologist, etc.) who coordinated research at the Hubbard Brook Experimental Forest in New Hampshire (Bormann and Likens 1979). The stability accrued to this research group occurred because the Forest Service was able to maintain the data collection during those periods when research funding temporally ceased for the university scientists. This coordinated research effort resulted in some major research findings that integrated the spatial and temporal scales integral to ecosystem analyses. This research also introduced, for the first time, the idea that management activities in forests may affect other ecosystem functions and had a dramatic affect on policymakers trying to decide how much to control forest management practices. The research results from the Hubbard Brook were used by policymakers to identify the impacts of clear-cutting on forest ecosystems, even though subsequent research showed that not all watersheds responded like the Hubbard Brook (see Section 2.1.1).

In 1980, the National Science Foundation, which is a major supporter of ecological research, established long-term funding for 18 permanent research sites in the United States representing several of the major ecosystem types in the world (Franklin et al. 1990). Four out of the original 17 sites were previously established Forest Service research sites (i.e., Hubbard Brook Experimental Forest in New Hampshire, Coweeta Hydrologic Laboratory in North Carolina, H.J. Andrews Experimental Forest in Oregon, and the Luquillo Experimental Forest in Puerto Rico) and have continued to take advantage of the infrastructure and stability generated by the Forest Service connection. The research groups associated with the Long-Term Ecological Research (LTER) sites are mainly comprised of more than 400 associated scientific personnel from universities, private research centers, and U.S. Forest Service scientists in several of the sites. The disciplinary representation on each LTER site is quite broad, and each LTER is capable of examining ecosystem level questions. Many LTER scientists concurrently conduct research in more than one LTER site. Similar to the IBP, all LTER sites are coordinating the methods used in data collection because one of its goals is to be able to compare data across all the sites. A significant part of the LTER research is to develop large data sets for the selected sites that would be available to all participating scientists and also scientists who are not part of the LTER program (see Chapter 4 on the importance of large data sets).

The LTER program also identified five core research areas to be included as part of all research (Franklin et al. 1990):

- "patterns and control of primary production;

- spatial and temporal distribution of populations selected to represent trophic structure;

- pattern and control of organic matter accumulation in surface layers and sediments;

- patterns of inorganic inputs and movements of nutrients through soils, groundwater, and surface waters;

- pattern and frequency of site disturbances."

These core areas are again common components of ecosystems, as discussed earlier in this chapter. The LTERs are explicitly analyzing the role of disturbances at different temporal and spatial scales on ecosystem functions, something that had not been addressed by most of the large research teams formed prior to this time. New LTER sites are still being added, while some have been deleted from the network. The LTER system and approach is quite active today and is one of the few infrastructures capable of conducting long-term ecological research.

In addition to the research being conducted in natural ecosystems, many federal laboratories have played important roles in trying to understand the effects of human-generated pollutants on ecosystems since the 1960s. Much of the early research specifically focused on determining the impact of one or two pollutants on seedling growth under controlled conditions. However, similar to how the rest of the ecological field has evolved, they have shifted to using a multitude of different approaches to study pollutants and global climate change effects on vegetative communities. Some of these approaches combine the use of models with very controlled laboratory experiments and field studies (Tingey 1994a,b). Much of this type of research would have been very difficult to conduct or implement with individual scientists at the university level. The complex interrelatedness of the research questions and the technical difficulty of setting up experiments that need to be maintained over longer time scales make the federal laboratories most suited for coordinating research addressing the global impacts of human activities. Much of the global change research being conducted by the United States Environmental Protection Agency (EPA) is coordinated by the EPA staff, with university level cooperators and other government laboratories contributing to the core research (Tingey 1994a,b). The urgency of understanding the impact of different human activities on our ecosystems precludes us from contracting research to only university scientists who conduct research as a part-time activity. It is critical for us to define what organizational structures or coordinated interdisciplinary groups work best at addressing the global environmental issues. Only a few examples exist of large interdisciplinary groups working effectively towards solving a common problem. Even the IBP was an early experiment in conducting interdisciplinary research in group settings.

Another more recent example of coordinated group research is the NAPAP program, which is discussed further in Chapter 4. The NAPAP program was launched by a federally mandated policy question (Russell 1992) to determine whether acid precipitation was decreasing the health of forest ecosystems in the United States. Much of the research program was largely driven by university research personnel in combination with researchers from some federal laboratories and private consultants. This program has been heavily criticized for its inability to give useful information to policymakers after being highly funded for a 10-year period (Russell 1992, Schindler 1992). Due to the critical global issues that have to be dealt with today, it is important to examine the management structure of these large programs and to determine how they can be made more effective for contributing to policy decisions (Russell 1992, Levin 1992). In addition to better policy integration of the research, the approaches used to decipher cause–effect relationships of pollutants at the scientific level are still in their infancy. Even Cowling (1992) stressed that the NAPAP did not identify mutually agreed-upon methods for analyzing the cause and effect relationships of one or multiple pollutants until the end of the program, a factor that limited its utility for contributing to policy decisions when a clear connection between a pollutant and damage to a forest cannot be shown.

Much of the research conducted by federal laboratories today is a mesh of core federal scientists from several agencies (i.e., Forest Service, Environmental Protection Agency, Department of Energy, etc.) conducting research in association with university scientists and graduate students. Again, the federal laboratories serve a useful function in being able to: (1) conduct full-time research on a particular topic and concentrate on producing timely research outputs; (2) effectively function as the stable focal point where interdisciplinary research teams from multiple agencies, universities, and private research organizations can be formed to focus on specific large-scale environmental issues; and (3) consolidate, maintain, and function as a central resource center for state-of-the-art equipment needed to address many of the global environmental questions. It is not clear how effectively universities can maintain research resource center accessible to outside scientists, even though many such centers have been established within the past 10 years.

The individual-researcher mode may not be as effective in addressing many of the ecosystem level questions, especially those that scale across large landscape units. The complexity of information that has to be filtered through to identify what is relevant in ecosystems and the fact that data from several different disciplines must be considered suggest an important role for large research groups. This does not imply that the financial support of individual researchers should be neglected, but for those questions that must be addressed at the ecosystem level, a need exists to consider interdisciplinary research teams and not the individual-scientist mode. There is a strong role for individual scientists to contribute to advancing our understanding of specific aspects of ecosystems (Lubchenco et al. 1991), but many aspects of ecosystems can only be addressed by interdisciplinary teams.

2.2   Roots of Management

The traditional uses and management of forest resources have been dictated by the demands of the public (e.g., fuel wood) and industry (e.g., iron production), or the needs of people in power (e.g., English kings; Fernow 1907, Perlin 1989, Williams 1989). Throughout history it has been the *perceived or real shortages* of forest products that have controlled and changed the intensity with which wood supplies were harvested from forest lands. Concurrent with these changes in the intensity of wood harvesting was the development of different forms of management. For example, in Germany the planting of spruce in monoculture plantations was initiated in response to the timber shortages that existed around the turn of the century. After World War II, Germany replanted spruce into previously forested areas that had been harvested as a means of obtaining sufficient funds to pay its war reparations to England. Similarly, in the United States the intensity of harvesting and management of forest resources were dictated by the increased demands made by the public for a sufficient supply of wood materials for building houses and railroads just prior to and after World War I and World War II. In response to the higher timber demands occurring during these times, government, private companies, and forest managers adjusted management practices to increase the supply of timber that could be made available to the public. During these times, the public supported the increased harvesting of timber to satisfy their demands for this resource and forest managers were perceived favorably by the public.

However, the social climate as to what was considered acceptable forest management practices began to change in the early 1970s with the development of the environmental movement and the public perception of damage to the environment due to common management practices (Table 2-4). The public began to perceive large-scale timber harvesting operations (i.e., clear-cuts) as being environmentally unacceptable because they were identified to result in the loss of species and environmental degradation. Forest managers were not as responsive to the change in the public mood as to what was considered acceptable. This converted the previously positive perceptions of forest managers to their being perceived as the villains of the day. The lag time between the response of management to the changing social climate was in part a product of the fact that trees

TABLE 2-4 Summary of ecological, management, and legal events leading up to ecosystem management

Time	Ecological	Management	Legal
Pre-1860		*Practice:* **resource exploitation and use**	**Transfer of public lands to private ownership**
		Social climate: **resource exploitation and use**	
1870–1890	1866: Hæckel defines œcology	*Practice:* **resource exploitation and use**	**Carelessly framed new laws and fraudulent application of old laws allowed continued liquidation of resources**
		Social climate: **preservation**	
1891–1904	Dominance of individual research and natural history approach to research	*Practice:* **custodial management**	**Mandate to improve and protect federal lands**
		Social climate: **preservation**	1891: Forest Reserves Act – provides mandate and halts sale of public lands
1905–W.W. II	**Mandated forest and ecological research begun** 1910: Forest Products Lab 1925: Lotka published *The Elements of Physical Biology*	*Practice:* **custodial mangement** *Social climate:* **manage by scientific principles "wise use"**	**Begininng of organized management agencies and laws** 1905: U.S. Forest Service 1916: National Park Service 1937: O&C Sustained Yield Act
Post-W.W. II–1960s	**Foundation of ecological principles formulated** 1953: Odum's *Fundamentals of Ecology* 1960s: programs on ecosystems in U.S. supported by U.S. Atomic Energy Commission 1969: Odum's "The Strategy of Ecosystem Development"	*Practice:* **range of practices, from scientific management (multi-use) to exploitation,** (e.g., clearcuts of bitteroot and Monongahela National Forests) *Social climate:* **multi-use and sustained yield,** birth of environmentalism	**Management and legal infrastructure well established** 1946: Bureau of Land Mgt. 1960: Multiple Use Act – for multiple use and sustained yield 1969: National Environ. Policy Act – requires EIS

take a long time to mature. What society demanded when trees were planted was different from what society demanded when the trees matured. Because forest do take a long time to mature, a 20 to 30 year lag

TABLE 2-4   *(Cont.)*

1970s	Development and expansion of the ecosystem concept 1971–72: Tropic model of ecosystems proposed by Wiegert & Owen 1979: Borman and Likens synthesis book, *Pattern and Processes of Forested Ecosystems*	*Practice:* multi-use, use, manage by single species 1972: U.S. Forest Service agrees to limit clear-cut size *Social climate:* preservation, with public participation in policy decisions, maturation of environmentalism, the greening of America counterculture	Numerous acts to correct past exploitative mgt practices 1971–72: Congressional hearing on clear-cutting. 1973: Endangered Species Act 1976: NFMA & FLPM required long-range management plans on federal lands
1980s	Numerous publications on management and the ecosystem concept Late 1980s ecosystem approach advocated by scientists & policy analysts	*Practice:* multi-use, use management scrutinized by the public, single species approach *Social climate:* sustainability	Extensive pollution control legislation and regulation, reauthorization of numerous environmental acts 1989: NFMA plans complete
1990s	Academic consensus toward ecosystem management	*Practice:* multi-use, use, single species approach *Social climate:* ecosystem management	Policy: ecosystem management 1992: USFS Chief announces new policy of multiple-use ecosystem management

can often occur between what forests are intended to provide and what the public demands. The social climate of the United States has changed much faster than the forests mature and much faster than management practices could adjust to those changes.

Historically, management practices have been more closely linked to the needs and social climate of the country rather than to any over-reaching policy or ecological knowledge base (see Table 2-4). The pace of change in management practices occurred slowly, taking longest early in our history and progressively occurring more rapidly up to the present. Correlated with this change is the low level of linking of the three fields that have affected resource management: (1) the management practices themselves, (2) ecological research, and (3) environmental policy. Before 1900, environmental policy dictated resource use by encouraging exploitation for the purpose of building the nation. However, beginning at the turn of the century, the role of policy changed. Policy no longer guided or determined management practices. Rather, it assumed a corrective, reactive stance, allowing other factors to influence or change management practices. This *corrective environmental policy* attempted to change practices by preserving, regulating, and restricting resource use. This was accomplished by establishing individual agencies and bureaus with their own specific mandates to control the land under their jurisdiction (see Appendix A and Table 2-4). As more all- encompassing policy was passed, management was restricted by legal and political boundaries as well as each agency's own mandate.

In contrast, ecological research has traditionally not been limited by any legal or political boundaries or restrictions. Academic and government researchers addressed ecological questions that developed a basis for active management. At the turn of the century, the idea of managing forests based on scientific principles began the race to acquire ecological knowledge that could guide management practices. However, until this decade research had not advanced far enough to effectively direct the implementation of resource management on a national basis (see Appendix A). Tracing the history of management, policy, and ecological research will help to illustrate these patterns and trends, and will be presented later. Such an examination will bring us to the present, when ecological research may guide management and policy for the first time in history and the social climate is such that policy can be presented that will allow a new form of management to be initiated.

2.2.1   Management, Policy, and Ecology Historical Review

Prior to and during the 19th century, resource use in the United States was driven by the need to develop arable lands and the need to use the vast timber resources for building the nation. Up until the middle to late 19th century, resource management practices were not a concern to the general public. Forests fed the growing nation by providing fuel and construction materials for cities and railroads. Laws such as the Preemption Act of 1841, the Homestead Act of 1862, and many others were effective in transferring public lands to private ownership and the subsequent use of the resources on those lands. Liquidation, utilization, and exploitation described the land-use philosophy, policy, and social climate of preindustrial America (see Table 2-4).

Changes in public attitudes toward how forests and natural resources were managed occurred when the results of the effects of the rapid use and exploitative management practices became apparent. In 1864, George Perkins Marsh, author of *Man and Nature*, suggested that wise management could mitigate the detrimental effects of exploitive land-use practices. Fears of a "timber famine" in the 1860s to 1870s changed attitudes and fueled the movement toward more balanced management of forest resources (see Table 2-4). The social climate at this time was also in a mood to preserve what was left of the forests, resulting in the 1891 the Forest Reserve Act, which proclaimed federal lands as public reservations. This act halted the sales of public lands and repealed the abused land grant acts (i.e., Preemption Act). However, exploitative management practices continued because policies enacted in the eastern parts of the United States were slow to have any effect throughout the vast nation.

Around the turn of the century, the idea of management based on scientific principles was being discussed. Gifford Pinchot and others began a push toward wise use of resources, protection from fires, and studying and applying the best scientific information. "You can either exploit them or you can manage them," Oberforster von Steuben told the Second American Forest Congress in 1905. The social climate at the time supported the management of the forest lands, providing President Theodore Roosevelt the public support needed to set aside more lands. Though opposition was locally strong, Roosevelt increased the forest reserve area to 40.5 million ha (100 million acres) by 1905. This large acreage fell under the jurisdiction of the newly established Forest Service. However, due to their vastness, custodial management was the best form of management the newly formed Forest Service could perform for the majority of these lands.

From 1905 up to World War II, this form of management focused on the important task of fire prevention (Clarke-McNary Act of 1924). Fire prevention was identified to be an important management role for the Forest Service because 8.1 to 20.3 million ha (20 to 50 million acres) burned annually prior to 1930. Fire was seen as one of the obstacles to profitable reproduction of timber, so that logged areas were not replanted with seedlings due to the risk of their loss due to fire (MacCleery 1993). It was not until the decline of the annual acreage burned in the 1930s and the establishment of the Civilian Conservation Corps (1933) that reforestation occurred on extensive tracts of the nation's forest lands. During this early period, ecological research was in its infancy and the development of scientific principles applicable to forest management was just beginning (see Appendix A). During this time and earlier, ecology and forest management were not disjunct from one another, even though ecologists were indirectly contributing to silviculture. The principles of silviculture being developed during this time were founded on a strong understanding of tree growth in ecological systems (Toumey 1928). It was also during this time that forests began to regrow in the eastern parts of the United States as people migrated to the cities and abandoned farms, established in forest areas in the previous century.

After World War II, the high demand for wood to build houses dictated the harvesting of timber on national forest lands. Forest managers met these demands by cutting timber. To ensure that these demands were met in the future, the concepts of multiple use and sustained yield developed. The Multiple Use Act of 1960 provided the mandate to develop and administer national forests for multiple use and sustained yield. Active management of forests increased concurrent with more forestry research as managers tried to catch up with the demand for wise use of forest lands. The infrastructure and administration for the active management of national resources was well in place.

In the 1960s ecological research began examining forests at larger scales and viewing ecological processes as occurring within a systems framework; the ecosystem concept was materializing (see Appendix A). At the same time, management practices covered the range of possibilities from those based on the social climate of the time (i.e., multiple use and sustained yield) to clear exploitation of forests in the form of large clearcuts. As a result of the perceived exploitation of some federal lands, a vocal rebelling public demanded a halt to the ecological destruction being caused by the government (i.e. clear-cutting on the Bitterroot and Monongalhela National Forests). The high demand for timber after World War II was over and a new generation of people perceived the forest resource differently than those of the past—environmentalism was born.

By the 1970s, environmentalism was maturing and the cry to preserve forests was once again heard. In an attempt to prevent the errors of the past, environmental impact statements and long-term management plans were required by law on federal lands. Social perception of acceptable practices on federal lands was changing rapidly, and the ability of managers to develop sound management practices compatible with these changes was a difficult task. Dr. R. Keith Arnold articulated the difficulty of the time in his address to the Sixth American Forest Congress when he said, " The accelerating momentum of change belies our ability to even describe the world as we know it today. We must invent something beyond periodic policy assessments. In the past, they have been sufficient. In the present they are clearly inadequate, and in the future they will be worthless. Policy formulation is a continuing, complex, and comprehensive task which has been accomplished on an *ad hoc* basis up to now. At the close of the Sixth American Forest Congress, it is obvious to me and I believe to you, that we need something new."

Management of federal lands in the 1970s and 1980s was driven by the multiple-use concept and the Endangered Species Act (see Appendix A). The resulting management scheme focused on single species or single resources for conservation or use. In the 1980s, ecological research moved rapidly forward in developing the ecosystem concept and pushed for a more holistic form of management. This contrasted policy debates and legislation, which focused on pollution controls and the reauthorization of past environmental acts. No new policies for forest management were considered during this time period (Vig and Kraft 1994). Sustainability was the buzz word of the 1980s, even though it was not exactly known what this meant or how to accomplish it.

In the 1990s the ecosystem concept gained wide attention from the academic community (see Appendix A). Numerous publications have been produced emphasizing the ecosystem approach as the way to manage the country's natural resources. The United States Forest Service announced a new policy of multiple-use ecosystem management in 1992. Currently, all federal agencies are seeking ways to implement this new management philosophy (GAO 1994).

Today, similar concepts and approaches are being simultaneously developed in ecology, management, and policy in an environment where consensus is possible (see Table 2-4). Historically changes occurring in one field were always disjunct from the other fields by at least 20 to 30 years. This resulted in a long transition period before policy or ecology were able to affect a change in management practices. In the early history of the United States (1800s), the social climate dictated the management practices (clearing and use) so that ecology did not contribute to the formulation of the policy. Then, in the early 1900s, policy sustained those land-use practices and ecology attempted to understand the impacts of prior land use practices. In the early half of the century, policies were being considered to implement "true" management practices but change was slow. Not until the enactment of the Multiple Use Act of 1960 did forest management practices begin to change. However, by the time that multiple use became the accepted practice in the 1980s, policy was already moving toward requiring environmental impact statements and long-term planning. At the same time, ecological research was moving ahead to develop the ecosystem approach to natural resource management. In the 1990s, the shortest transition time and narrowest gap exists between the three fields affecting natural resource management. For the first time, ecology may contribute to driving policy, with management ready and waiting for the mandate to carry it forward.

In the past, the large size and extensive availability of natural resources in the United States strongly contributed to how resource use and management developed. The early perception was that errors in management were not critical because there was always another forest where one could go to extract timber. The reduction of available old-growth timber, whether due to liquidation of existing stands or the changes in policy that made the remaining stands inaccessible, has changed forestry from an industry that has always been able to harvest the resource from region to region, to one that has to treat the resource everywhere as strictly an agricultural crop. It is this history that makes forest management in the United States fundamentally different from most European and Scandinavian nations. Unlike our European counterparts, who have had to deal for centuries with the risk of uncertainty with regard to wood supply due to the smallness of their resource base, the United States has just begun to face that uncertainty and ecological risk associated with a restricted supply of natural resources.

This shift creates circumstances under which timber growers and resources managers need to ensure the long-term maintenance of timber supply from the existing land base. No longer can timber be viewed as a resource to be mined, and it is unethical to do so when the methods, technology, and financial resources are available to properly manage the lands already utilized for timber production. The need to ensure long-term timber supplies from existing forest lands fundamentally changes the way natural resources are managed in the United States and drives the current impetus to utilize a new management approach, in this case, ecosystem management.

2.3   Ecosystem Management

2.3.1   Development of the Concept

The term *ecosystem management* has been used and defined by various groups in several different ways and, because of somewhat vague definitions, has the potential to cause confusion among managers. If the ecosystem management definition is kept sufficiently vague, it will be very easy for individuals with their own agenda to begin to associate undesirable characteristics with these words, with the end result being that the term becomes ineffective and is abandoned. It is very important that this does not happen because there are core attributes that form the ecosystem management paradigm that should not be lost from our management system.

Other terms that are sometimes utilized synonymously with ecosystem management are *ecosystem health, ecological integrity*, and *sustainability* (two of these terms are further discussed in Chapter 4). Terms like *ecosystem management, ecosystem health*, and *ecological integrity* have been used to connote the broad aims of environmental management, that is, to suggest managers should protect ecosystem health or integrity through ecosystem management. The terms *health* and *integrity* have also been used in association with theories of ecosystem development and have been suggested to correlate with quantitative indices of ecosystem conditions (Haskell et al. 1992, Bourgeron and Jensen 1993). If there is no strict theory or mechanistic understanding of what constitutes ecosystem health or integrity (see

Chapter 4), the development of quantitative indices will become based on value judgments to identify what to include in the analysis. Whenever science and values mix in analytical endeavors, there is a tendency to be less comfortable with conclusions that are based on these analyses. But, as part of ecosystem management, it will be necessary to explicitly incorporate scientific information with social values as an integral part of the formula for management.

Ecosystem management similarly suffers from sounding like something good but so vaguely described that managers are left trying to implement something they cannot define. Phrases like *ecosystem management* and *ecosystem health* are terms that have been selected in reaction to a dissatisfaction with previous approaches. These approaches are selected because they sound good and they appear to replace methods that have been classified as "undesirable". These terms have also become associated with all management regimes, causing a lack of credibility in the eyes of individuals not comfortable with this approach and a lack of clarity for those hoping to adopt the new approach.

Ecosystem management is currently an ill-defined management approach, adopted in an attempt to move away from previously identified unsatisfying management schemes. Because it has been an attempt to move away from something unwanted rather than a move toward clearly outlined management goals and strategies, it is easier to define what it is not rather than what it is. Ecosystem management is:

1. **Not** *multiple-use*, in which everyone was offered everything with no one having to sacrifice anything;

2. **Not** a *single species approach*, which emphasizes the particular species people think are important, and often involves crisis management, in which species are targeted for conservation only when they become very close to extinction; and

3. **Not** grounded firmly in either *biotechnologist* views, which suggest that nature can be improved by the works of humans, or *bioconservative* ideals, which treasure and seek to preserve the biological and ecological status quo (Sagoff 1992).

Defining what something is not, however, offers little guidance to those charged with implementing ecosystem management. Difficulty in defining ecosystem management to a wide audience stems, in part, from the fact that ecosystem ecology is in itself a relatively recently accepted concept in ecology (see Appendix A). Ecosystem management is a very new idea for many federal agencies who have embraced it as their management paradigm. For example, in 1992 the U.S. Forest Service was directed to manage under the ecosystem management concept, focusing on ecosystem sustainability (Robertson 1992, Overbay 1992). There exist numerous managers at federal agencies and academicians at universities who have described ecosystem management in a variety of ways. Some of the definitions are given below to highlight what variables were perceived by each to be inclusive as part of ecosystem management.

U.S. Forest Service

"... means that we must blend the needs of people and environmental values in such a way that the National Forests and Grasslands present diverse, healthy, productive, and sustainable ecosystems" ... "protecting or restoring the integrity of its soils, air, waters, biological diversity, and ecological processes" (Robertson 1992).

"Ecosystem management as an approach to the management of natural resources that strives to maintain or restore the sustainability of ecosystems and to provide present and future generations a continuous flow of multiple benefits in a manner harmonious with ecosystem sustainability (Unger 1994)."

"... a system of making, implementing, and evaluating decisions based on the ecosystem approach, which recognizes that ecosystems and society are always changing." ... "the degree of overlap between what is ecologically possible and what is desired by the current generation. We advocate that the desires of future generations be protected by maintaining options for unexpected future ecosystem goods, services, and states" (Bormann et al. 1993).

U.S. Multi-agencies—Forest Service, Bureau of Land Management

"A strategy or plan to manage ecosystems to provide for all associated organisms, as opposed to a strategy or plan for managing individual species" (FEMAT 1993).

U.S. Department of Interior, Bureau of Land Management

"Ecosystem management is a process that considers the total environment. It requires the skillful use of ecological, social, and managerial principles in managing ecosystems to produce, restore, or sustain ecosystem integrity and desired conditions, uses, products, values, and services over the long term... Ecosystem management recognizes that people and their social and economic needs are an integral part of ecological systems." (USDOI BLM 1993)

U.S. Environmental Protection Agency

"To restore and maintain the health, sustainability, and biological diversity of ecosystems while supporting sustainable economies and communities" (Lackey 1995).

Professional Society—Society of American Foresters

"Ecosystem management focuses on the conditions of the [ecosystem], with goals of maintaining soil productivity, gene conservation, biodiversity, landscape patterns, and the array of ecological processes" (SAF Task Force 1992).

Academicians

"Ecosystem management involves regulating internal ecosystem structure and function, plus inputs and outputs, to achieve socially desirable conditions" (Johnson and Agee 1988).

"... is partly a matter of redefining management units and partly a matter of building on the best ecosystem science... The goal is to provide a framework and a research agenda that will facilitate the joint achievement of environmental protection and economic development through modified planning, management, policy, and decision-making activities" (Slocombe 1993).

"We must emphasize the placing of forest landscape objectives in a larger context in both space and time" (Mladenoff and Pastor 1993).

"Every system definable in biological and physical terms connects to and interacts with a network of human values, uses, institutions, and other social structures... tracks transactions across defined system boundaries and moves the boundaries themselves when necessary" (Gordon 1993).

Consultant

"... ecosystem management, focuses on the maintenance of an ecosystem's natural flows, structure, and cycles, displacing the traditional emphasis on the protection of such individual elements as popular species or natural features" (Goldstein 1992).

Interestingly, all these definitions published by different agencies and academicians repeat certain words, even though they do not all include the same words in their definitions. Many of them explicitly present ecosystem management as being a balance between sustaining natural resources at the same time as sustaining the social and economic systems dependent on the natural resources. Some directly address the fact that ecosystem management means a movement away from the protection of individual species to protecting the whole ecosystem. There are no conflicting statements among these different definitions. They all say about the same thing, but at such a broad level that it is impossible to characterize what ecosystem management should really manage for. Only the definition from the Society of American Foresters identified the specific functions in ecosystems that should be sustained, such as "soil productivity, gene conservation, biodiversity, landscape patterns, and the array of ecological processes" (SAF Task Force 1992). Many of the attributes of ecosystems that we discussed earlier are mentioned, such as ecosystem structure and function, ecosystem health, ecosystem sustainability, biological diversity of ecosystems, and sustaining economies and communities.

2.3.2   Operational Definition of Ecosystem Management

Components of Ecosystem Management

Ecosystem management has been defined in many ways in the few years that the term has been in use. To date, no single, unifying definition has been accepted by either scientists or managers, but certain principles and concepts are present in virtually all literature discussing ecosystem management. The principles and concepts relevant to ecosystem management are articulated as follows:

1. Ecosystem management is necessarily a melding of natural science tools and data, and bureaucratic and social science techniques. A balance must be struck between the physical and biological facts of ecosystems and the equally real human factors.

2. Ecosystem management requires the active management of both the natural system (or the internal dynamics) and the anthropogenic effects or external influences impacting that system (Saunders et al. 1991).

3. Ecosystem function should be measured using two parameters, biodiversity and productive capacity. While biodiversity is more easily monitored and quantified, the ecosystem management idea demands that the functions and processes of the ecosystem be considered in both analysis and decisionmaking.

4. Ecosystem management necessitates the identification of *thresholds,* levels of degradation below which ecosystems cannot drop without losing certain vital attributes or functions. A major role of the ecosystem scientist and manager is the development of tools to identify these thresholds, to identify different threshold levels within an ecosystems and the communication of these data to decisionmakers (NCASI 1992).

5. Ecosystem management necessitates the systematic, scientific study of human uses and effects on the ecosystem. Ecosystem management requires a balancing of both sides of the equation. This principle is less developed in the existing literature but is of crucial importance.

6. There is no "free lunch," that is, some loss is inevitable as expanding human uses meet existing ecosystem functions. Thus ecosystem management is ultimately about presenting the choices and trade-offs, and estimating and monitoring the costs and benefits of making these choices. An understanding and acceptance of loss is part of ecosystem management.

7. The scale of ecosystem management must be flexible enough to respond to management goals and objectives. No one spatial scale by itself is adequate to manage ecosystems. Similarly, the temporal scale

must be adaptable enough to allow for regeneration of an ecosystem through a full cycle or a catastrophic disturbance.

8. Adaptive management is an essential part of ecosystem management. The rules and criteria must be flexible enough to adapt to changing biophysical conditions, human behavior and objectives, and scientific advances. Ecosystem management requires a system that learns from its own mistakes, that is not rigid but has feedback loops.

Throughout the development of the ecosystem concept, the issue of ecosystem complexity has prompted debate on whether an ecosystem can have properties at larger scales that can be understood from the interactions occurring at smaller scales or whether the systems' expression of properties (i.e., organization and regulation) may be deduced from their parts (Ricklefs 1990). Consequently, there have been many theories surrounding ecosystem processes, ranging from Lotka's 1925 thermodynamic perspective, to Lindeman's 1942 energy-transformation processes, to Odum's ecosystem energetics, and lastly to systems ecology (Ricklefs 1990). These questions are still being asked today with no clear resolution of how to scale information in complex systems.

Within an ecosystem, there are many hierarchical levels of organization that are linked to one another (i.e., nutrient cycling, trophic levels, water flow, etc.). To adequately ensure that the best management decisions are being made with the best available *scientific* data, changes in the functional processes of an ecosystem must be measured on the physical, biological, and chemical levels using various indicator parameters. These may range from monitoring species, populations, nutrient cycling, decomposers, and/or net primary production. Because there are different levels of organization, a number of parameters must be measured at all times. Although this sounds time consuming and expensive, it should be pointed out that many areas already have multiple management objectives—wildlife, clean water, etc.—that may be integrated into an ecosystem monitoring program. It will require additional coordination of information collection, but will also serve to inform managers where they have information gaps.

In the vast majority of the existing ecosystem management literature, the functions and viability of the ecosystems have been monitored using either a species level approach or an ecosystem level approach (i.e.,

NPP). In the past, the focus of managers and environmentalists has been principally on biodiversity and the prevention of loss of genetic diversity, especially species. There are a number of reasons for this focus. Biodiversity, compared with productive capacity, is relatively easy to measure, assess, and monitor. When genetic diversity is lost from the system, it is assumed to to be very final with few recourses for re-establishing that diversity, and there is a sense of urgency that this is occurring very rapidly due to human activities. In addition, biodiversity issues, particularly of charismatic megafauna, are attractive from both a fund-raising and public consciousness-raising standpoint. For these reasons, a lot more scientific attention has been paid to understanding biodiversity issues than ecosystem level indicators such as NPP.

Net primary production estimates measure the capability of ecosystems to do work (utilize, transform, and store energy), to evolve structurally, and to withstand disturbances. In general, NPP, unlike biodiversity, is a measure of the overall functions of the system rather than a counting of its individual parts. NPP estimates reflect the activities of the whole ecosystem, which is why it better reflects ecosystem function compared with biodiversity estimates (except for keystone species, see Section 2.1.2).

One of the great limitations currently for using NPP measurements in ecosystem management is the limited understanding of productive capacity and what controls within-plant carbon allocation patterns (see Section 3.6.1). Forest managers have traditionally not measured NPP in their stands but have concentrated on estimates of above-ground wood volumes. This would necessitate managers collecting additional information on their sites because past approaches cannot be easily converted to NPP values. Similar to NPP, nutrient and carbon cycle models are also in the early stages of development, especially for management purposes. Yet it seems clear that NPP estimates are an essential component of ecosystem management and that biodiversity by itself is an insufficient tool. To overcome this knowledge gap, greater attention will have to be placed on measuring NPP in ecosystems. Productive capacity will have to overcome its obvious public relations and comprehension disadvantage to biodiversity. How can soil, leaf mass, and unseen nutrients be made as charismatic as grizzly bears? Some of the work of ecosystem management will be this education process.

The two measures of ecosystem activity (i.e., biodiversity and NPP) are not decoupled from one another, so they should not be analyzed in isolation from each other. For instance, if the system loses biodiversity, it loses the capacity to do work on an evolutionary level. Depending on the nature and extent of the loss, it may also lose resistance, resiliency, and/or productivity as measured by photosynthetic efficiency. Likewise, a system with degraded productive capacity will probably be less able to support a broad range of flora and fauna.

Looking at species using a systems perspective means concentrating less on the *number* of species in a system but rather examining the interactions between species and their functional roles, especially as related to "perturbations" to the system. In the context of perturbations, a systems view alerts us to the role of species or species assemblages as *modulators* of various stressors by virtue of their function in the system. Thus there is less of tendency to proclaim that biodiversity reflects the state of ecosystems, or that high diversity necessarily provides "resilience." A systems perspective acknowledges the role of diversity *as it is relates to other factors that influence the system,* such as disturbance cycles, novelty of stressors, energy and nutrient pools and cycles, hydrologic regimes, and adaptive strategies of the species. Rather than a universal question, such as "Does diversity promote stability or enhance functions in ecosystems?", a systems perspective asks "What would happen *in a particular ecosystem* if certain species are lost?", or *"What might happen to particular functions that allow the continued output of some desired system attribute?"* or, "If a number of species are lost, what else (in terms of ecosystem function or structure) is lost?" In short, a systems view lets us consider that the effects of species loss may depend on more than the just the number of species that are lost.

Because productive capacity has been defined as the ability of the ecosystem to do work (i.e., whether it be to produce human-desired products or adapt to natural variations within the system), it is possible to use consumptive measures to monitor the functional processes. To date, this has been the most common indicator of ecosystem processes, such as declines in fish runs, difficulties in regenerating previously forested landscapes, desertification, etc., which are unmistakable signs of an imbalance between human desires and the ecosystem's ability to sustain its rate of production. Yet changes in productive capacity also serve as an accurate assessment of societal demands on the system. Ecological problems were not the sole impetus for the forestry "reform" movement (Brunson

1993); public ideals of land use have changed and non-extractive uses, such as outdoor recreation and enjoyment of wildlife, were gaining in constituents. Changes in the public support base calling for land-use "reform" were in actuality calling for more emphasis on the productive capacity of the ecosystem to produce their desired objectives.

Ecosystems and Humans

Because ecosystem management requires a balancing of human and natural system needs or demands, the conceptual framework for ecosystem management must include ecosystem functions, human uses, and the interplay between the two. This is represented in Figure 2-1 as a system in which a finite pool of resources is divided between ecosystem functions and human uses. The threshold region represents the area in which further expansion of human uses will almost certainly compromise or endanger ecosystem functions. While there are many systems and situations in which human activity and ecosystem functions can coexist, for every system there is some threshold beyond which the system cannot shift without resulting in a degradation of some structures and/or functions.

Human uses are defined here as all human effects on the ecosystem, whether physical, social, economic, or aesthetic. One of the limitations of ecology in the past as an ecosystem management tool has been that it tends to focus on nonhuman "natural" systems. Human activity tended to be separated and treated as the domain of the social sciences. This frequently resulted in human uses not being examined with the same scientific methodology and tools that were applied to the rest of the system.

Ecosystem management requires the understanding, assessment, and management of human activities (Brunson 1993). Ecosystem science can analyze and predict human behavior in ecosystems just as easily as the behavior of other fauna and flora. It can examine the actual and likely impact of interactions between humans and nature. For instance, in managing grizzly bear habitat, it is just as important to understand what humans will do in the presence of a bear as what grizzlies will do when they encounter a person (Mattson 1994). This kind of analysis is essential because ecosystem management necessarily involves balancing and making choices between ecosystem functions and human uses. Ecosystem managers need to understand and predict human population growth and migratory activity, as well as human consumption, resource utilization and destruction, and waste production. They also need the analytical tools to understand the impact of management actions on human behavior.

Thresholds and the Inevitable Trade-offs

Ecosystem management necessitates the identification of *thresholds*, levels of degradation below which ecosystems cannot move without losing certain vital attributes or functions, or perhaps even experiencing a collapse of some part of the ecosystem (see Figure 1-2, NRC 1986). In many cases, the threshold will represent a trade-off between ecosystem functions and human uses. There will not be a clear line but instead a zone in which increasing human uses begin to compromise ecosystem functions. Nonetheless, this is the area in which most ecosystem management in fact operates. Like the 55 mile per hour speed limit or the 50/500 rule in conservation biology, judgments in ecosystem management are made on the basis of the best available science and information. For instance, the risk of death or injury increases along a gradient with automobile speed. The speed limit is a determination of what society considers an unacceptable level of risk. Because the exact threshold cannot be known or is dependent on stochastic or catastrophic events, a comfortable threshold will always include a margin of error. The relaxation (November 1995) of speed limits in the United States will have costs associated with the benefits of being able to legally drive faster.

Whether thresholds can be defined in every case is subject to debate, as is whether theoretical thresholds hold up in the real world. An approach is presented in Chapter 4 that is worthwhile pursuing for identifying thresholds in ecosystems. From a management perspective, thresholds are necessary to set policy; without them managers do not know where to draw the line between competing ecosystem uses. Thresholds also ultimately provide the enforcement mechanisms available to ecosystem management.

A basic concept in ecosystem management is that some loss is inevitable as expanding human uses meet existing ecosystem functions. In a system with finite resources, there is no ability for all resources to be accessed at maximal levels. Thus ecosystem management is ultimately about presenting choices and trade-offs, as well as estimating and monitoring the costs and benefits of making these choices. An understanding, estimation, and identification of the choices that will occur with certain management decisions is integral to ecosystem management.

The quality of decisions ultimately depends on the quality of the data. Ecosystem scientists are charged with providing the best possible information about ecosystem functions and human uses, and their present and potential impacts on the system. The goal of ecosystem management is to present choices in such a way that the effects of the chosen policy or activity are understood, foreseen, and accepted. Loss is inevitable, but ecosystem management can delineate what is retained and what is lost.

2.3.3   Ecosystem Scale and Management

Spatial Scale

Choosing a spatial scale for ecosystem management will be difficult if one continues to regard an ecosystem as a individual unit, isolated from the surrounding landscape. It is better to think of the ecosystem management unit as a system composed of physical-chemical-biological processes active within a space-time unit (Tansley 1935, Lindeman 1942) that has physical boundaries but not functional ones. Therefore, choosing a spatial scale is to define the physical area with which one is working but not to design to isolate the functional processes.

The issue of scale becomes especially important these days as many natural resources managers realize that current management units may not be large enough. Concerns that reserves may not be large enough to protect some of the larger vertebrates and other taxa (Soulé 1980, Frankel and Soulé 1984, Salwasser and Schonewald-Cox 1987) have prompted calls for increasing the scale at which wildlife is managed by incorporating corridors and buffer zones into other management units, instead of trying to produce viable populations within the confines of parks and wildlife refuges (Servheen 1986). Depending on management objectives, watershed or landscape level approaches are more appropriate management units. This holds particularly true for managers whose resources are migratory or mobile.

Natural resources disregard political boundaries. Inputs into an ecosystem may flow across several different human-designated jurisdictions before becoming retained, depending on the scale at which one is measuring. Therefore, one can see the scaling of ecosystem management must be flexible in order to respond to changes in knowledge of ecosystem function, and management goals and objectives. Scaling is need-driven and may go up or down depending on the scope of the project. The issue of transboundary management is becoming more important and will require greater cooperation between private landholders and state and federal agencies. Yet, how does one ensure continuity amongst parties and practices?

One approach to addressing these scaling issues is to use *nesting*. Whatever the scale chosen, one must link it with the larger and smaller scales around it. One way of thinking about this is to imagine the management unit as a city. While semi-autonomous, the city is still responsible for coordinating with county, regional, state, and ultimately, federal governments. Certain federal laws must be observed on the city level, and yet the city chooses how it will do so. The same may be done to measure productive capacity using such indicator parameters as previously discussed. While not all parameters need match, processes must function in enough of a similar fashion to protect the integrity of the ecosystem and to maintain some larger degree of functional continuity. Buffer or transition zones are one example of a nesting effect, with gradual changes over a spatial and temporal period.

For decades, demands to either preserve or develop resources have been the focus of many environmental conflicts. A typical example of this is the Greater Yellowstone Ecosystem (GYE). The GYE area has also been traditionally dependent on the bust-and-boom industries of logging, grazing, and mining (Power 1991). These extractive or impactive effects may change the visual appearance of the GYE landscape, which conflicts with the competing demands that have been occurring with people's desire to utilize this land area for conserving large mammals and for recreational uses. Increasingly, it becomes more difficult to set aside lands for conservation purposes and still sustain the private industries seeking to use public lands. Such conflicts between users and use of public lands will continue as resources become more scarce, resulting in increasing competition for their use. However, it is certain that federal land holdings are not sufficient to function as a comprehensive national biodiversity repository unless devoted solely to that function (Soulé 1991). Even then, there are biomes that would be inadequately represented. In this respect, nesting is vital to maintain the levels of biodiversity the public may desire.

Nesting will also help manage across public and private lands by providing a framework and identifiable goals. It will be easier to arrange for coordination once specific targets have been designated, although obviously some conflicts will persist. Currently, some states such as Washington State, have or are considering adopting SEPAs or State Environmental Protection Acts. Similar to NEPA, such acts seek to assess environmental impacts, which may eventually provide statewide data on the status of the ecosystems. Ultimately, a NEMA or National Ecosystem Management Act is envisioned to give legal and policy support to the idea of nesting.

Temporal Scale

The question of temporal scales is more difficult to integrate into ecosystem management. Ecosystem management must occur on a temporal scale that allows ecosystem functions to be maintained through a complete natural cycle. Quite often there are reoccurring triggers within an ecosystem (i.e., fires, floods, hurricanes, or cyclical drought periods) that establish a pattern within the ecosystem functions. For the grasslands of Brazil it may be the yearly floods or for the Yellowstone region it may be periodic forest fires (Knight 1991). Without accurate data and careful attention, it will be hard to determine the reoccurrence or return interval of many of these phenomena. The appropriate temporal scale would be to allow the ecosystem to return to the predisturbance or threshold levels, depending on which one was closer to protecting its functional integrity.

In addition, temporal scales raise the issues of intergenerational equity and long-term and short-term planning. Often, we fail to look far enough into the future when considering the impacts of our actions and rarely, if ever, assess the cumulative impacts (Wargo 1996). This is due, in part, to our inability to quantify or even identify current changes or impacts. However, management decisions affect future generations, and it is the responsibility of current generations to make judicious choices that will not severely impair the ability to make resource use decisions in later years. It is worthy of serious consideration, especially as population levels and resource consumption continue to grow and more demands are placed on the existing supplies. Consequently, resource allocation should be considered in both long- and short-term frameworks.

2.3.4   Summary

Figure 2-1 is a useful way for conceptualizing how managers should perceive the role of species in ecosystems. It (the functional spiral) graphically represents the functional group and functional redundancy concepts, and how they relate to ecosystem response and resilience to perturbation. Each circle represents a number of functional components within a given ecosystem; the larger the circle, the more functional components (more slices of the pie) and the more redundancy within each functional component (larger pie slice). In response to perturbation, the system may lose functional components as well as redundancy, causing the circle to contract. A degraded ecosystem is thus depicted by a smaller circle. Theoretically, a manager could therefore identify the appropriate tier and then manage for the functional components in that circle or any larger circle. However, it is a bit more complicated, as ecosystem functional components may vary in both space and time, so that the circle may expand and contract naturally, moving up and down the spiral. As a result, managers must identify a threshold below which the system does not naturally drop unless it is degraded.

Ecosystem management will require more than an expanded scientific knowledge base. It necessitates an approach that integrates the study and understanding of both ecosystem functions and human uses. The major scientific challenge is to define the "threshold" level and to communicate its value and importance to the decisionmakers and the public at large. The major societal challenge is to define what is retained and lost as part of management, and to make choices depending on values assigned to various practices. However, the public must also realize that it is impossible to continue the current level of consumption without loss. Ecosystem management is not only management for the people and by the people, but also of the people.

It is relevant to define what good management is and not worry as much about the definition of the terms. Today this is being defined as ecosystem management, but the terms being used are not as important as the activity that is generated by a new approach. Ideas about good management might be defined more easily by what they are not: *not* a single-species approach, *not* a multiple-use approach, etc. But good management can also be defined as a perspective that emphasizes management units nested in a larger system or landscape matrix, and that acknowledges the limits to what the system can provide over the long term.

Principles of good management will consider and incorporate the following:

1. *Spatial and temporal scales are critical.* Problems of scaling and transfer of information between levels exist (the management unit may be at a very different scale from the scale in which feedbacks are impacting on the management unit).

2. *Limits to system capacity are considered.* Biological/ecological thresholds are acknowledged and investigated. Good environmental management recognizes the *limited productive capacity of ecosystems.* Productivity in a system (and thus the amount of system output that can be harvested or harnessed by people) is constrained by limited available energy and nutrients, periodic disturbance cycles, environmental factors (temperature variations, wind, etc.), and limited energy transfer efficiency.

3. *Ecosystem managers look for the root of the problem, rather than treating one symptom at a time.*

4. *Social constraints are considered.* Often social constraints drive and "run over" biological constraints. Ecosystem management involves looking at both social constraints (and benefits rendered by human activity) and ecological constraints on what people can do (harvesting, etc.) in the system. Trade-offs are made explicit. Ecosystem management is not based on a precautionary principle nor on the premise that impacts of human behavior are bad. Humans do change ecological systems but not always in ways that irreversibly change ecological processes. With a systems view, human activity can be viewed as causing change or as influencing ecological processes, and *then* evaluated with respect to its good or bad points. Ecosystem management acknowledges the important role of humans in shaping ecosystem structure and function, and so necessarily must consider the social constraints that influence human behavior.

5. *Science is used to inform decisions*, to let us know both what is known and how much is not known about the system and human influences on it.

6. *System function as well as structure are investigated* (maintenance as well as output). More than a species management approach, ecosystem-based management involves consideration of ecosystems of which individual species are but one component. Individual species may have a greater or lesser effect on system function, depending on the functional role of the species in the system. Characterizing species dominance, functional role, and functional redundancy within the system aids in getting a richer picture of the role of species in ecosystems.

Also, systems-based management allows us to look at the forces that influence human behavior, and so obtain a better picture of why environmental problems arise. For example, knowing about the economic incentives that influence the use of natural resources gives one perspective on ecological problems and how they should be managed. Then it can be asked whether economic incentives that lead to destructive activities can be changed, and how economic incentives can be used to both obtain the natural resources/amenities desired by humans and to maintain the function of ecosystems in a condition that is also desirable.

It is also important to understand that there are social system attributes that are related to the intensity of natural resource use. One of these is the diversity of natural resources that provide an economic basis for a society. Dependence on a single natural resource promotes intensive use of that resource, as well as political pressure to resist regulation of its use. For example, logging towns dependent mainly on timber as their economic base will demand compensation for their economic losses due to government restrictions on logging. With a systems view, this situation can be seen as a lack of redundancy in the social system when the local economy is built on a single resource. The impacts of restrictions on use of that resource are fed back quickly through the social system. In contrast, a system that is heterogeneous and consists of subsystems that make use of different resources is buffered from shocks that come from the loss of any single one of the resources.

One of the key aspects of good management is to design the management activity such that it is flexible, especially in complex systems where uncertainty is ever present. A systems perspective suggests examining the positive and negative feedback cycles that connect system components. By doing this, an idea about what is driving and constraining the system can be obtained. With this information, good understanding about trends can be obtained and used to design management with what is known. At the same time, an appreciation of the complexity of the system suggests the importance of building flexibility into management.

Ecosystem management does not provide a panacea. Environmental policy and resource management will remain fraught with conflict, conflict that may become heightened with tightening budgets and pressing social problems that compete for attention and funds. Potential problems may accompany ecosystem management, both existing problems that are expected to persist and new ones specific to ecosystem management. For example, using a new name (i.e., ecosystem management) does not guarantee better management. Changing the guiding phrases without changing what is conducted operationally will not necessarily guide management to better practices. The term may also cause great confusion among practitioners because the ecosystem concept and ecosystem management have not been readily comprehensible to a wide audience (Norse 1991). Ecosystem management, although a relatively new concept, already means different things to different people and to different groups. Both the lack of understanding and variety of interpretations have the potential to cause confusion about what is being managed and how management should proceed.

Because managers have to work with physical units, there may be a tendency to try to delineate "ecosystems" geographically and to assume that changing the management boundaries (generally enlarging them) changes the management approach and solves problems. Instead, the ecosystem concept is most useful as an approach to management that emphasizes driving variables and constraints on ecosystem structure and function, and the relationships between management activity and system structure and function. A more successful approach might be to manage environmental problems at the appropriate scale. Because various problems occur at different scales (even within a single ecosystem or management unit), they must be managed accordingly. Therefore, no single set of ecosystem borders will automatically be appropriate for management of an array of regional issues. When managing for air quality (a problem that involves sources and sinks spatially distant), borders will be quite different than when managing timber yields. A more realistic approach might be to bound environmental problems by the biogeochemical processes that dictate the fate and effects of human activities.

Also, problems have to be bound by the stakeholders involved in the problem. When stakeholders can be identified, the management unit can be defined. The delineation of stakeholders (including future generations and those distantly affected by environmental problems) can be as difficult as the delineation of ecologically sensible boundaries. This is especially difficult where common property resources are involved, such as with marine issues. Here it may be difficult to identify the injured parties when overfishing depletes fish stocks and the fishing industry declines, or the parties injured by a disastrous oil spill at sea.

Because ecosystems can be bounded using any process or structure-related criteria, they do not coincide with political borders. Implementation of comprehensive management regimes that are based on ecological processes will then be a problem in ecosystem management. Again, locating stakeholders and negotiating problem amelioration is a key issue. Ecosystem management requires integrated management, bringing together economic, social, and environmental aspects of management issues (Norse 1991), which is difficult and time consuming. However, this is the only way to get beyond the traditional piecemeal problem approach. The traditional approach to environmental problems entails reacting to symptoms of problems, usually when they have developed to a crisis stage. This has resulted in different laws and sets of regulations aimed at specific symptoms—one set of regulations for the adverse effects of building and development, one set for water pollution, one for air pollution, and another for limiting resource overexploitation. This symptom-by-symptom approach had led to fragmentary environmental management decisions and resulted in agencies or divisions within agencies working at cross-purposes (Norse 1991).

3 | Tools and Knowledge Base Presently Available to Do Ecosystem Management and to Assess Its Success

3.1 Overview

Managers and researchers are currently defining and setting objectives for ecosystem management; however, it is not clear what ecological information is needed as a basis for setting specific management goals and priorities, or to implement ecosystem-based management in the field. The focus of this chapter is to describe how and what type of data have been collected in the past to assess ecosystems. The objective of this section is to identify what tools and bodies of knowledge are currently available to guide the planning and implementation of ecosystem-based management.

As discussed in Chapter 2, the adoption of an ecosystem perspective is a necessary first step toward planning for the management of ecosystems and is central to its implementation. The next phase in the development of an ecosystem management plan should be an ecological assessment of the specific system being managed. Advancing the ecosystem perspective further requires the application of more in-depth ecological concepts and theories in ecological assessments to inform management planning and implementation. A body of ecological theory and knowledge will be discussed that, when applied to specific systems, may aid managers in viewing both the systems they manage and the information currently available about those systems from an ecosystem perspective. When components of this body of theory and knowledge are used to guide ecosystem management, these can be referred to as *conceptual tools*. The conceptual tools discussed in this chapter include knowledge and theory about ecosystem structure and function, biodiversity and resiliency of ecosystems to perturbations, scale and hierarchy, productivity, and ecological indicators. The actual tools used to assess ecosystems are called *resource tools* and consist of the field methods and analytical techniques employed to compile and collect data to assess ecosystem states. The conceptual tools discussed in this chapter should guide the use of the resource tools. The intent of this chapter is to take the conceptual tools out of the academic realm and to show how they can be employed in the management of ecosystems, a merger of science, technology, and management.

Employing the ecosystem perspective entails a process of examining a specific system in light of current ecological theories and knowledge (i.e., using the conceptual tools) and considering what resource tools will best aid in organizing existing data and collecting new data. Collective use of conceptual and resource tools may

1. Organize existing information: help managers and scientists to organize existing information in more useful or accessible configurations

2. Identify new information: help managers and scientists to identify what new information is needed to inform management decisions

3. Set management priorities: aid managers in setting and ordering management priorities

As it is not practically feasible to obtain information about all aspects of ecosystem function, the purpose of the conceptual tools is to help identify which resource tools are most appropriate to utilize, and to guide scientists and managers in decisions about what new information is needed.

In this chapter, a view of how ecosystems are now studied and have been studied in the past will be presented, specifically focusing on the primary tools and approaches used. These will be discussed in the context of the spatial and temporal scales considered relevant for the questions being addressed. Both the positive aspects of these methods and their limitations will be discussed, with the purpose of identifying their utility to the ecosystem manager. This chapter begins with a review of spatial and temporal scales relevant for implementing ecosystem management, followed by the individual techniques used to assess species, physiological, and ecosystem attributes of systems.

3.2   Factors Contributing to Ecosystem Scales

Several problems with scaling information across different spatial scales can be identified, many of which are related to the fact that the quantity and type of information needed to analyze a system varies by the scale of analysis. For example, analyses at larger regional scales require less data to predict what is happening at that scale than analyses conducted at a stand or microsite level (Gosz 1992; Table 3-1). Problems also arise because different stresses may be expressed at one scale but not at the other scales (and not always at the smallest scale). Many environmental questions are being asked at the regional or global scale, where scientists are attempting to identify regional impacts on different ecosystems due to climate change (i.e., global warming, higher CO_2 levels) (e.g., Baldocchi 1993). A problem that arises is that most of the information available to assess how the systems are responding has been collected at the physiological levels, where more data are needed to predict plant responses. This contrasts with the larger regional level analyses, which appear to be able to predict plant responses to climate change using mainly climatic variables. This results in a predicament when attempting to answer questions at larger spatial scales, where less information is needed from data collected at smaller scales, where a significantly higher number of feedback loops control the type of responses measured (see Table 3-1). How much information is lost in this scaling process and if the predictions made at these larger scales have any relevance to any specific site or to particular species is difficult to presently state. The large-scale data analyses discussed in Section 4.2.7 suggested that many hardwood species and also many of the pine species were not sensitive to climatic variables such as precipitation. In such cases,

TABLE 3-1 The type of data that is needed to predict ecosystem processes occurring at the different scales of analyses

Scale

Number of constraints

Biome

Climate	Topography	Vegetative types (i.e., grasses, trees, etc.)	Time scale: frequently not relevant

Landscape

Climate	Topography	Community and ecosystem types	Soil physical characteristics	Spatial distribution of ecosystem types	Time scale: comparisons between years

Ecosystem

Climate + microclimate	Topography + micro-topography	Species composition and dominance	Soil physical, chemical characteristics	Consumer levels	Plant tissue turnover rates and their decomposition
Spatial distribution of live and dead organic materials, of soil types and textures	Plant morphological adaptations for acquiring nutrients and water	Symbionts	Nutrient and water availabilities	Time scale: annual, several years	

Population and plant

Climate + microclimate	Topography + micro-topography	Microsite soil physical, chemical characteristics	Physiological adaptations for carbon fixation, nutrient acquisition	Consumer levels	Plant carbon allocation patterns: above- (leaf) and below-ground (fine roots)
Plant genetics	Symbionts	Nutrient and water availabilities	Time scale: hourly diurnal, annual		

other variables would have to be identified that could be used to assess the impact of ecosystems to climate change that can be scaled from these smaller to the larger levels without losing their sensitivity of analysis.

The use of seedlings to study ecological processes highlights how scaling is problematic in conventional studies. Many of the studies trying to understand the mechanisms behind woody plant responses to a changing environment have used seedlings because of the ease of testing and experimentally manipulating small plants. This creates problems because of the difficulty of transferring data from seedlings to mature trees; the seedling data are more relevant to addressing issues related to plant regeneration d do not reflect how mature plants will respond. The physiological differ ces between seedlings and mature plants are highlighted in Table 3-2 discussed further in Section 3.4.1.

so has been assumed that it is necessary to develop techniques to olate information from a small scale (i.e., typical of most ecological es) to address questions at the larger ecosystem or landscape level. This assumption has generated much research attempting to address how to transfer information between scales, especially when the processes are not linear between each scale. One of the rationales for hierarchy theory is the need to acknowledge the existence of different levels of biological organization (O'Neill et al. 1986, 1989). Hierarchy theory is useful to pursue, but it must be acknowledged that it may not always be necessary to transfer information between scales to manage ecosystems. There is a need to identify how the matrix in which a particular analysis unit is embedded affects the patterns and processes observed for that unit and to focus on the scale that most sensitively reflects how that system is responding. In some cases, the scale at which a process is important to examine may vary depending on the disturbance type and frequency.

Changes in species or ecosystem function in response to a disturbance are not detected with the same efficiency at all spatial scales. This was reported by Hansen et al. (1993) while searching for the scale that would best predict the abundance of the spotted owl in a watershed in western Oregon. In that article, a greater proportion of the variability in the abundance of the spotted owl was explained at the landscape level, which integrated the scale of habitat use by the spotted owl (i.e., old-growth area, fragmentation index). Other scales and their associated characteristics were poor at explaining the abundance of owls (i.e., stand or patch level, which in-

TABLE 3-2   Structural and functional characteristics of seedlings and mature trees

Structure/function	Seedling	Mature trees
Carbohydrate storage capacity	Low	Higher
Within-plant carbon allocation annually	None to reproduction, higher to foliage than roots	Higher proportion to reproduction, 5–70% to roots, 5–30% to foliage, 1–15% carbohydrate storage
Drought resistance	Lower	Higher
Leaf conductance	Higher	Lower
Nutrient uptake requirement	Higher	Lower
Translocation of nutrients limiting plant growth	Insignificant	30–70%
Percent fine roots suberized	Few in young plants	~ 100% except during periods of fine-root growth
Proportion of fine roots to total root biomass	~ 80–100%	< 5%
Proportion of photosynthetic tissue biomass to total plant tissue biomass	> 1/2	~ 1/5
Proportion of respiration to photosynthesis	Lower	Higher
Rate of net photosynthesis	Higher	Lower
Secondary defensive chemical production	Low	Higher

corporated habitat structures such as variation in tree size and number of snags, and regional level parameters, which reflected elevation and latitude). This study shows the importance of focusing on the particular scale of analysis that most strongly reflects the parameter of interest.

In another review, the importance of one environmental variable in predicting fine-root biomass or fine-root net primary production varied quite dramatically depending on the spatial scale of analysis (i.e., comparing forest climatic types—boreal to the tropical zones—or comparing deciduous or evergreen behavior of tree species; see Section 4.2.7 for further discussion). This latter review highlighted the importance of identifying if and what functional groups exist within each scale because there not exist a ubiquitous response to a disturbance by the different ive communities comprising each scale. This data review also em- ed that some species or functional groups do not respond in the ted manner to a disturbance or some do not respond at all to a nge in an abiotic variable. Therefore, it is not only important to focus on a particular spatial scale of analysis, but there may be a necessity to further group within each scale, depending on the variable being studied.

Studies to detect ecosystem state changes at specific scales are further complicated by the fact that ecosystems may be impacted by one to several different disturbances or changes in abiotic resources at any given time period. These multiple disturbances impinging on one system at the same time are further confounded by the fact that studies conducted at smaller scales have to deal with more variables affecting how the system responds. It has been extremely difficult to design an experimental system that duplicates the multiple interacting environmental factors that can simultaneously occur in the field. This explains the popularity of greenhouse or growth chamber studies, where environmental conditions can be controlled with great precision so that disturbance effects can be isolated from one another. Some of these controlled studies have varied a couple of factors at one time, however, it is an enormous challenge to identify the temporal scales at which these disturbances should be imposed to try to mimic natural cycles in the field. At larger scales, where less information (i.e., mainly climatic) is technically needed to predict ecosystem change, the existence of multiple interacting factors complicates the ability to predictably state that any one factor is of greater significance than another in controlling the response of the vegetative community. The few variables needed at this scale means that they may not be sensitive to a particular disturbance unless the change is a climatic one.

So the problem that has to be addressed in ecosystem management is, *How do we measure the impact of multiple interacting factors impinging on ecosystem function at the different scales pertinent to management of ecosystems?* McLaughlin and Downing (1995) proposed a regression model approach that they stated allowed them to separate the effects of two interacting factors (i.e., ozone and climatic factors) on the growth rate of *Pinus taeda* L. in southeastern United States. As part of their analysis, they combined direct measurement of seasonal changes in stem growth with a regression model approach that described separately the impacts of ozone from soil moisture on growth rates. This study highlights two of the problems that exist in scaling physiological data to the ecosystem level: (1) identifying what parameters or measurements best integrate the individual species response to an environment with multiple interacting factors functioning simultaneously and (2) justifying the use of seedling studies to implicate the effect of a variable (in this case, ozone) on the growth of plants when these relationships have not been verified at the ecosystem level. The regression model presented by McLaughlin and Downing (1995) has not been directly tested or verified in the field due to the difficulty of designing an experiment capable of separating the impacts of two interacting factors on stem growth of mature trees. In this case, using a regression model approach is probably the only realistic approach to examine the possible range of impacts due to two interacting factors and to analyze each at the temporal scale at which they exhibit a stronger control on stem growth.

There is a further need to expand our understanding of how an individual structure or function occurring at a specific scale is impacted by processes occurring at other scales, and how it, in turn, feeds back to affect the functioning of the core system at another scale. It is important to realize that structures or functions at even small scales can mitigate processes occurring at larger scales. For example, a piece of coarse, woody debris located at the ground level but where there exists an opening in the forest canopy (i.e., gap) can change the water and nutrient availabilities at the microsite scale but also control nutrient cycling rates at the stand level (Vogt et al. 1995a). Similarly, an individual plant that appears to be a minor component of the system (less than 20% of the stand biomass) may control the cycling rate of phosphorus (P) at a stand level because it is an aluminum (Al) accumulator—it increases the amount of Al being cycled and therefore how much P is complexed in an unavailable form (Vogt et al. 1987, Dahlgren et al. 1991).

In an another example in northern California, USA, the presence of a dominant tree species (i.e., *Pinus muricata*) did not change along a coastal terrace, even though extreme edaphic gradients existed along this same terrace. In this example, *P. muricata* changed N cycling rates and forms in those portions of the gradient that were N-poor in contrast to more N-rich areas; therefore, changes in N cycling corresponded to the soil chemical characteristics along this edaphic gradient (Northrup et al. 1995). In those ⌐ctions of the gradient that were highly N limited, *P. muricata* appeared to ⌐tain organic N as the dominant soil-available N source by its ability to ⌐ polyphenolic complexes (Northrup et al. 1995). By controlling the ⌐nant N form in some portions of the gradient, *P. muricata* minimized competition from other organisms for abiotic resources, because many plants are unable to utilize organic N forms (Northrup et al. 1995). Because mycorrhizal fungi are very capable of taking up organic N forms and *P. muricata* has well-developed mycorrhizal associations, it was hypothesized that *P. muricata* was dependent on these symbionts for accessing N from decomposing litter (Northrup et al. 1995).

Some ecosystem structures (i.e., coarse woody debris, inoculum of symbionts, etc.) existing at smaller scales may also function as "legacies," because they may buffer components of ecosystems after disturbances or between different stand developmental stages or during succession (Franklin et al. 1981). Coarse wood in a large gap may be important as a storage site for fine tree roots, and mycorrhizal fungi that survive when the soil environment may be nonconducive for their growth. For example, 6 months after a gap was created in an old-growth *Pseudotsuga menziesii* forest in Washington, USA, most of the live fine conifer and mycorrhizal roots were found in coarse wood and not in the soil (Vogt et al. 1995a). The role of the coarse wood also appeared to vary depending on where it was located because coarse wood under forest canopies did not function as strongly as a reservoir for live fine roots and mycorrhizas (Vogt et al. 1995a). In the dry forest studied by Harvey et al. (1976), coarse wood was important for allowing mycorrhizal roots to remain active during very dry periods. This means that there are microsite structures that may modulate how the total plant community maintains itself during particular times of the year or in response to disturbances that change canopy cover in forests. To really understand what controls ecosystem function and/or structure under a variety of different disturbances, it is necessary to determine how small-scale processes feed back to the larger scale and vice versa, and how the mosaic patterns change the growth environment for plants.

Scaling issues also arise at the individual plant level because the scale of the above-ground parts of the plant is very different from the below-ground parts. The design of field experiments with experimental manipulations has to consider that measuring environmental impacts at the plant root level will require a larger spatial scale than examining the canopy of a tree. Many research studies in the past established a 10-m wide plot to study the response of a stand to a perturbation. However, when the root system of one tree is capable of growing horizontally for more than 30 m from its base, this plot size may not effectively sample the space being utilized by individual trees or the stand. In other cases, if a plant primary root extends more than 30 m vertically into the soil from its base (Stone and Kalisz 1991), traditional research designs are not accounting for all the abiotic resource pools being accessed by the individual plant (in these cases stable isotopes should be used, and the reader is referred to Section 4.2.4 for more discussion on this topic). In the past, field studies were not designed to account for the different rooting zones of the plant because the perception existed that the above-ground space would be equivalent to the below-ground space (Vogt et al. 1993).

Some deep-rooted plants may also have a strong impact on the soil environment perceived by adjacent plants, again not reflecting the spatial scale that they appear to occupy. For example, some deep-rooted plants increased the amount of water available to other plants, with more shallow roots growing adjacent to them (Richards and Caldwell 1987, Caldwell and Richards 1989). In management, it is important to identify when deep-rooted plants exist in an area and to understand their roles as potential keystone habitat modifiers (see Chapter 2). Managers must also be aware that the planting of deep-rooted trees as part of plantations may create problems if the species are too effective at accessing resources from deeper depths. This has recently become a problem with *Eucalyptus* species, which have been utilized as a plantation species in many parts of the world. *Eucalyptus* has been shown to have some of the deepest rooting (up to 60 m) of any plants based on a worldwide literature review of plant rooting depths (Stone and Kalisz 1991). The concern that is being raised with *Eucalyptus* is that it is overutilizing water resources from deeper depths and lowering water tables at rates beyond the site's capacity to replenish them (S. Nambiar, personal communication).

3.3   Spatial Boundary Considerations

Several factors have to be considered when trying to define the boundaries at which ecosystem processes are to be measured in the field. An ecosystem study by definition incorporates the movement of energy and materials ᵕcluding atmospheric pollutants etc.) into and out of its boundaries. the plant level, some ecosystem edges are clearly identified by drastic ᵕanges in the vegetative community type (e.g., a forest compared with an agricultural field). Because ecosystem borders are identified by definition to be where one detects a significant change in the rate of exchange of materials or energy (Westman 1985), a change in species composition is by itself insufficient to state that one has shifted into another ecosystem. For example, a significant shift in tree species composition can occur without concomitant changes in the energy flow; therefore, even though species have changed, this is still classified as one ecosystem type. Furthermore, different components of the ecosystem will have different spatial scales at which their activities or impacts are confined. Identifying the vegetative boundaries of an ecosystem may be insufficient when managing for some animals or when studying the impact of a pollutant. For example, where the dominant vegetative community borders have been demarcated, some animal species will have reached the boundary of their migration (e.g., burrowing rodents), while others will continue beyond the stand border (e.g., caribou) (Hunter 1990). The spatial scale of analysis for each animals will be quite different. Similarly, if one is examining the effect of waterborne pollutants, some pollutants may be retained within the watershed boundaries if they are transferred to the riparian zones; however, there is also a strong likelihood that they may be transferred beyond the watershed boundaries by the river draining the watershed or as part of groundwater flow.

   The problems in defining ecosystem boundaries and ecosystem size are highlighted when considering how something generated in another country or part of the world can impact ecosystem processes occurring at a local or regional level. There are many examples involving pollutants, both natural and human generated (i.e., volcanic activity, Sahara dust being deposited in Caribbean forests, acid deposition, higher CO_2 levels, etc.), which can significantly decrease or increase nutrient cycling rates, change plant species composition, and change the net primary production of a managed ecosystem. There are many examples throughout the world where pollutants (especially nitrogen- and sulfur-based compounds) have affected forest ecosystem structure and/or function. For example, acid precipitation from Europe has been reported to be responsible for the decrease in soil base saturation measured in Sweden (Abrahamsen 1980). In the United States, pollutant inputs originating in Ohio have changed the nitrogen economies in forests located in New Hampshire and Vermont, and have contributed to the high mortality of spruce that occurred in these states in the 1980s (Johnson and Siccama 1989). Many coniferous ecosystems are not adapted to utilizing the higher nitrogen inputs occurring with pollution, resulting in less "healthy" forests (Waring 1987; see later discussion on fertilizer effects on forest stands in this chapter). In Holland, high use of N-based fertilizers has resulted in changes in the grass species dominating the landscape (Van Breemen and van Dijk 1988).

Plant adaptive strategies can also define or limit the distribution of plant species and therefore define distributional boundaries. There exists a range of climatic tolerances (temperature, rainfall, wind speed) for plants that will determine where plants can or cannot exist (Woodward 1987, Mooney and Winner 1991). For example, tropical plants will not grow at temperatures below 0°C but have a tolerance range that varies from 0° to 42°C, while boreal plants have a tolerance range that varies from −80° to +42°C. The absence of a plant in a region does not mean, however, that it is not capable of growing in that area, as evidenced by the introduction of plant species by humans that have become weeds or have outcompeted indigenous species (i.e., *Tamarix* is a weed in the southwest United States riparian zones; nitrogen-fixing trees are outcompeting native species in Hawaii; Westman 1985, Vitousek and Walker 1989). These types of analyses tend to produce very broad distributional boundaries for species but can also determine regions (i.e., ecotones) where species appear to be quite sensitive to any environmental change because they are not in their core habitat (Gosz 1992).

3.4 Spatial Scales Overview

Studies of ecosystem responses to disturbances have used several different scales, ranging from greenhouse or growth chamber studies using seedlings, herbaceous species, or grasses (Naeem et al. 1994); to the stand level approach, with experimental manipulations in the field (Vitousek et al. 1979, Gower et al. 1992); to the watershed level approach (Bormann and Likens 1979, Hornbeck and Swank 1992). Both the watershed and stand level approaches have attempted to experimentally manipulate the site by changing the water or nutrient relationships, and to determine how the system responded to this manipulation by comparing the system with a control. The watershed approach as an experimental system requires that one can verify that all the inputs into a watershed can be measured and that all outputs can be quantified at one location (Swanson and Franklin 1992). Frequently the watershed approach has been presented as the scale at which to do ecosystem management; however, the use of this term is unclear. For some, a *watershed* means an easily identifiable land base that can be used to demarcate a management unit, while for others it is an ecological definition that specifically incorporates the scale at which all the water inputs and outputs are self-contained for a particular land base. If the issue of concern is how nutrient and hydrologic cycles are changing within this land area, this approach cannot be used in all areas because the site has to have an impermeable bedrock or hydrologic budgets cannot be developed (see Section 3.4.3). The watershed approach is very important for examining fisheries issues and trying to understand when the integrity of a stream is changing in a direction in which the viability of organisms using that resource are being detrimentally affected (Hornbeck and Swank 1992, Swanson and Franklin 1992).

The scale of study may also determine the variables and the amount of information that will have to be collected to assess the state of the ecosystem. In most cases, large-scale (i.e., landscapes or biomes) analyses of systems have low data needs, and only a few specific parameters appear to give high predictive capabilities for identifying how the system will change in response to an alteration in the environment (see Table 3-1). Smaller scale analyses have greater data needs and a higher diversity of parameters must be monitored, mainly due to the fact that there are a greater number of variables and feedbacks controlling ecosystem functions (Gosz 1992). These differences in the type and amount of data that need to be collected at the small and large scales suggest great difficulty in using large spatial scale analyses based on climatic variables to predict how the system will respond at any given local ecosystem scale. However, if climatic variables can be used to predict the most important controlling processes occurring at a smaller scale, the ability to predict across scales will be greatly facilitated (see Section 4.2.7 for a discussion of large data sets).

Ecosystem research has been approached by scientists at several different scales (see Section 4.2.6). The spatial scale of a research study has varied from using individual plants grown in pots in greenhouses or controlled environmental chambers, to stand level studies, to ecotones between different plant communities, to watershed scales, and recently to landscape scales. In practical terms, ecosystems have generally been defined as being a stand or a watershed. Many questions can be addressed at the stand level, while interaction among stands involves at least a drainage basin. Questions involving wildlife usually encompass larger regions (landscape) to evaluate the significance of wildlife in ecosystems. Originally ecosystem studies were confined to easily defined land areas, such as a lake, a drainage basin, or a forest plantation. However, today it may not be necessary to specifically define ecosystem boundaries when examining issues concerning insect outbreaks or the use of fire or cutting practices to manage wildlife populations. The different scales used to study ecosystems are now discussed individually and are further discussed in Section 4.2.6.

3.4.1   Individual plants

Individual plant level studies have mostly focused on seedlings as a means to experimentally examine tree growth and to study the mechanisms underlying tree responses to disturbances and stress. Seedlings give researchers the ability to control the variables in their system so that they are able to specifically identify the impact of one or two variables on plant growth. In the past, it was believed that results from seedling studies directly transferred to the field and could be used as a tool to identify how mature plants would respond. Today, it is understood that seedling studies under controlled conditions should be used to *identify the mechanisms* by which plants react to a changing environment, but the results from these studies *do not directly* transfer to the field. The relationships obtained under controlled conditions should be used as a *filter* to select those variables that best reflect how a plant is responding to a change in the environment. Seedling studies have a very important role in identifying which variables might hold promise for detecting plant responses. Seedling studies have emphasized using physiological measurements to detect plant responses to a variety of disturbances, specifically focusing on a plant's ability to fix carbon as part of photosynthesis and measuring nutrient uptake and allocation relationships of plants (Table 3-2). These physiological measurements are the tools that will allow for the identification of plant responses to stress at the mechanistic level and have tremendous potential for being early warning signals of a changing plant state.

There are several reasons why results from seedlings will not directly transfer to explaining how mature trees are responding to stress or disturbance: (1) Seedlings were usually grown under noncompetitive conditions—one plant per pot; (2) seedling physiology and response to stress is different from mature trees (see Table 3-2); (3) seedlings frequently are not inoculated with mycorrhizas, even though it is unusual to find plants without mycorrhizas in the field and mycorrhizas may "filter" or "buffer" how plants respond to their environments (Vogt et al. 1996a); and (4) seedling growth rates are affected by the size of the containers used to grow them. This especially becomes a problem at the root level, where growth can become restricted by the pot size, such that deep-rooting adaptations will not be expressed or unnatural growth occurs along the edges of the pots. As these problems became apparent to researchers, researchers have responded by increasing the size of the growth containers, inoculating seedlings with mycorrhizal fungi, and growing a mixture of plant species in each pot.

It becomes obvious from comparing the structural and functional differences between seedlings and mature plants that they will not respond to a disturbance in a similar manner (see Table 3-3). Seedling studies are highly relevant and directly transfer to address questions about the regeneration of plants on a site. However, when past studies used seedlings to address questions of tree growth at the ecosystem level, they inadequately examined trophic relationships and how these might modify plant responses to a stress. This changed recently when Naeem et al.

TABLE 3-3 Ecosystem and physiological parameters that are useful to measure for detecting ecosystem responses to stress

Ecosystem and physiological level parameters	Low stress condition	High stress condition
Annual growth rates	+ + +	+
Photosynthetic rates	+ + +	+
Foliage longevity	+	+ + +
Nutrient uptake	+ + +	+
Nutrient- and water-use-efficiencies	+	+ + +
Lifespan	+	+ + +
Allocation to foliage biomass	+ + +	+ +
Mycorrhizal dependency	+	+ + +
Percent NPP[a] in roots	+	+ + +
Allocation in root biomass	+	+ + +
Allocation to secondary defensive chemicals	+	+ + +
Nutrient leaching	+	+ + +
Total NPP[a]	+ + +	+
Importance of decomposers	+	+ + +
Activity of pathogens	+	+ + +
Surface litter accumulation	+	+ + +

+ low importance; ++ medium importance; + + + high importance.
[a] net primary production.

(1994) specifically established a study where trophic relationships were controlled in an attempt to examine the relationships between species diversity and ecosystem function. This study utilized controlled conditions but had several herbaceous plant species competing with one another, and had a complement of consumers and predators that were introduced into the system. This type of approach has been criticized in the past for creating artificial trophic interactions because species relationships evolve with time and adding all the components of an ecosystem into a container at one time may not cause species to respond in the way that they would in nature.

The temporal scale at which species respond to disturbances is difficult to mimic in individual tree studies because these studies are typically of shorter duration, monitored for a period of 1 to 2 years. These controlled studies are also ineffective in detecting changes in some parts of the ecosystem due to varying buffering capacities of the individual parts resulting in different temporal scales of response. For example, changes in soil characteristics in response to a disturbance may take several years to be detected, while population level responses will occur much faster. Therefore, phytotron or controlled environmental chamber studies are not effective for detecting the cause–effect relationships between highly buffered (i.e., soil) and less buffered (i.e., consumer populations) parts of the ecosystems when exposed to a chronic disturbance. Phytotron or controlled environmental chambers studies would not have detected the relationships between declining populations of passerine birds and acid precipitation described in Chapter 1 (Graveland et al. 1994). It is very difficult to isolate the cause–effect relationships or to link processes occurring at very different spatial and temporal scales using controlled environmental chambers. Furthermore, indirect impacts of a stress may be difficult to detect with controlled environmental chambers because the experiments are designed to control many variables in the experimental system so that a change may be perceived as being an experimental error and corrected.

Small pot studies are also unable to effectively test several plant responses that occur at the root level. A mature plant growing in a forest has several root morphological adaptations that significantly modify how that plant responds to a disturbance. For example, if a plant's adaptation to the site consists of producing a tap root system that extends some 30 m vertically or horizontally to the surface of the ground, or if root grafting is common between individuals in the community (Graham and Bormann 1966, Basnet et al. 1993, Vogt et al. 1993), growing plants in pots will not allow these adaptations to be expressed. This ability to access carbon or nutrient resources from your neighboring plants can buffer an individual plant's response to perturbations so that it may be difficult to detect a response at the individual plant level.

When the National Atmospheric Precipitation Assessment Program (NAPAP) was initiated in the 1980s to assess the "health" of forest ecosystems in the United States being impacted by acid deposition, more than 80% of the studies conducted were under controlled conditions using seedlings (see NAPAP discussion in Section 4.2.9). Most of these studies also focused on determining the sensitivity of plants at the species level to one or a combination of two pollutants. Most of these experiments obtained mixed results and were unable to unequivocally state that acid precipitation was negatively affecting forest growth; the only exceptions were the red spruce forests in New England and several point source pollution gradients in California. Growth reductions for red spruce were determined from field studies that used experimental manipulations of branches enclosed in chambers (Johnson and Siccama 1989). Branch studies were perceived to be the next step for scaling seedling studies to mature trees. However, branches do not represent the whole canopy of a large tree because physiological processes fluctuate dramatically within different parts of the canopy (Sprugel et al. 1994). The branch chambers also imposed other artifacts because the charcoal filters used to produce control chambers also changed the chemical environment around the branch by removing natural atmospheric constituents; therefore, these chambers were not good controls.

In the late 1980s it became apparent that the timing of the acid precipitation inputs into forests in Europe and the United States occurred at the same time as the growth of these forests was also being affected by an extended drought period (Johnson and Siccama 1989). Even though drought was identified to be a significant contributing factor to tree mortality, its relative importance in causing tree growth reductions could not be isolated from the impacts due just to acid precipitation. Because the occurrence of this extended drought was not evident until the end of the NAPAP study, the combined effects of the pollutants and drought did not receive much attention. This is not a criticism of the type of studies conducted as part of NAPAP but again highlights the fact that our knowledge base is dynamic; ecosystem analysis systems have to be designed that are capable of incorporating relevant new information in a timely fashion. However, it should be kept in mind that it is always easier on hindsight to complain about the approaches used in past studies, ignoring the context of when that research was conducted.

Even though there may be problems in using seedling studies to address questions related to the response of a forest to a particular stress, some experiments are technically impossible to conduct on mature plants and require the use of seedlings. For example, seedling studies have been useful to identify where and how a plant may respond to higher carbon dioxide levels (Norby et al. 1986). Seedlings studies have the ability to examine factorial experiments where the carbon dioxide levels can be varied at the same time as the nutrient availability is modified. Establishing this type of experiment in the field is extremely difficult. Carbon dioxide applications systems (i.e., FACE systems; Hendrey 1992) have been successfully used with agricultural plants in the field where computers were used to maintain the CO_2 at desired levels. However, trees are much more complicated to study, and it is difficult to establish experimental systems that will maintain consistent CO_2 levels around tree canopies.

3.4.2 Stand Level Approach

A stand is typically defined as an area characterized by having sufficient homogeneity of vegetative structure, similar stand ages, soil types, topography, microclimate, and present and past disturbance histories to be treated as a single unit (Kimmins 1987). The typical spatial extent of a stand can vary from 0.05 to 1 hectares in size. There are several assumptions and attributes of stands that affect the type and amount of data that need to be collected at this spatial scale:

1. The smaller scale of a stand may require more data to be collected due to the increased complexity and higher number of variables driving ecosystem functions (see Table 3-1). Therefore, one has to select what is important to analyze.

2. Climate is assumed to be constant within the defined ecosystem.

3. A higher number of variables have to be studied to understand the controls and feedbacks on processes.

4. Defining the sampling unit is difficult because we have to identify a relatively uniform ecosystem structure to study (typically avoiding site heterogeneity and assuming that it does not contribute to driving ecosystem functions).

5. It is difficult to transfer site-specific analyses to address larger scale issues (i.e., the stand or plot level can be a unique site that does not represent the region or landscape patterns because they only examine a part of that heterogeneity).

6. The scale is effective for studying nutrient and water relationships using experimental manipulations of an ecosystem, and identifying

the specific cause–effect relationships associated with these variables; pollution impacts on plant growth within an ecosystem framework; functional diversity relationships in ecosystems; and the historical and present land-use effects on systems.

7. The scale is not appropriate to study the role of large animals in ecosystems, except those that are not too mobile or have small habitat requirements.

3.4.3   Watershed Level Approach

The classical definition of a watershed is that it is a topographically defined area where all the precipitation falling into that area leaves in a single stream. A watershed study is basically an inputs–outputs study in which the ecosystems may be manipulated within the watershed but are not directly measured as part of the study. This means that the number of different variables that need to be studied in a watershed can be lower than in a stand level study. The size of a watershed can be quite variable. For example, the watershed of the Mississippi River includes roughly one third of the United States but in general most watersheds are typically less than 100 ha in size (Waring and Schlesinger 1985, Hornbeck and Swank 1992). Watershed studies cannot be automatically established everywhere because the ability to monitor the total hydrologic cycle is an important part of the input–output budgets and many areas do not have a bedrock sealing or restricting water outflow to one stream. If the bedrock is fractured where an undetectable portion of the water is able to leave, it will be impossible to obtain total water budgets for a system. A watershed can be a drainage basin or catchement (land area in which precipitation is distributed into components of the hydrologic cycle; see Hornbeck and Swank 1992). Today the term *watershed* is also used by people to identify a given land area, typically a drainage basin, without any attempt to study the hydrologic and/or nutrient cycles within that area. It is merely being used as a convenient land area to manage, especially when mixed ownership is occurring within this land base.

The following assumptions and attributes are relevant to traditional watershed studies:

1. Inputs from outside its boundary are minor (i.e., pollution, animals etc.).

2. Vegetation has a consistent level of water and/or nutrient uptake.

3. The ecosystem is not leaky and there are no losses of water or nutrients through the bedrock that cannot be accounted for.

4. More than one watershed has to be compared, especially because the entire watershed is being monitored as one unit; however, watersheds are frequently not of uniform size or occupied by similar vegetation or soil chemistries, making it difficult to identify "control" watersheds.

5. They are inherently long term, due to variability in year-to-year weather and in wet and dry deposition (data sets exist that span 30 to 50 years for a few sites, such as the Hubbard Brook Experimental Forest in New Hampshire and Coweeta Hydrologic Laboratory in North Carolina).

6. They do not effectively study the interaction of large animals or birds within the ecosystem.

7. They are a good scale to study water pollution, acidic deposition, harvesting practices, water quality, and stream or river relationships to the land.

8. Watershed ecosystem analysis establishes cause and effect relationships and ecosystem responses at the landscape level.

3.4.4 Landscape Level Approach

A landscape incorporates many stands within its spatial scale or can consist of a watershed. The use of the landscape level approach has the following assumptions and attributes:

1. Landscape analyses tend to assume clearly defined boundaries for the different habitats distributed within their area (Meentemeyer and Box 1987) and in the past have not dealt with the heterogeneity of ecotones existing within a habitat [i.e., a forest landscape has many different-sized gaps existing within its boundary, but these gaps do not have a discrete clear edge to the forest; the size of the ecotones existing between the gap and forest will vary and can be very indistinct (this is

very similar to Swiss cheese that has been left out of the refrigerator, allowing fungi to grow along the edges of the holes, giving them a "fuzzy," indistinct appearance due to the fungal growth; Publicover and Vogt 1991)],

2. This approach assumes that one can sum up the component parts of the landscape and understand how the whole system is functioning.

3. This approach in the past has been mainly based on visually assessing boundaries of different forest types or vegetative communities, with the analysis itself being a mapping operation using GIS; modeling has been used to help integrate information across the landscape.

4. This is the best scale to monitor and assess animal activities in ecosystems, and this is the scale that can integrate information across different ownership classes and different habitat types.

3.4.5   Biome Level Approach

A biome consists of a distinct vegetation community that is characterized by having similar qualities, for example, canopy height, number of strata of vegetation, etc. (Kimmins 1987). Biomes are typically delineated by using mean annual precipitation and mean annual temperature to demarcate vegetative communities (the Holdridge System of Vegetation Classification is an excellent example of this). For example, the biome classification system breaks the tropics, which is itself defined by a mean annual temperature above 24°C, into the following categories based on precipitations and evapotranspiration rates: rain forest, seasonal forest, woodland, thorn scrub, and desert. The biome level has been frequently used to scale information from several sites to develop large-scale averages. Which are then used to analyze different possible scenarios of change in soil or vegetation carbon pools due to deforestation or other land uses (Brown and Lugo 1982, Brown and Iverson 1992).

Biome or climatic vegetation types have the following characteristics:

1. The dominant factors controlling vegetative distribution are assumed to be climatic.

2. Temporal patterns in precipitation are not effectively incorporated. In

some ecosystems the total precipitation values are not able to predict the functions of particular ecosystem components (Vogt et al. 1996b).

3. Other parts of the system (i.e., soils) are not recognized to have overriding control on the functional processes occurring in a system, as climatic data are insufficient by themselves.

4. The effect of altitudinal changes on species distribution is ignored.

5. One cannot go to a specific site and ever duplicate the climate-vegetation relationships produced at the biome level.

3.4.6   Ecotone Level Approach

An ecotone is the boundary or transition zone between either individual plants or communities of plants (Gosz 1992). An ecotone can be considered as a differentially permeable membrane whose boundary is highly plastic. Ecotones can also have edges within edges, with one edge being embedded within a matrix of different communities. They can be thought of as boundaries between zones or communities that occur at distinctive locations along environmental gradients and are expected to be especially sensitive to environmental change (Gosz 1992). Boundaries are a continuum of change where some places are more sensitive than others, and therefore they are highly dynamic and will move because interannual variations in abiotic variables (climate) modify patch edges of species. It is important to manage for ecotone movement to maintain high productivity and diversity. Boundaries frequently have an increased density of individuals and an increased number of species because three groups of individuals may be utilizing or relying on different resources (Hunter 1990).

Ecotones have the following characteristics (Gosz 1992):

1. Boundaries are assumed to be like membranes in organismal and physical systems. They have varying permeabilities.

2. Gradient analysis is used, which is a landscape level approach, but it focuses on ecotones or transition zones as a unit of analysis. It can be done at a regional level by comparing biomes, at an individual plant level by comparing the zone between plants, or at the community level by comparing zones between communities.

3. It may be very sensitive in detecting small changes in systems due to climatic conditions because it can examine several different boundaries. The key is subtle changes occurring simultaneously at several boundaries.

4. There does not exist a single edge effect, as it varies by the vegetative communities that form the edge. Dynamics in ecotones in landscapes are probably nonlinear, perhaps chaotic, and not an average of adjacent resource patches.

5. Plot studies become valuable if studied along points of an environmental gradient, as gradient analysis allows for the extrapolation between sites. One can examine regional variations in vegetative communities based on latitude and elevation, as demonstrated by Gosz (1992), who examined patterns of community composition in the Rocky Mountains, USA.

3.5   Disturbances as Temporal Agents of Ecosystem Change

Because ecosystem function is often greatly influenced by both the structure and productive capacity of the ecosystem (King 1993), it is important to understand what factors or feedbacks determine the present structure of a system and which may change its productive capacity. Disturbance, both natural and anthropogenic, drives ecosystem structure, in that it affects plant architecture and community composition, sets the seral stage, and may shape dominant land forms. Disturbances may also affect the productive capacity of an ecosystem by (1) changing spatial and temporal patterns of nutrient availability and cycling, (2) adding or removing biomass, (3) changing the rate of plant succession (i.e., mainly due to consumers changing plant competitive interactions), and (4) changing the ratio of live to dead material in the system (Pickett and White 1985, McNaughton et al. 1988, Pastor et al. 1988).

Ecosystem management has to incorporate into its analyses the fact that ecosystems are very dynamic, that is, there does not exist a steady-state condition for an ecosystem. This makes it imperative to understand the difficulty of maintaining an ecosystem in its present state in perpetuity without actually causing the system to change. This requires an understanding of the natural successional processes in an ecosystem and how disturbances influence these processes over longer time scales. It is also important to remember that the ecosystem response to a disturbance is linked to how the management unit fits into the landscape. For example, if a forest exists in a matrix landscape in which it is surrounded by fields or a network of roads, the impact of a disturbance will be expressed in a different manner (Franklin and Forman 1987, Saunders et al. 1991).

In many cases management actions attempt to mimic natural disturbances. For example, silvicultural thinning of a forest stand by girdling trees closely mimics the stage in stand development when trees are fully occupying the site and competition between trees increases natural tree mortality (Oliver and Larson 1990). However, in other cases, management actions may stray from the natural disturbance regime of an area. For example, by cutting forest stands before a natural rotation is completed, humans increase the frequency of disturbance and potentially the degree of disturbance if many trees are removed and little effort is made to restore the site or to facilitate natural regeneration. Humans also disturb ecosystem in a uniquely human manner by directly or indirectly introducing, extracting, or altering biotic and/or abiotic resources (Mooney and Drake 1986). Disturbance can also be reflected in systems in many different ways, and sometimes there may not be any loss of biomass but there may be a loss in the ability of some function to be maintained. The introduction of many competitive exotic plants (i.e., kudzu in Florida, USA) has resulted in their dominating total system biomass; ultimately functional redundancy and perhaps functional capacity is reduced due to the disappearance of many native species (Ewel 1986).

In trying to identify the type and frequency of disturbance in an ecosystem, the manager should look to both historical records of natural disturbance and ongoing studies of precursor events to disturbance. Such records are found in national and international databases, in academic studies of plant life histories and disturbance events, and in ecosystems themselves. Soil cores, tree cores, and chronosequence studies may provide numerous examples of disturbance over time (Johnson and Siccama 1989, Shortle et al. 1995). Prior indications of a high incidence of disease or mortality, an increase in the proportions of r-strategists, or a decrease in the abundance or distribution of large organisms, may well indicate past disturbances (Odum 1969). Abiotic indicators, such as the presence of high concentrations of certain pollutants or unusual variation in soil horizons (e.g., the presence of a flood or Ap horizon; Marsh and Siccama 1996) may also provide a historical perspective on disturbance. Historical records and data may help the manager identify the probability of the occurrence of certain types of disturbance, for example, flooding or hurricanes.

An understanding of natural disturbance processes will allow the manager to begin to understand the potential impact of current and future human impacts on the system. Often the combination of natural and anthropogenic disturbances results in unexpected degradation of ecosystem processes. Therefore, managers need to account for the possible cumulative effect of different types of disturbances. Frequency analysis can be used to predict the probability of the occurrence of certain natural disturbances within a given amount of time. Although the means for understanding future human impacts on the ecosystem are beyond the scope of this book, managers should certainly acknowledge political, cultural, and economic pressures on the ecosystem.

One disturbance that has had a significant impact on forest ecosystems in many parts of the world is fire (Agee 1991, Specht 1991). In many cases, fire may have been a natural part of the ecosystem and the dominant species are adapted to the fire cycles. However, in many places the frequency and intensity of fires mostly of human origin have increased dramatically within the last 100 years (Specht 1991). In Australia, eucalypt forests are very resilient to fire and are able to regenerate rapidly after a fire however, these forests have been experiencing localized areas of degradation due to several land uses ("timber extraction, selective clearing, cattle grazing, road-making, and rubbish disposal") and the increased incidence of repeat fires (Specht 1991). Once these areas begin to degrade, insect pests, root pathogens, and weeds having become more competitive and have outcompeted the eucalypt trees. There is discussion of using controlled fires as a tool to reverse this land degradation occurring in the eucalypt forests (Specht 1991), but this also assumes that an understanding exists of how to use fires to control invasive species and to reduce the susceptibility of systems to pests and pathogens.

Others have also prescribed the use of controlled fires to restore coniferous ecosystems in the southwestern and northwestern parts of the United States (Agee 1991, Sackett et al. 1993). Due to several management activities (i.e., heavy grazing, selective timber extraction, fire exclusion within the last 100 years), these ecosystems have had their structures and functions modified to the extent that some of them are not considered "healthy" (Sackett et al. 1993, Covington et al. 1994). The typical symptoms used to identify whether ecosystems are classified as healthy is the increased incidence of pests and pathogen, and the increased size and frequency of wildfires (a prime example of this was the large-scale fire in the Yellowstone National Park in 1988) (Knight 1991). It has been suggested that fire is one tool that can be used to re-establish vegetative community compositions of the past and to reduce the incidence of pest and pathogens. This would require developing a historical information base on fires that could be used to develop reference conditions of what the spatial and temporal scales of fires were in the past (Graham 1994). Because fire has not been the only change that has occurred in these ecosystems, fire by itself cannot be used to restore ecosystems to more "healthy" conditions.

Ecosystems can similarly incorporate, that is, dampen the effect of or recover from, anthropogenic disturbances. Such effects are necessary to consider when establishing the spatiotemporal scale of the ecosystem to be managed. If disturbances, such as timber harvesting, are in the management plan of the ecosystem, a large enough scale for the managed ecosystem must be chosen so that the disturbance is incorporated.

3.6 Parameters Used to Study Ecosystems

Many studies have focused on several variables in trying to assess plant growth and what regulated its growth using a combination of techniques developed in physiology and ecosystem ecology: plant photosynthetic rates; plant biomass; plant net primary production (NPP; most pre-1980 research only measured above-ground NPP); plant allocation to canopy components (especially leaves), to carbohydrate storage, and to tree boles and branches; plant morphological characteristics; plant uptake and internal use of water and nutrients; and nutrient cycling in litterfall (especially because above-ground litterfall was assumed to indicate site nutrient status). Many of these variables vary in response to changing stress levels in ecosystems and therefore hold promise for being useful indicators that the system state may be changing (Table 3-3). Several of the parameters listed in the Table 3-3 are further examined.

TABLE 3-4   Net primary production (NPP; mg/ha/yr) for different vegetative communities[a]

Forests	6–23
Tropical humid	23
Tropical seasonal	16
Mangrove	10
Temperate	13–15
Boreal	6–9
Forest plantation	18
Savanna	13–23
Temperate grassland	5–12
Tundra	< 1–3
Desert	1–2
Lakes and Streams	4
Swamps and Marshes	25–40
Bogs	1

[a] Adapted from Kimmins (1987), with permission.

3.6.1   Net Primary Production

Estimating net primary production allows researchers to determine the efficiency with which energy enters and is passed through the primary producers (plants). Because primary producers provide energy for all other components of the ecosystem, NPP is fundamentally important to ecosystem function. In terrestrial plant communities, production is typically measured as the increase in total weight (biomass) or quantity of organic material accumulating on a given land area over a defined period of time (usually 1 year). Net primary production is expressed as either the weight, volume, or energy content of the primary producers per hectare per year. Net primary production is a useful measure of how a plant community is growing on a site and how that community responds to a disturbance. This is a parameter that responds to changes in the environment in a very detectable manner and is very sensitive to the abiotic and biotic environment.

Most past studies (pre-1977) estimating NPP in ecosystems only examined above-ground production because the assumption was that roots could be de-emphasized in studies without loosing an ability to understand how the system was responding. This decreased emphasis on roots was due to the perception that (1) both the above-ground and below-ground parts of plants would respond in a similar manner to manipulations and (2) only a small part of the plant was in roots because coarse-root biomass typically comprised less than 30% of the total biomass of a plant (Bray 1963). These perceptions, combined with the difficulties of studying something that could not be observed without a major effort, meant that most research was confined to above-ground plant parts. Today it is apparent that the NPP of fine roots is important to understand in comprehending ecosystem function. For example, even though fine roots may contribute less than 2% of the total plant biomass, up to 40% of the annual growth may occur in this root size fraction (Vogt et al. 1996c). However, in ecosystems it is very useful to obtain estimates of both above- (i.e., especially leaf) and below-ground (i.e., fine-root) productivity because both are important indicators of how the plant is responding to its environment and each portion is sensitive to a different set of variables.

Terminology and Ecosystem Energy Flow Determinations

There are several typical approaches that have been used to determine NPP, two of which are given below in an equation format:

1. *Net primary production = gross primary production (GPP) − plant respiration*

   Net primary production represents the gross amount of energy fixed into chemical form by plants (GPP) minus the energy respired (R) by plants. This is a value that has explicit spatial and temporal scales, typically a 1 year period and over a 1 ha land area.

2. *Net primary production = change in annual plant biomass + losses of that year's biomass during the year [detrital inputs (litterfall) + consumption by animals + other losses (organic acids, leaf exudates, root exudates, mycorrhizas)].*

   In this case NPP is the annual change in plant biomass with the addition of those losses occurring during that annual cycle due to animal herbivory and litterfall. Above-ground litterfall is relatively easy to measure and the consumption of foliage by consumers can be monitored, but below-ground senescence of root tissues and grazing by soil animals is much more problematic to measure (Ingham et al. 1985).

In actuality, the second equation is the most frequently used by researchers because it is almost impossible to obtain realistic values for GPP and plant respiration. Plant respiration by itself has been very difficult to measure and has been found to vary significantly within the canopy of a tree (Sprugel et al. 1994). It is much easier to measure changes in tree biomass using tree diameter measurements, to assess litterfall inputs, and to assume that grazing by consumers is minor and therefore could be ignored (grazing by animals or insects has typically been found to be less than 10% and therefore is within the error terms of the plant biomass measurements; Bray and Gorham 1964). The greatest losses of annual NPP in forests occurs as litterfall; above-ground litterfall has been shown to account for 57% to 85% of the above-ground NPP (Kimmins 1987).

Any management activity or change in the environment (i.e., global warming) will affect NPP in an ecosystem because it either (1) causes higher respiration rates, which decrease the NPP measured on the site; (2) increases or decreases the leaf area of a stand, which affects the amount of carbon fixed by plants and indirectly the amount of nutrients taken up by a plant as allocation to roots changes; or (3) increases the consumption of leaf or root tissues by insects or pathogenic fungi, which decreases the plant's ability to fix carbon and also to take up nutrients. Insect outbreaks can result in significant transfers of energy to the grazing and detrital portions of the ecosystem as part of insect frass (feces), dead insect parts, and dead trees. Initially, insect outbreaks may cause short-term reductions in NPP for several years (Romme et al. 1986). However, herbivory can be a positive feedback in the long term by converting low productive and degraded (e.g., "doghair" lodgepole pine stands) ecosystems to stands capable of achieving higher productivities compared with predisturbance levels, mainly by changing the competitive environment between trees, with surviving trees not being killed by the herbivores having increased access to abiotic resources (Waring and Pitman 1983, Romme et al. 1986, Alfaro and MacDonald 1988). In such a case, NPP estimates are quite useful for monitoring how the ecosystem is responding positively or negatively to its environment.

Tree *biomass* values by themselves are not useful measures to obtain in ecosystems unless there is an interest in estimating how much carbon has been sequestered in an ecosystem, because biomass has no temporal scale. Because total tree biomass accumulates over longer time scales, and part of it is also dead (e.g., heartwood), it is less likely to reflect how a plant is responding to its environment in the short term. Biomass is a buffer for plants over longer term time scales but frequently short-term impacts are typically more interesting to assess in ecosystems.

It is important to remember also that the desire to achieve or strive for high total NPP levels in ecosystems is a *human-generated value* derived from managers attempting to maximize the outputs of specific products that do not really consider the ability of the site to sustain these high NPP levels without modifying other attributes of the ecosystem (i.e., *nutrient retention capacity, species composition, susceptibility to insects and diseases,* etc.). This desire to achieve high NPP levels has resulted in a significant amount of research by forest managers and ecologists to determine what the optimal fertilizer regimes are to enhance the growth of the above-ground tree stems (Axelsson 1985, Ingestad and Ågren 1992). However, the use of fertilizers in forests since the 1950s has resulted in many problems at the ecosystem level because typically only one nutrient (i.e., N or P) was applied on a site under conditions where the limiting nutrient had not been clearly identified and usually a nutrient was applied in large quantities because the purpose was to achieve a growth response at the tree level (Ballard 1979). Despite the desire to increase tree growth rates with fertilization, the ability of trees to access the applied fertilizers has been low (ranging from 3% to 53%; Ballard 1979), suggesting that many other parts of the ecosystem are being impacted by these changing nutrient levels other than the trees. This low recovery of a nutrient in trees also shows the difficulty of identifying what and how much nutrients to add to obtain a tree growth response on different sites without converting the fertilizer treatment into a disturbance effect similar to pollutants (see Section 3.6.3).

It is important to determine what the appropriate fertilizer application levels are for forests so that excess nutrients are not applied as part of management. There are regional fertilizer programs in almost every country managing forest resources. Some of these programs, like those in Sweden, have been quite sophisticated and have used a combination of laboratory-based experiments conducted under controlled conditions (Ingestad and Ågren 1992) and extensive field experiments (Andersson and Persson 1988, Andersson 1989) to assess plant growth in response to different availabilities and frequencies of fertilizer applications. Part of the purpose of the Swedish research is to determine what the minimum fertilizer application levels should be and their seasonal time of application to maximize trees acquiring these applied nutrients. The relationship between N and plant growth is important to assess because this is a nutrient added to ecosystems due to industrial activities and as part of fertilizing agricultural fields and grasslands. There exist several examples of a changing N economy in ecosystems due to pollution inputs. For example, excess fertilizer applications or maintenance of a high population of N-fixing plants (i.e., *Alnus rubra*) have impacted plant growth rates, increased the incidence of pests and pathogens, changed soil chemistries, and resulted in losses of species (Johnson and Todd 1983, Van Miegroet and Cole 1984, Waring et al. 1985, Van Breemen and van Dijk 1988, Vogt et al. 1990).

Total *ecosystem metabolism* has also been estimated for many ecosystems, even though this measurement is very difficult to determine and one should have low confidence in its use. The following equation is typically used for this estimate:

3. *Net ecosystem productivity = gross primary production − heterotrophic respiration or NEP = GPP − R_c.*

This includes all trophic levels, where R_c is community respiration by all plants, animals, and saprophytes.

There is tremendous interest in measuring NEP because global climate change is not only affecting the primary producers in an ecosystem but is having significant effects on animals and saprophytes. Where NEP has been estimated in an ecosystem, it typically is a low percentage of GPP (2% in old-growth Douglas fir forests in Oregon; Grier and Logan 1977). In the latter study, NEP value was estimated indirectly and may be realistic for an old-growth forest. How high NEP values might be in young forests would be important to determine because the old-growth forest value is so low that it is not worthwhile measuring; in fact, it would be undetectable as part of an ecosystem carbon budget because of the error associated with some of the other biomass measures.

Utility of NPP Measurements in Ecosystems

Because commercial forestry is concerned with optimizing forest productivity and maximizing the conversion of photosynthate into marketable products, most silvicultural practices have been directed towards this end (e.g., thinning to optimize canopy size and structure, fertilization to increase leaf area and improve photosynthetic efficiency of leaves; Kimmins 1987). Even though forest managers are ultimately interested in NPP, they have not traditionally collected data on NPP values but have focused on assessing the part of the tree that has traditionally been most important for management purposes (i.e., trunk or stem volumes). This focus on one part of the tree is quite obvious when one is interested in maximizing that part of a tree. Because an understanding of the remaining portions of the trees is more relevant for ecologists, they have been instrumental in collecting NPP data in forest ecosystems. Because of the time-consuming nature of collecting NPP data for a system, the actual number of forest sites where total NPP data are available is less than 200 (Vogt 1991, Vogt et al. 1996c).

It is not uncommon to observe summarized NPP values reported for different forest ecosystems (Kimmins 1987) that were developed from the review published by Whittaker (1975). It is valuable to compare past NPP values published for forest ecosystems to recently synthesized values because much of the early research consisted of only above-ground NPP estimates (they did not include the below-ground parts) and the previous NPP estimates were derived from very few sites (Vogt et al. 1996b; Table 3-4). Despite the weaknesses of these previous values, they have been used to determine how effective model results were at representing NPP values at the biome level and to assess the importance of forests in contributing to annual carbon sequestration in the world. The importance of having large data sets to produce average values for different forest types is quite apparent when examining the older data set and the recent more comprehensive data set.

A comparison of two tables (Tables 3-4 and 3-5) indicates how the wrong opinion can be obtained from making large-scale comparisons across vegetative types when using data from a few sites to generalize for a region. For example, in the temperate zone, the NPP values derived from the earlier data were generally in the range obtained in the more recent analyses if only examining the above-ground NPP values. Even in the comparison of just above-ground NPP, the earlier results showed a much tighter range of values than the larger data set now available for both the cold temperate (3.3 to 26.1 mg/ha/yr) and the warm temperate (12.1 to 20.9 mg/ha/yr) zones. The more recent analyses highlight the fact

TABLE 3-5   Recent estimates of above-ground and total net primary production (NPP) values for different forest communities[a]

Climatic forest type	Above-ground NPP (mg/ha/yr)[b]	Total NPP (mg/ha/yr)[b]
Boreal broadleaf deciduous	3.5	5.9
Cold temperate broadleaf deciduous	9.7–12.4	12.3–15.8
Cold temperate needleleaf evergreen	3.3–26.1	10.6–30.2
Warm temperate broadleaf deciduous	12.1–20.9	16.8–30.9
Warm temperate needleleaf evergreen	15.9–19.5	20.2–21.8
Subtropical broadleaf evergreen	10.6	23.8
Subtropical needleleaf evergreen	14.9	16.0
Tropical broadleaf deciduous	12.6	17.5
Tropical broadleaf evergreen	7.6	11.4

[a] From Vogt et al. (1996c), with permission.
[b] The variations in the NPP values are related to their grouping by soil order; total NPP values by forest climatic type had a mean value with a deviation that was 30–80% of the mean.

that there exist wide variations in the soil availability of resources that are ineffectively integrated using the climatic groupings. The inclusion of the below-ground data as part NPP showed how much higher the total NPP values were in the temperate zone than what the original data would have suggested. This contrasted sharply with earlier data recorded for the tropics which suggested the tropics to have some of the highest NPP values in the world (see Table 3-4); however, recent data do not approach the high values reported earlier (see Table 3-5). Recent data show that total NPP values recorded in the tropics were much lower than in the temperate zone, again, in contrast to earlier published information.

There are some very good reasons why total NPP values have not been obtained for more ecosystems, even though total NPP values are such a good integrator of how an ecosystem is functioning. Serious methodological difficulties and the heavy time commitment needed to obtain below-ground NPP estimates have hampered people's ability to incorporate this half of the ecosystem into their analyses (Vogt and Persson 1992). Because part of an individual plant's and a stand's response to a changing environment include shifts in the amount of carbon allocated between the above- and below-ground biomass, the ability to more easily detect within-plant carbon allocation shifts is a useful tool that needs to be developed.

As demonstrated by this discussion, measures of NPP may serve as particularly useful integrative indices of how an ecosystem is functioning and how it may respond to changes in environmental constraints. NPP may also provide a comparative measure or baseline between different stages of development within a particular ecosystem state, between different ecosystem states along a gradient of potential ecosystem states, and potentially between different types of ecosystems. For example, process-based modeling has incorporated correlations between production and certain environmental parameters in order to simulate those processes of production that are affected by the environmental factors (Kimmins 1988). Finally, hybrid forest production simulation models (e.g., FOREST-BGC; Running and Coughlan 1988) are capable of taking into account multiple growth factors, such as (1) soil characteristics (e.g., moisture, nutrients), (2) changes in climate and atmospheric temperatures, (3) nutrient cycles, (4) decomposition processes, and (5) plant interactions (e.g., competition for nutrients and light), and thus have much greater predictive capacity (Kimmins 1988). However, a common limitation of all these models is that they require extensive empirical data on the specific sites that are to be modeled (including biomass and productivity data, and plant growth under different conditions), which often requires destructive sampling.

Some modeling approaches have attempted to utilized key ecophysiological processes and relationships to derive estimates of ecosystem production in an effort to reduce the number of necessary input variables and to eliminate the need for allometric equations for estimating biomass production. For example, the model PnET (Aber and Federer 1992) is based on two relationships (i.e., photosynthesis as a function of foliar nitrogen content and water-use efficiency as a function of vapor pressure deficit), which are then combined with aggregated climatic data. The PnET model was validated against 10 well-studied temperate and boreal forest ecosystems and was shown to be relatively sensitive. However, the model is limited by the fact that different ecosystems may exhibit a diversity of responses to identical alterations in climatic variables (Aber and Federer 1992). Furthermore, the limited capacity of models such as PnET and other similar models to specify the mechanisms behind an ecosystem's response to change or a shift in ecosystem health demands that data be collected and experiments conducted on a site-specific basis.

A major limitation to current estimates of forest productivity is the incomplete understanding of the environmental controls on carbon allocation. For example, it is not clear what controls carbon allocation between above- and below-ground biomass, and respiration is not accurately incorporated into the models (Gower et al. 1995, 1996). The inability to predict respiration rates is a real problem because this is a process that can consume more than half of the annual carbon fixed (Ryan et al. 1994). Using nitrogen fertilization experiments in conifer forests located in four different climate regions, Gower et al. (1996) showed that leaf area index (LAI) and annual allocation to below-ground biomass were inversely correlated; however, the specific patterns observed varied among the different tree species and climatic regions. Allocation between current twigs and leaves, stem wood, and bark production has also been shown to vary significantly among different co-occurring species and also varied by canopy position and life form (Andersson 1970). Thus, the capacity to develop generic models for carbon allocation shifts in response to environmental change is strongly limited by differences among species and sites. Furthermore, the need for destructive sampling to develop allometric regressions for estimating stand productivity has also restricted this type of data collection to a few selected intensive study sites.

Estimates of NPP are useful for determining the range of productivity values common to an ecosystem through its developmental history, for determining the development stage of a particular ecosystem, and for identifying when an ecosystem may be approaching a threshold based on deviations from expected NPP values. The analyses of NPP for different developmental stages of an ecosystem can be further strengthened by obtaining NPP estimates for different states of a particular ecosystem. However, general models relating leaf area and climatic variables are limited by their inability to accurately incorporate other potentially important factors affecting ecosystem state, such as the interactions between nutrient availability and cycling, and between tree growth and carbon allocation (Vogt et al. 1996c). Furthermore, differences among species' adaptive strategies, such as deep rooting, mycorrhizal associations, and the amount of coarse, woody debris in an ecosystem, may compensate for environmental constraints and confound expected patterns predicted from large-scale models. Additionally, models that rely on leaf area measurements to estimate stand productivity do not adequately account for complex canopy dynamics that affect carbon assimilation, such as ambient humidity, temperature, and turbulence (Field 1991). Factors that are limiting or play a predominant role in affecting plant growth must be incorporated into monitoring designs and their mechanistic relationship to larger scale patterns, as they may prove to be particular sensitive indicators of ecosystem change.

Another primary concern in assessing the productivity of entire forest ecosystems is the high degree of variability that results from different patterns of disturbance and successional stages and that should be incorporated into ecosystem monitoring programs. This may require stratifying the inherent spatial biophysical variability (Kimmins et al. 1990). For example, many forest ecosystems support both deciduous and evergreen trees either as codominants during different successional stages or associated with different topographic position and/or soil types. Because productivity and nutrient cycling patterns have been shown to vary among evergreen and deciduous shrubs and trees (Mooney and Kummerow 1971, Vitousek 1982, Chapin and Kedrowski 1983, Waring and Schlesinger 1985, Son and Gower 1991, Vogt et al. 1995b), and not necessarily in predictable patterns, data should be analyzed separately by climatic forest type and species grouping (Vogt et al. 1996c). In addition, disturbance patterns will also significantly influence the scale at which an ecosystem should be monitored to assess its integrity. For example, Bond et al. (1988) attributed low species diversity on small islands of fynbos shrublands compared with mainland communities to the change in disturbance frequency. Islands have fewer fires and thus lose species dependent on frequent fires. Unless an adequate scale, which includes the full range of disturbance processes that directly influence an ecosystem's structure and function, is covered, the assessment will be inaccurate.

Alternative Approaches to Estimate NPP

The labor-intensive field work and large amount of time required to obtain direct measures of ecosystem NPP has focused attention on alternative methods for estimating NPP. One of the most promising alternatives is the use of satellite imagery and remote sensing combined with knowledge about ecophysiological relationships, such as between leaf area index (LAI) and above-ground productivity (DeAngelis et al. 1981, Gholz 1982, Waring and Schlesinger 1985), or leaf area and site soil water potentials (Grier and Running 1976). These estimates can be made more accurate by taking into account the length of the growing season, which is calculated as the leaf area duration (LAD = LAI × months in which the canopy may conduct photosynthesis) (Waring and Schlesinger 1985). Thus, LAI has been shown to provide a fairly good indication of photosynthetic capacity across broad scales and has been effectively used to estimated above-ground NPP.

The use of models also holds great promise for estimating NPP. Earlier models for ecosystem level processes were based on the "big leaf" approach (Sellers et al. 1986), in which the canopy was represented as one large leaf that responded to the environment according to simple physical principles. According to the *functional convergence hypothesis* from which these models were derived, the biochemical capacity for CO_2 fixation should be highly correlated with gross primary production (Field 1991). To the extent that biochemical capacity for CO_2 fixation is dependent upon local resource availability, the photosynthetic response of the canopy should depend on the quantity of light absorbed and the availability of other resources required for growth (Field 1991).

Models that simulate ecosystem processes are becoming more advanced and can incorporate multiple factors. For example, the FOREST-BGC model (Running and Coughlan 1988, Running and Hunt 1993) is based on a carbon budget approach that integrates carbon, nitrogen, and water characteristics and runs simulations for 100 year periods. When productivity data were collected for forest stands during periods of markedly different climatic patterns and were used to validate the predictions made by the model, 65% of the variation in annual growth increments was explained (Hunt et al. 1991). An important assumption of the FOREST-BGC model is that a number of key physiological responses (e.g., stomatal control, photosynthesis) can be treated generically (Running and Hunt 1993). Existing knowledge, such as the control of maximum photosynthesis rate by leaf nitrogen concentration (Field and Mooney 1986), provides a critical connection between the carbon and nitrogen cycles and is important for building models (Running and Hunt 1993). These models hold great promise for addressing large-scale questions related to changing forest conditions and how these feed back to affect forest NPP.

Although simulation models provide a powerful tool for understanding large-scale processes, several limitations must be recognized. Scaling from the leaf to the canopy requires the integration of both a biological component (consisting of the canopy architecture and plant physiological responses) and a fluid-dynamical component (considers the environmental gradients that occur between the vegetation elements and the bulk environment above the canopy; Norman 1993). The complex natures of both components are difficult to integrate in space and time and to scale effectively (Norman 1993). For example, complexities associated with canopy structure (i.e., turbulent transport of mass and energy) may cause differences in production capacity (Field 1991). Furthermore, individuals within a canopy may differ in exposure to stress and their ability to access resources other than light (Field 1991). Consequently, spatial heterogeneity in both resource variability and species composition results in a mosaic of dissimilar patches having differential influences on the overall ecosystem (Bazzaz 1993). In addition, generalizations or assumptions are often quite broad in these models. For example, nitrogen retranslocation from tissues prior to litterfall of leaves and roots is defined as a constant 50% (Running and Hunt 1993), so that these models are often not very sensitive to potentially important species and community differences in nutrient use. Validation and enhancement of these models requires new methodologies for measuring ecological processes over large spatial scales and integrating smaller scale parameters directly into modeling approaches. An important opportunity for moving in this direction may lie in remote sensing technology (see Section 4.2.5).

These productivity models can function as powerful tools for scaling between different levels of ecosystem functioning because of the existence of strong relationships between leaf nitrogen levels and maximum photosynthetic capacity, plant nitrogen content and photon-flux density, and APAR and stand GPP (Field 1991, Baldocchi 1993). However, recognized sources of variation exist for each of these relationships, which means that the models must incorporate that variation or they will be of limited predictive capacity. For example, the efficiency of nitrogen use varies among sclerophyllous and nonsclerophyllous plants (Field and Mooney 1986), between sun and shade plants (Seemann et al. 1987), and between C3 and C4 plants (Sage and Pearcy 1987). Interspecific differences in how plants access nitrogen (Field 1991) and seasonally inactive evergreen vegetation (Running and Nemani 1988) may also confound the interpretation of these relationships. Models driven by LAI and light (Waring and Schlesinger 1985) would not work effectively in those regions where the relationships between moisture and productivity are highly variable due to small scale temporal and/or spatial differences in climate, edaphic, or topographic factors.

Furthermore, although aggregation of information at different scales is often a necessary process in most modeling exercises in order to avoid the cumulative error associated with the estimation of a large number of parameters, aggregation also produces errors due to the variation among the aggregated components (Rastetter et al. 1992). Thus, the coarse-scale model that is derived from fine-scale relationships may be inaccurate, even when the underlying, fine-scale processes are well understood and modeled (Rastetter et al. 1992). In addition, the use of spectral analysis is particularly limited by the fact that as spatial scales increase, spectral variation decreases nonlinearly due to the averaging of components (i.e., vegetation, soils, atmospheric conditions; Ustin et al. 1993). These restrictions, combined with the difficulties in collecting the site-specific data needed to develop accurate predictive models for particular ecosystems, demand the development of indices that are relatively easy to monitor and that reflect relative changes occurring on larger scales, even in the absence of large-scale data.

3.6.2   Soils

To understand the impact of different land uses or different management activities on the short- and long-term integrity of soil structure and function, measurements have been collected on the biological, physical, and chemical characteristics of soils and/or their biological activity (Brady 1990). Integral to managing soils is the desire to maintain plant growth at similar levels and at the same time not to degrade the ecosystem. A fundamental question that this raises is, *What is and how does one measure high soil "quality" and soil "health"?* See the later discussion in Chapter 4 (Section 4.1) on the problems of characterizing *health* in ecosystems and how the definition of this word varies depending on the values of the person assessing the condition of the system. In the soil literature, *health* is used synonymously with *soil quality* (Warkentin 1995). The Soil Survey Division (1995), in a National Committee Report, has proposed that soil quality is the capacity of the soil to sustain plant and biological productivity, to maintain environmental quality, and to promote plant and animal health. These general functions are described within ecosystem boundaries by the following soil functions (Soil Survey Division 1995):

- "providing a physical and chemical environment for gas, nutrient, water, and heat exchange for living organisms;

- regulating the distribution of surface water to runoff or to infiltration, storage, and deep drainage;

- providing mechanical support for living organisms and their structures;

- buffering the life support system against thermal, chemical, gaseous, or other stresses; and

- acting as a source, sink, and filter reducing contaminants that affect water and other resource quality."

Furthermore, the Soil Survey Division (1995) has proposed that soil health is the evaluation of soil quality over time. The measured values obtained at the beginning of a soil health monitoring program would function as the base level values for future soil health comparisons. Soil quality and soil health terms, therefore, incorporate both spatial and temporal scales, respectively, that are used to report the physical, chemical, and biological condition of a soil area.

New strategies to protect the soil resource and to measure changes in soil quality have been proposed by the Natural Resources Conservation Service (Soil Survey Division 1995) as follows:

"The development of methods to quantify changes in soil quality will require measurable indicators that are relatively easy to sample and not subject to extreme variation in time or space. This research effort should include:

- identification of the soil attributes that can serve as indicators of change in all soil functions and development of simplified models that relate changes in the selected attributes to changes in soil quality;

- standard field and laboratory methodologies to serve as indicators to measure changes in soil quality;...

Soil quality indicators and models should be used to set threshold levels of soil quality that can be used as quantitative guides to soil management."

The identification of the *soil quality indicators* referred to by the National Research Council becomes a very important step in determining the health

TABLE 3-6   Listing of potential soil health indicators identified by the Soil Survey Division (1995).

Visual indicators of soil health	Physical soil health indicators	Chemical indicators of soil health	Biological indicators of soil health
1) Adjacent deposition	1) Aeration	1) Acidity/alkalinity	1) Active carbon
2) Bare patches	2) Aggregate stability	2) Cation exchange capacity	2) Active nitrogen
3) Cloddiness	3) Amount of coarse fragments	3) Carbon a) Cycling b) Labile carbon	3) Biological activity
4) Crusting	4) Available water capacity	4) CO_2 evolution	4) Disease resistance
5) Dunes	5) Bulk density	5) Contaminates	5) Ergosterol content
6) Erosion a) Class b) No. of rills and gullies c) Phase d) Runoff	6) Crusting	6) Exchangeable calcium	6) Hyphal length
7) Evidence of deposition	7) Erosion potential-wind and water	7) Exchangeable magnesium	7) Invertebrate population – abundance, biomass, density, richness
8) Plant growth a) Crop color b) Uneven growth	8) Hydraulic conductivity	8) Exchangeable potassium	8) Microbial biomass
9) Slick spots	9) Infiltration	9) Exchangeable sodium	9) Organisms–nematodes, earthworms, ants, ground beetles
10) Soil cover	10) Macropores	10) Exchangeable sulfur	10) Residue decomposition
11) Soil use	11) Mechanical strength	11) Ion exchange capacity	11) Respiration
12) Subsoil exposure	12) Penetration resistance	12) Metal oxide content	
	13) Permeability	13) Microtox bioassay	
	14) Pore size distribution	14) Mineralization	
	15) Porosity	15) Nitrogen a) Cycling b) Mineralizable nitrogen	

TABLE 3-6 (*Cont.*)

Visual indicators of soil health	Physical soil health indicators	Chemical indicators of soil health	Biological indicators of soil health
		c) Ratio of mineralizable nitrogen to carbon	
		d) Status	
		e) Total N	
	16) Reconstituted bulk	16) Nutrient	
		a) Availability	
		b) Leaching	
		c) Status	
	17) Residual porosity	17) Organic matter	
		a) Distribution	
		b) Quality	
		c) Quantity	
		d) Rate of organic matter breakdown	
	18) Rooting limitation	18) Phosphorus	
		a) Exchangeable	
		b) Organic phosphorus	
		c) Total	
	19) Soil water content	19) Plant bioassay	
	20) Structure	20) Salinity	
	21) Structure form–porosity, pore size distribution,	21) Type and amount of clay mineral	
	22) Structure resiliency		
	23) Subsoil aeration porosity		
	24) Temperature		
	25) Texture		
	26) Thickness of topsoil		
	27) Water erosion potential		
	28) Water-holding capacity		
	29) Water supplying capability		
	30) Wind erosion potential		

of an ecosystem. A list of indicators identified by the Soil Survey Division (1995) as being important to consider as potential indicators of changing health conditions in the soil is given in Table 3-6. Obviously data will not be collected on all of these indicators for each system, but the manager should be aware of the type of information that will result from collecting these data. Prior understanding of the ecosystem being studied and the type of disturbance being imposed on the system will help to identify which of these indicators will most sensitively reflect the change that is occurring in the soil. It is useful to know all the potential indicators, and then to eliminate those considered to have a low probability of influencing the system (e.g., soil salinity in an ecosystem with acidic soil).

Once the most sensitive soil indicators have been selected, they should be quantified using standard field and laboratory methodologies so that comparisons can be made between other studies (e.g., Page et al. 1982, Klute 1986). It is not uncommon for different studies to use different chemical or physical analysis techniques that produce different results so that some standardization of methods is necessary if comparisons are to be made between studies. It is also quite useful to measure these indicators of soil quality over time on the same site; then rates of change in individual indicators of soil quality can be documented and the site indexed as to its soil health. This type of information will also indicate possible management effects on the ecosystem.

If several indicators are being quantified at one time, contradictory results may be obtained, which should not be unexpected. If this occurs, each indicator will have to be judged or weighed as to which one most significantly reflects the changes that are occurring in the ecosystem or one may have to reassess the indicators that were initially selected for analysis. It may be necessary to select other indicators because the most sensitive ones may not have been initially chosen for analysis. It is important to always incorporate several indicators into one's experimental design because our understanding of ecosystem interactions is still developing.

One procedure that has been developed to evaluate several indicators at once is called the Multiple-Variable Indicator Kriging (Smith et al. 1993). This procedure was designed to provide a means of integrating soil quality parameters into an index to produce soil quality maps on a landscape basis. These maps would indicate areas on a landscape that have a high probability of having good soil quality according to predetermined criteria. It would also identify indicator parameters responsible for zones of low soil quality, thus allowing specific management plans or land-use policies to be developed for these areas. This type of approach is very useful for scaling information between small and larger spatial scales, and for integrating soils information into ecosystem management. For these to be useful, there has to be a strong mechanistic understanding of the variables being examined; otherwise, this becomes a mapping operation that is not readily transferable anywhere else.

3.6.3   Abiotic Resources

Nutrient Pools and Fluxes

Understanding nutrient cycles in ecosystems has been an important research area for a long time, with much of this research focusing on nutrient uptake and use by individual plants. Plants require a total of 21 different essential elements, with the required amount varying by element and plant species, in order to successfully establish, grow, and complete their life cycles. Along with water, nutrients are also a dominant factor limiting tree growth. Plants acquire nutrients from the ecosystem from several different pools: soil exchange sites, soil weathering, above- and below-ground litter that is decomposed by soil fauna and flora, and internal retranslocations within the plant tissues. In most forest ecosystems tree growth is constrained by one or more of the following nutrients: nitrogen, phosphorus, potassium, calcium, or magnesium (Lassoie and Hinckley 1991). Nutrient availability may also be directly influenced by soil moisture, as water is required for decomposition and weathering processes, as well as for the transport and dissolution of nutrients within the soil matrix.

Because of the direct link between productivity and nutrient availability, forests and agricultural lands have been managed in an attempt to increase their productivity over short and long time scales by using nutrient additions. Much of that research has hinged upon trying to increase the output of products from these managed lands in an efficient and least costly fashion. Many of our past and current land uses have also directly disrupted or altered natural nutrient cycles, typically manifested as an excess input of nutrients. Many of the acid deposition research programs have focused on understanding the impacts of pollutants on terrestrial and aquatic ecosystems, and many of the pollutants in question are nutrients. There has also been considerable concern over how nutrient cycles are affected by whole tree harvesting because nutrients are also removed from the site during tree harvesting (Johnson and Todd 1987). Even the use of fire as a management activity has to consider the quantity of nutrients lost from the system due to different fire intensities (Boerner 1982). Nutrients have also been excellent parameters to monitor when attempting to assess the impact of a management activity on ecosystem resistance and resilience (see Section 2.1.1 on watersheds).

Documenting nutrient uptake and cycling in ecosystems is a very valuable tool for managers. Nutrient uptake and use by plants can be measured using several different approaches. The fastest and easiest approach is based on visual appearance and the identification of symptoms indicating nutrient deficiencies. However, the symptoms of multiple deficiencies of nutrients may be difficult to detect and distinguish, because many other factors besides nutrition may be involved, such as fungal diseases, insect attacks, drought, frost, pesticides, or air pollution (Waring and Schlesinger 1985). Chemical analysis of leaf tissue provides a more precise and quantitative means of determining plant mineral nutrition. However, several factors must be considered in selecting leaves for tissue analysis because they significantly affect what nutrient contents are measured. These include the position of the leaves within the canopy, leaf age, tree age, and time of year of sampling. Environmental variation due to microsite differences in soils, climate, wind exposure, and moisture may also alter nutrient relations among trees (Kimmins 1987). In controlled greenhouse experiments, quantitative relationships can be developed between the concentration of an element in the plant tissue and the growth or yield of the plant. However, application of the results to the field are often difficult due to complex interactions between different nutrients, genetic differences between individuals, and environmental influences that confound relationships between nutrition and plant growth.

Another common approach to studying nutrient relations in ecosystems is to assess the availability of nutrients in the soil. An important limitation of soil analyses to assessing plant mineral nutrition is that soil nutrient content does not always accurately reflect the amount of nutrients that is actually accessed and taken up by the plant. This is particularly true for phosphorus because plant–mycorrhizal associations may significantly enhance phosphorus uptake, these associations often differ substantially among species, and their efficacy varies under different environmental conditions. A common approach to determining which nutrients are limiting to a plant is to conduct controlled nutrient addition experiments either in the lab or field. This consists of applying fertilizers to an individual plant or a stand at different application frequencies, and with different combinations and concentrations of individual nutrients. These experiments have provided substantial insight into the role of mineral nutrition in controlling ecosystem productivity and the potential of management to influence nutrient relations.

Studies on plant nutrition have shown that nitrogen is often the most limiting nutrient in many ecosystems. It is required by plants in relatively large amounts and is not contained in the parent material. Typically the input of N as part of atmospheric inputs is quite low and is insufficient to satisfy plant requirements, unless the areas has been exposed to atmospheric pollution or fertilizer inputs (see previous discussions). Nitrogen, therefore, is tightly cycled within the ecosystem, with two activities playing significant roles in contributing to the whole ecosystem availability—internal vegetative recycling and decomposition of litter material (Cole and Rapp 1981). For example, in a 20-year-old loblolly pine plantation, 16% of the plant N requirement was obtained from precipitation, 5% from canopy leaching, 40% from the decay of litter material in the forest floor, and 39% from internal redistribution within the trees (Kimmins 1987). In some ecosystems (especially early successional or highly disturbed systems), the presence of N-fixing plants or associative N fixation by bacteria provides important sources of N input. Management has to consider the role of these N-fixing plants in contributing to maintaining the N status of previously forested ecosystems and to ensure that they are not eliminated as part of the managed succession of vegetative communities. Nitrogen limitations to plant growth are particularly common in the temperate regions, where N-fixers are absent or where fire is frequent. Many dry forests are limited by both low water availability and also low N availability at different times of the year (Gower et al. 1992).

Because the productivity of forest ecosystems is closely linked to plant N uptake and use, a strong positive correlation is exhibited between a leaf's photosynthetic capacity and its nitrogen content (Field and Mooney 1986). This relationship persists across species and growth forms, and is relatively robust to variation in leaf age, nutrient availability, or light availability (see Field 1991 for review). Sparks and Ehleringer (1995) showed that leaf carbon isotope discrimination and N content were highly correlated, suggesting that leaves with higher photosynthetic capacities had lower intercellular carbon dioxide concentrations. Results from long-term N fertilization experiments in Swedish forests reported that the effect of N fertilization on carbon isotope discrimination was strongly correlated with production and was weakly correlated with foliar biomass during a dry year, but there was no correlation during a wet year (Hogberg et al. 1995). These studies indicate the potential for integrating nitrogen isotope and carbon isotope analyses in order to evaluate the relationships between ecosystem-level processes, such as net primary production, with smaller scale phenomenon occurring at the individual tree level measured directly using isotope techniques.

The use of indices related to nutrient use and cycling may be particularly appropriate in sites where nutrients limit plant growth, and thus strongly influence carbon allocation patterns (Ingestad and Ågren 1992). One example is seen in the observed relationships between N and production in forest ecosystems. For example, Pastor et al. (1984) found that net above-ground production across a series of edaphic climax forests was highly correlated with soil N mineralization. Nitrogen availability indices have also been effective in predicting root biomass and productivity for some ecosystems (Aber et al. 1985, Vogt et al. 1986, Gower et al. 1992, Vogt et al. 1996c). Furthermore, leaf nitrogen content may also influence herbivory and pest interactions (Jones and Coleman 1991).

Nitrogen is also a valuable indicator to monitor to detect changes in ecosystem processes that are reflected at the level of nutrient cycles and decomposition. For example, trees respond to N deficiencies by retaining a greater proportion of internal N in their biomass, which reduces the cycling rate of N in the ecosystem because plant materials added to decomposition are chemically of poorer quality and require a longer time to decompose. Plants respond to N deficiencies by increasing tissue longevity and the residence time of N in tissues, increasing the amount of N retranslocated out of tissues prior to senescence, resulting in a litter material with relatively high C/N ratios being transferred to the decomposition system (Bloomfield et al. 1993, Gower et al. 1992, 1996). Because N plays a controlling factor in decomposition processes, and thus the rate of nutrient cycling in forest ecosystems, litter with lower N concentrations (i.e., high C/N) has slower rates of decomposition (Berg 1984). This litter decomposes slowly and available N may be taken up by the microorganisms, further reducing N availability to plants, and creating a positive feedback to increasing N limitations of primary production (Vitousek and Howarth 1991). The influence of N on decomposition reflects its role in regulating rates of nutrient cycling within forest ecosystems (Swift et al. 1979).

Consequently, changes in N availability due to stress or disturbance will likely have direct implications for processes occurring at many different scales, providing a useful measure for transferring information between scales. Because N integrates as a dominant factor in system functioning across so many different levels, easily measurable and sensitive parameters that are mechanistically related to N effects may be especially useful in understanding how ecological processes at different scales are interconnected.

In forest ecosystems where N is a limiting nutrient, such as boreal and subalpine coniferous forests, N deposition due to atmospheric pollution may exceed the growth requirement of trees and reach toxic levels (Waring 1987). The physiological basis for nitrate effects on boreal and subalpine conifers has been relatively well established, such that forest health affected by nitrate deposition can be monitored based on key stress responses (including the accumulation of amino acids in foliage and in the xylem sap) (Waring 1987). For example, air pollution may cause a decrease in the C/N ratios in leaf tissue due to the induced mobilization of N in plants and increased accumulation of amino acids and other N rich polypeptides (Bryan et al. 1983, 1985, Jones and Coleman 1991). Thus, monitoring changes in the N content of these key compounds may provide information about the effects of pollution stress on the ecosystem. As with plant storage of starch, N storage may also serve an indicator of stress. Rubisco can perform an important storage role in plants, particularly on less fertile sites (Chapin 1990), and changes in stored N may be linked to stress response. Specialized storage proteins and amino acids may also store N (Chapin 1990) and serve as indicators to changing environmental conditions.

However, it is important to recognize that N interacts with several other environmental constraints and that other limiting factors may also strongly influence how plants respond to stress. For example, Lajtha and

Whitford (1989) found that net photosynthesis was positively correlated with leaf N only when the plants receive supplemental water. In tropical broadleaf evergreen forests, where both N and P can limit tree growth (Cuevas and Medina 1986), the ratio of N/P in above-ground litterfall explained 99% of the variation in fine-root NPP and provided a stronger index than N alone (Vogt et al. 1996c). Reich and Schoettle (1988) also found that for white pine seedlings grown in forest soils with different N and P availabilities, the proportion of leaf P/N correlated better with maximum net photosynthesis than foliar P or foliar N alone. Thus, in some ecosystems it may be necessary to determine the interactions between N and P or some other element in influencing system behavior, and to incorporate multiple components within an index for monitoring ecosystem health.

Phosphorus is also required by plants in relatively large amounts. Phosphorus commonly limits plant growth in regions characterized by old, highly weathered soils, such as the tropics, southeastern United States, and parts of Australia (Moran 1987, Brady 1990). Furthermore, because P is not highly soluble in water and is often tightly fixed as unavailable compounds within the soil matrix, P may often be limiting to plants, even though the total amount of P in the soil may be high (Brady 1990). Similar to N, annual P requirements of plants are mainly satisfied by internal transfer within plants and from the decomposition of litter material transferred to the forest floor. For example, 2% of the annual P requirement of a 20-year-old Loblolly pine plantation was obtained from the mineral soil, 6% from precipitation, 9% from canopy leaching, 23% from litter decomposition and 60% from redistribution within the plant (Kimmins 1987).

The evaluation of mineral nutrition and nutrient uptake of individual trees may provide useful information to the manager because once the limiting nutrient or nutrients are identified, and the extent to which they limit productivity is better understood, management interventions may be designed by increasing or modifying nutrient availability in the system. However, most previous studies on nutrient relations in managed ecosystems have focused either on agricultural crops or forest plantations, and have focused almost exclusively on the effects of nutrients on above-ground biomass production in isolation from other aspects of tree growth and the integrity of the ecosystem as a whole (Vogt et al. 1996a). As a result, crucial interactions between mineral nutrition and other factors such as disease and pests, carbon allocation, and defense mechanisms are poorly understood, yet these interactions may be critical in determining our capacity to successfully manage natural ecosystems and plantations. It is necessary to begin to move beyond the "quick fix" approach of fertilizer applications, which merely replace lost nutrients, towards focusing on a greater understanding of how individual plants and ecosystems regulate and maintain a balanced mineral nutrition. For example, in natural ecosystems, the biological activities of plants and microorganisms significantly alter nutrient availability. Symbiotic associations (i.e., mycorrhizal, N-fixing, etc.) have a major role in changing the nutrient environment perceived by a plant (Harley and Smith 1983, Allen 1991). Even plants themselves appear to be well adapted to respond to changing nutrient availabilities by shifting carbon allocation between the shoot and root tissues (Vogt et al. 1990). Plants and the associated mycorrhizal associations are also very effective at changing nutrient availability at the rhizosphere level by secreting organic compounds or modifying the activity of microbes involved in nutrient mineralization.

Understanding the physiological basis for how different plants acquire and utilize nutrients, and how these processes are altered in response to natural or anthropogenic disturbances, may be particularly useful in addressing many management issues. Key changes in the patterns of nutrient use by vegetative communities following a particular disturbance may be identified and may directly affect our capacity to manage the ecosystem. For example, lands abandoned from human land use generally support early successional vegetation (with higher rates of nutrient absorption and photosynthesis), which can result in the depletion of internal nutrient reserves and greater incidence of disease (Chapin 1983). Vitousek (1990) suggested that different nutrient cycling characteristics of exotic plants that successfully invade disturbed habitats may drastically alter patterns and processes at the ecosystem level. For example, the biological invastion of *Myrica faya* Ait (an exotic actinorrhizal N-fixer) in N-deficient volcanic regions of Hawaii has significantly changed N budgets and, in turn, successional processes of the native vegetation (Vitousek 1990). Finally, understanding the capacity of different species growing under different climatic conditions to utilize and allocate nutrients for production can be useful in making management decisions about species–site interactions and in predicting future productivity potentials on a global level, as demonstrated for pines (Gower et al. 1995). These examples demonstrate that information about how individual plants access and utilize nutrients at different scales may provide valuable insight into ecosystem functions and facilitate the design of more realistic management interventions for achieving particular objectives.

Water Uptake and Use

Over much of the Earth's surface, water is the primary factor controlling the distribution and growth of trees (Kramer and Kozlowski 1960, Gholz et al. 1990, Lassoie and Hinkley 1991). Water influences plant growth and development directly through its effects on turgor pressures and transport processes, and indirectly through effects on processes, such as protein synthesis and photosynthesis (Lassoie and Hinkley 1991). In order to maintain well-hydrated leaf tissues that can support high levels of gas exchange, water must be continuously moved from the soil to the leaves through the plant vascular system. The flow of water up the plant is driven by solar energy, which causes the evaporation of water from the leaves, creating a difference in the energy potential within the plant (Lassoie and Hinkley 1991, Pallardy et al. 1995). There are a variety of adaptive strategies that plants use to tolerate or avoid water stress, including:

1. Decrease in canopy leaf area

2. Stomatal closure

3. Increased leaf angle

4. Decrease in leaf absorptivity (e.g., leaf hairs, salt glands, waxes)

5. Maintenance of high turgor pressures

6. Increased root-to-shoot biomass ratio

Managers may be concerned about measuring the water status of trees for several reasons. First, because water plays an integral role in photosynthesis and primary production, knowledge about the effects of water stress on tree functions may be useful in making predictions about tree response to changes in microclimate conditions or major disturbances, such as drought or flooding. Furthermore, an understanding of how different species respond to water stress may be important in selecting species for plantation systems or in prescribing treatments to forest stands.

Water relations in forest ecosystems can be measured directly by determining the water content of either the plant tissues or of the soil, expressed as the fraction or percent of water in the plant tissue or soil dry weight, or the relative water content of the plant tissue compared with the total water present in fully hydrated tissues. The technique most commonly used to directly measure plant water stress is water potential, which is measured using a pressure chamber (for review of techniques, see Pallardy et al. 1991). Because of the high diurnal variation exhibited by plant water potentials, predawn leaf water potential is often used as a baseline for estimating the water potential of the soil (based on the equilibration of the hydraulic system that occurs overnight because of stomatal closure). Several other methods are available for measuring water potential but are generally more complicated to use (Pallardy et al. 1991).

In many cases it may be easier to use indirect measures of plant water use and uptake that do not require extensive equipment and also provide more integrative information. For example, measures of plant anatomical characteristics of leaves, such as cuticle and blade thickness, stomatal density, and thickness of the palisade mesophyll, may also provide information on water-use efficiency and photosynthetic capacity. In a comparison of three different oak species, drought tolerance was shown to be inversely correlated with stomatal area per unit area of leaf and positively correlated with leaf anatomical plasticity, suggesting stomatal characteristics may provide indices of adaptive strategies related to water uptake and use (Ashton and Berlyn 1992). Similar to the relationships obtained between leaf area and photosynthesis, good relationships can also be derived for leaf area and both the hydraulic conductivity of the wood and transpiration rate (Pallardy et al. 1991).

One application of using physiological measurements to assess management impacts on ecosystems is the case of grazing impacts on California blue oak (*Quercus douglasii*) savanna ecosystems in the western United States. As a result of previous land-use history under which large numbers of livestock were grazed in the region following European settlement, plant species composition changed with annuals largely replacing perennial grass species (Bartolome et al. 1987), with the former having greater leaf areas and depleting the soil water faster (Gordon et al. 1989). Welker and Menke (1990) studied the interaction of simulated browsing and drought stress on blue oak seedlings in which seedlings were treated with different levels of defoliation and water stress. Measurements of plant water potential were used to assess plant response to the treatments. The study showed that severely defoliated seedlings exposed to slow rates of induced soil water stress maintained greater water potential, which was associated with greater survivorship the second year; however, these patterns were not as strong under rapid rates of soil water depletion. Plant soil water potential served as a fairly reliable indicator for predicting plant response to grazing and water stress interactions, which were not apparent in observed survivorship until the following year. The physiological measures also provided valuable information about the mechanisms for vegetation response to grazing that may be applicable to managing grazing lands for specific objectives or products.

In some cases, physiological attributes can provide a more accurate assessment of how plants integrate their environmental conditions than measurements of microclimate alone can provide. Physiological measurements can also provide information before plant performance is visible through biomass accumulation and growth. For example, Crunkilton et al. (1992) studied the response of northern red oak (*Quercus rubra* L.) seedlings to two different types of silvicultural management practices: clear-cutting and shelterwood treatments. They used physiological measurements of plant water potential to document that seedlings planted into the clear-cut site maintained better water status than those planted into the shelterwood during the first year, despite the greater temperature and evaporative demand in the former. Gas exchange levels were also greater for seedlings in the clear-cut site. By the second year, the water statuses of seedlings in the two sites were no longer significantly different, which was attributed to greater competition for moisture with understory herbs in the clear-cut site. The physiological measures revealed the importance of moisture in regulating regeneration success and their importance in the design of silvicultural techniques.

Physiological measurements may be particularly useful in management for verifying relationships between plant response and more easily measured biophysical characteristics. Once the physiological mechanisms underlying these relationships are understood, monitoring and assessment may rely primarily on non-physiological parameters. For example, Meiners and Smith (1984) used a pre-existing site quality index based on topographic parameters (e.g., aspect, slope inclination, slope position) to select sites along a xeric to mesic gradient. They then verified the accuracy of the site quality index in predicting soil moisture and plant moisture stress by measuring soil water potential and predawn plant water potential along the moisture gradient. Their results showed a direct relationship and a strong seasonal trend between soil water potential and plant water potential. However, discrepancies between some of the sites selected according to the site quality index and the corresponding soil and plant moisture relations suggested that certain modifications to the index could be made to improve its accuracy. As site quality indices are a relatively rapid and easy measure for evaluating site quality, initial investment in refining indices based on physiological and biophysical measurements may be particularly valuable in long-term management.

Abiotic Plant Resource–Use Efficiency

The efficiency with which plants use resources (i.e., light, nutrients, water, carbon) has traditionally been expressed by calculating ratios of productivity per unit of resource; or, in the case of carbon, relative growth rate (Sheriff 1995). Most often, efficiency values are used as a basis for comparing differences among species or individuals, or the magnitude of the efficiency value in relation to growth and survival. The degree of efficiency with which a particular resource is used reflects the mechanisms that enable a plant to grow in an environment in which that resource is limiting, and will depend on the extent to which the different resources are limiting and their interactions with each other. Differences in the ratio of nutrients that plants obtain or release (i.e., as reflected by nutrient-use efficiency) are also important at the ecosystem level, because they determine the ratio of elements that are found in live plant tissues, and ultimately in dead tissues, that are the substrate of the decomposer community (Vitousek 1983).

For example, nutrient-use efficiency has important implications for decomposition and nutrient cycling process by changing the quality of the litter (Aber and Melillo 1982). However, because resource-use efficiencies are ratios, they do not necessarily relate to absolute levels of either productivity or resource use, but merely show the relationship that exists between them (Sheriff et al. 1986). Furthermore, a high resource-use efficiency may increase productivity only if that resource is limiting, while compensation for limiting resources can occur through enhanced uptake of another resource or through particular adaptive strategies (Sheriff et al. 1995). For example, the efficiency of use of nitrogen or water was not strongly associated with productive capacity (Field et al. 1983). Thus, values of resource-use efficiency must be used with a clear understanding of both the constraints and advantages associated with them.

Previous studies have shown that the use efficiency of a resource generally increases as the resource becomes scarcer (Vitousek 1982, Ehleringer and Cooper 1988, Schlesinger et al. 1989). For example, several studies have shown that nutrient-use efficiency decreases with increasing availability of a nutrient (Shaver and Melillo 1984, Aerts and Caluwe 1994). Boerner (1984) demonstrated that nutrient-use efficiency and resorption of N and P for four deciduous tree species along a site fertility gradient was consistently greater on the less fertile soils, which also indicated a relatively high physiological plasticity within a species for nutrient-use efficiency. However, trade-offs between the use of different resources, for example, nitrogen use for increased photosynthesis versus water loss via transpiration, creates complex interactions between use efficiencies of different resources that must be considered (Field et al. 1983, Lajtha and Whitford 1989). Water and nutrient availability are closely interconnected at the level, where (1) water controls the availability of nutrients at the root surface and controls decomposition processes, weathering, and nutrient fluxes in the soil; and (2) gas diffuses into stomates on leaf surfaces concomitantly with water vapor losses occurring at the same location. This evaporation of water from the leaf surface helps to cool leaf surfaces and allows for the photosynthetic process to continue (Waring and Schlesinger 1985).

Water-use efficiency is related to gas-exchange processes occurring during photosynthesis. Movement of CO_2 from the atmosphere and into the plant occurs through openings or pores called *stomata*. This occurs because the fixation of carbon dioxide at the cell surfaces within the leaf reduces the internal CO_2 concentration gradient. The same cell surfaces are constantly moist, which results in the evaporation of water from that surface. Thus water loss to the atmosphere is an inevitable part of the CO_2 uptake. Transpiration causes the constant movement of water to the leaf to replace evaporation. If the plant is unable to acquire sufficient water, water stress develops, which eventually leads to stomatal closure. This results in a water cost/carbon gain process that can be presented as follows: water-use efficiency = weight of carbon gained per unit weight of water lost.

This water-use efficiency varies between species. Because sun foliage are thicker and smaller, and transpiration is closely linked to total leaf area, leaves with higher rates of photosynthesis per unit area (high N) also have lower rates of transpiration per unit area. This suggests a trade-off between water-use efficiency, which would be higher in the thicker, high N leaves, and N use efficiency, which would be higher in thinner, low N leaves (Field et al. 1983). Furthermore, this also suggests that there may be limitations to the extent to which the efficiency of one resource use can be increased without reducing the use efficiency of another resource.

Nutrient-use efficiency is related to the capacity of a plant to produce carbohydrates per unit of a particular nutrient utilized. Measures of nutrient-use efficiency must take into account both the mean residence time of the nutrient in the plant (e.g., leaf longevity and retranslocation) and the instantaneous rate of carbon fixation per unit nutrient (Berendse and Aerts 1987). As demonstrated earlier, there are trade-offs in the conflicting needs to conserve nutrients on poor sites and to produce the high nutrient content foliage required for high photosynthesis and nutrient uptake. Increasing nutrient-use efficiency for allocating more carbon to wood instead of leaves may prove not to be very adaptive if it seriously reduces leaf area and total photosynthetic capacity. Another approach is to maintain foliage for more than 1 year. If foliage is retained for 2 years, then a significant increase in carbon gain may be realized per unit of nutrient input to leaf tissues. The "evergreen" habit is considered a primary adaptation to low nutrient sites in forests (tropical to temperate), where the benefits of increased nutrient-use efficiency offset the respiratory cost of maintaining foliage through the winter (Sprugel 1989). These same allocation characteristics partially explain the preference of forest managers for evergreen species. Even on rich sites, planted pines and other conifers will allocate less to foliage and more to wood production than deciduous species, which need to replace the entire foliage canopy each year.

Efficiency measures have been used to evaluate individual plant productivity and photosynthetic capacity. For example, Waring and Pitman (1985) assessed changes in the susceptibility of lodgepole pine stands to mountain pine beetles in response to changes in nutrition, light, and moisture. They developed an index of carbohydrate stress (the tree growth efficiency index, based on the ratio of wood production, one of the last priorities for carbohydrates) with the amount of photosynthetic tissue. Waring and Pitman (1985) determined that dense stands had a relatively low growth efficiency and a greater susceptibility to insects than less dense stands, resulting in a possible risk rating system for assessing stand condition. Such an approach must include both the intraspecific range of plant adaptive strategies related to phenotypic plasticity (which serve to adapt to local variation or change in resource availability) and the interspecific range (which enables species to replace each other and to continue to perform key functions in response to environmental change).

There are several problems with measuring resource-use efficiencies that must be considered. Resource-use efficiencies are often determined using instantaneous measures of gas exchange and therefore are difficult to relate to the uptake of a resource that occurs on a longer time scale. These instantaneous measures of gas exchange are also highly variable due to fluctuations in environmental conditions to which stomata are highly sensitive. Furthermore, the allocation of photosynthate within the whole plant may change in response to the availability of a resource and may affect overall plant productivity without being reflected in the efficiency measure.

There are several measures of resource-use efficiency available that are more integrative in nature than the instantaneous measures discussed earlier. Integrative measures of nutrient use have reported varying results, suggesting that species' nutrient use characteristics are due to complex species–site interactions. For example, some studies have used nutrient reallocation techniques (Fahey and Birk 1991) to compare temperate deciduous with evergreen forests. These studies have suggested that deciduous trees have greater nutrient uptake and nutrient loss in leaf litterfall (Gosz et al. 1981, Vogt et al. 1986) and lower nutrient retranslocation (Chapin and Kedrowski 1983, Schlesinger et al. 1989) than nondeciduous species. In contrast, others have shown that when trees are grown under similar conditions, deciduous species may translocate the same or greater amount of nutrients than evergreen trees (Son and Gower 1991, Reich et al. 1992). Integrative measures of plant water use have been obtained using carbon isotope analysis to determine the mean intercellular carbon dioxide concentration (Farquhar et al. 1982; see Section 4.2.4). Because carbon isotope discrimination is affected by relative changes in stomatal conductance, it provides an integrative measure of how plants balance transpiration with carbon assimilation, and therefore can be interpreted to reflect plant water-use efficiency (Ehleringer 1990). Plants adapted to moisture-limiting conditions tend to have lower discrimination values than plants from moist habitats (Ehleringer and Cooper 1988). Studies have also suggested that variation in carbon isotope discrimination is associated with life form (plants with high relative growth rates have higher discrimination values, Ehleringer et al. 1993) and with leaf duration (evergreens have lower discrimination values than deciduous species; Valentini et al. 1992). Evaluating carbon isotope discrimination in conjunction with plant nutrient use can provide information about the complex interactions of plant adaptive strategies with multiple factors.

3.6.4   Physiological Measurements

Physiological measurements are generally conducted because they provide quantitative estimates of how individual plants are responding to their environment, which in turn can be used to understand ecological processes and landscape patterns. Physiological measurements can also be used to determine the effects of particular management activities on desired outcomes (e.g., increased productivity, decreased pests or disease incidence). Physiological studies on the individual plant level may be important in management for several reasons:

1. More precise information about plant health and productivity can often be obtained as compared with information collected at larger scales.

2. Information can be obtained relatively quickly and, in many cases, at lower expense.

3. Experiments can be designed using individual plants, which allows for more controlled manipulations of experimental conditions and thus provides more robust results.

4. Physiological measurements are often directly linked to primary management concerns, such as productivity rates, plant mineral nutrition, and plant water relations, and thus may yield valuable information on how to improve management activities.

Major constraints of using physiological measurements on the individual plant level are largely due to the highly specific nature of the data, which is usually limited to only a few number of plant species and/or individuals, and thus often cannot adequately account for the high interspecific and intraspecific variability. Furthermore, because physiological measurements generally measure a process or rate at a particular point in time, they do not incorporate the short-term temporal variability associated with physiological responses to changing environmental conditions as well as internal plant diurnal and seasonal fluctuations. Most physiological measurements are taken at the seedling level, where it is easier to grow seedlings under controlled conditions, and less at the larger scales (i.e., whole tree or stands) because of the logistical problems of size and the associated problems of controlling the environment. With our current technology and understanding of plant physiology, it is difficult to scale up from individual plant physiological measurements to processes and functions occurring within the ecosystem as a whole. For example, the important question that arose out of the NAPAP studies was how to transfer the high intensive sampling occurring at the seedling level to address questions concerning the forest level (see Section 4.2.9). However, it is often not necessary nor desirable to directly transfer information between scales. For some management questions, information obtained about physiological responses at the individual plant level may be the most appropriate and efficient approach.

Physiological measurement parameters should be selected to reflect those strategies or attributes that contribute most strongly to plant survival and growth within a particular set of environmental constraints and/or management operations (Table 3-7). Plant adaptive strategies are broadly defined as plant attributes that perform specific functions that increase the overall fitness of a plant and enable it to moderate the effects of the abiotic and biotic constraints operating within its environment (Solbrig 1993). It is necessary to identify those adaptive strategies that play a dominant role in regulating ecosystem processes within a particular site. Often those strategies directly related to stress response may be particularly relevant, especially if the stress or stressors and their corresponding response variables by the individual plant can be identified. Although the key plant strategies and the parameters measured will depend greatly on the particular ecosystem and the specific management questions being addressed,

TABLE 3-7   Above- and below-ground plant variables that can change in response to stress

Above-ground changes
Chlorophyll and protein production
Foliage biomass and leaf area
Net primary production (NPP)
Nutrient-use efficiency
Photosynthetic efficiency
Secondary chemicals, changes in sink strengths
Stem wood
Stored carbohydrates
Below-ground changes
Coarse-root biomass and net primary production
Fine-root biomass and net primary production
Fine-root distribution
Mycorrhizal colonization of roots
Mycorrhizal species composition
Nutrient and trace element ratios
Root morphology
Secondary chemicals
Stored carbohydrates

several basic physiological measures can be identified that provide a useful framework for making assessments across different ecosystems. These will be discussed in more detail later.

Photosynthetic Measurements

Ultimately, all living organisms and their associated processes and functions required to sustain them are dependent upon photosynthesis in which green plants use light energy to covert carbon dioxide and water into chemical energy in the form of carbohydrates. During photosynthesis, carbon dioxide is fixed and oxygen is released by the leaf. Approximately 18% of the photosynthetically active radiation that is absorbed by the leaf is converted to chemical energy, while the remaining energy is converted to heat (> 80%) or to fluorescence and luminescence (2% to 3%) (Leverenz and Hallgren 1991). Photosynthesis can be measured using techniques that determine the rates of oxygen and carbon dioxide exchange, or by measuring the amount of energy converted to heat or lost by fluorescence or luminescence of chlorophyll on intact leaves of a plant (Leverenz and Hallgren 1991). Because wood production requires energy produced by photosynthesis, measuring photosynthesis is often of particular importance to both forest scientists and managers.

Although various techniques are available for measuring photosynthesis, the most commonly used instrument is the infrared gas analyzer (see Leverenz and Hallgren 1991 for review of techniques). Measurements can be taken either in the field or laboratory; however, in both cases part of the foliage must be enclosed in an assimilation chamber to measure CO_2 uptake or loss. Therefore, these conditions differ from the highly variable and heterogeneous conditions experienced by the plant under natural conditions. One of the greatest limitations in applying photosynthesis measurements to the management of ecosystems involves issues of scale. At present, photosynthesis measurements using the gas chamber technique must be made on an individual leaf basis. The high degree of variability in photosynthesis occurring between different leaves, as affected by location in the canopy, changes in environmental conditions in time and space, and physiological differences due to leaf size, thickness, or shape, requires that intensive sampling be conducted within the canopy of an entire tree to collect information that can be scaled up to a stand level. Both the time and expense required to obtain this information would generally preclude the widespread application of photosynthetic measurements in ecosystem management.

However, such measurements may be particularly useful for studying specific responses of individual leaves or shoots to a stress or perturbations, which may serve as a useful predictor of tree or stand level response. Of particular relevance to management would be the establishment of links between different scales, for example, between photosynthesis measures at the leaf level and productivity measures of the entire tree or stand (see Section 3.6.1).

Because of the high degree of fluctuation and sensitivity of photosynthesis measures, and the specific equipment and time required, in some instances it may be more useful to focus on other indirect but related measures of photosynthesis that integrate longer time scales. For example, leaf nitrogen content, leaf area index, leaf chlorophyll content, and carbon allocation patterns have all been shown to be correlated with photosynthetic capacity, and are highly adapted to the particular environment in which the plant grows. For example, there exists a close relationship between the amount of chlorophyll and the amounts of proteins and enzymes required to carryout the biochemical reactions of photosynthesis (Chapin et al. 1987). Leaf life span has also been shown to be closely associated with photosynthetic performance. Consequently, these strategies are often sensitive to changes in environmental conditions or management treatments. Furthermore, carbon allocation patterns within the plant are often strongly influenced by changes in the environment or in the health of the plant, and thus measurements of carbon allocation patterns may also provide valuable information about the photosynthetic performance of the plant (see later).

Another useful indirect measure of plant photosynthetic capacity can be obtained from measurements of leaf area index (LAI), which can be collected at the individual or stand level (Gower et al. 1992, 1995). Leaf area index can be defined as the projected total area of foliage of a tree or a stand over given area of land. Leaf area index strongly reflects how plants are responding to their environment and has been conceptualized by the following equation:

Leaf area index = f (site water balance, nutrients, light, species)

(Waring and Schlesinger 1985). Plant leaf areas change significantly in response to many abiotic variables, and there exists a strong correlation between site water balance and leaf areas (Grier and Running 1976). The variables causing changes in plant LAIs are also quite sensitive, and changes caused by many human land-use activities, making this a useful parameter to monitor. Leaf areas of forest communities also change during stand development (Kimmins 1987), and one would have to distinguish between normal and atypical changes in leaf areas. This is an extremely useful technique in management because tools exist by which this parameter can be relatively easily collected in the field (Bolstad and Gower 1990).

The range of leaf areas that have been compiled for different forest types and agricultural crops are given in Table 3-8. The wide variability in forest types reflects how this parameter varies significantly in response to the abiotic environment in which the plant is growing. Leaf area index for the individual tree has been shown to provide a close estimation of the above-ground NPP of a stand and to be closely correlated with its

TABLE 3-8   Leaf area indices (m^2/m^2) for different forest ecosystems and agricultural crops[a]

Agricultural crops	3–6
Wet tropical forest	7–11
Broadleaf deciduous forests	5–10
Rocky Mountain coniferous forests	6–12
Pacific Northwest U.S. confierous forests	6–22[b]
Pine forests	5–8
Spruce forests	6–11

[a] Adapted from Jarvis and Leverenz (1983), and Kimmins (1987), with permission.
[b] Originally high leaf area indices but have been changed to ~ 12 by current estimates.

photosynthetic rate (Gower et al. 1995). This relationship between the amount of foliage maintained by a vegetative community and the achievable above-ground NPP has also been shown to vary across a broad range of vegetative communities, from coniferous forests to grasslands and deserts (Webb et al. 1983). Again, these relationships were developed with above-ground NPP only and do not include the below-ground parts; therefore, total NPP is not integrated into this estimation. Because carbon allocation to foliage and fine roots is highly variable and presently not well understood (see later), it may be impossible to estimate total NPP from measurements of leaf areas alone.

Carbon Allocation

Closely associated with photosynthesis is the distribution of the carbon produced as a result of photosynthesis within the plant. For example, carbon allocation between the below- and above-ground components is a particularly important stress-response mechanism (Vogt et al. 1990). Changes in fine-root allocation may produce an early warning signal of stress because fine roots have a lesser priority for carbon allocation than foliage and shoot growth (Jarvis and Leverenz 1983). Therefore, any stress that decreases carbon fixation should be quickly reflected at the fine-root level and measured as a decrease in root turnover rates or the colonization of fine roots by mycorrhizal fungi (Vogt et al. 1993). Another important stress response in plants is reflected in carbon allocation patterns to defensive compounds, which are directly related to trade-offs determined by spatial and temporal variation in resource availability imposed by environmental constraints (Bazzaz and Swipe 1987). The quantity and biochemistry of defense compounds produced by plants may provide an index of ecosystem stress (Vogt et al. 1993).

Patterns of carbon partitioning and storage have also been shown to be sensitive to changes in air quality. For example, some air pollutants (i.e., sulfur dioxide, nitrogen oxides) appear to alter nutrient balances and to reduce carbon transport to roots, possibly accounting for the observed increases in wood accumulation to the upper trunks and less accumulation to the base of spruce trees in declining European forests (Schutt and Cowling 1985). Reproductive partitioning of carbon is another stress response in plants that may be useful in assessing ecosystems. Major buffering mechanisms against stress include overinitiation of reproductive structures, prolonged flowering, and plasticity of yield (Chiariello and Gulmon 1991), all which may reflect more long-term trends in population dynamics on the ecosystem level.

3.6.5   Biodiversity, Species

Overview and Definitions

Biodiversity is currently a topic of concern in the popular and scientific literature (Myers 1988), and has been proposed as a fundamental aspect of conservation policy (see Section 2.1.2, Falk 1990). The concept is valid as a management guideline for two reasons. As a moral issue, most humans feel strongly that it is not a good idea to cause the elimination or loss of other species. A second reason for conscious efforts to preserve species has to do with the long-term maintenance of life on Earth. Over the long run, biological diversity is perceived to produce a higher number of options for maintaining viable ecosystems under conditions of high uncertainty in our understanding of the role of biodiversity in systems and how humans are altering their environments. Maintaining higher diversity in ecosystems allows for more adaptations to be expressed to a changing environment and allows those species or genes to survive whose role in the ecosystem has not been currently demonstrated. Species have been evolving since the beginning of life on Earth and preserving biological diversity maintains the evolutionary "anvil" that allows for continued adaptation by species to their habitats. However, the role of biological diversity in ecosystem function has not been fully explored and is not well understood (see Chapter 2). This section explores the role of biodiversity in ecosystem management.

Biological diversity, like most observable phenomena in ecology, can be observed at a number of hierarchical levels of biological organization (Norse 1993). While most people equate biological diversity (or biodiversity) with the number of different species in an ecosystem, there are other important aspects of biological diversity that need to be considered. Heywood et al. (1995) separated biodiversity into three distinctive groups: (1) ecological diversity, (2) genetic diversity, and (3) organismal diversity. The most accepted and utilized definition of biodiversity is at the *species* level (defined as a "group of similar organisms that interbreed or share a common lineage of descent"), which is just one of the components of the *organismal biodiversity* group of Heywood et al. (1995). This attention to biodiversity at the species level has been quite logical because it is the systematics group that originally identified the loss of species as being at a crisis stage, and they were also the ones who have actively documented this loss of species (Savage 1995).

Several levels of biological diversity can be identified based on nested sets that have been organized for conservation purposes. The following groups are useful to categorize the biological diversity of ecological units (adapted from Soulé 1991, Heywood et al. 1995, Frankel et al. 1995):

1. *Genetic diversity* refers to the total genetic information of an individual species (this is considered the genetic basis of plant adaptations to environmental change). For example, there exists more than 15,000 distinct land races for maize that are maintained by the International Maize Germ Plasm Bank (Silen 1982).

2. *Species diversity* refers to the number of different groups of populations that exchange genetic material via breeding. While species delineations are at times not sharply defined (as when dogs and wolves interbreed), this definition roughly defines the different types of sexually reproducing organisms.

3. *Whole-system diversity* or *ecosystem diversity* refers to the range of different distinctive community types that consist of all the biological, physical,

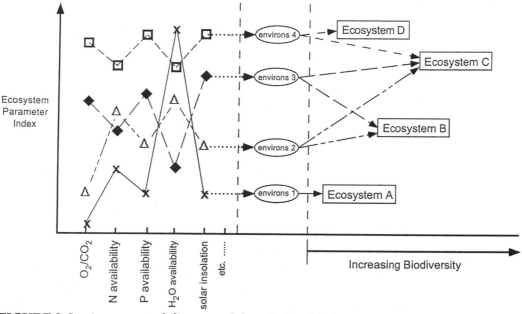

FIGURE 3-1   A conceptual diagram of the relationship between ecosystem parameters, the heterogeneity of environments, and biodiversity in ecosystems.

and chemical characteristics that form a system (see Figure 3-1). Different ecosystems show not only different collections of species and physical structures, but also different rates of nutrient and carbon fixation, cycling, and fluxes. The biodiversity of the ecosystem reflects the number of different environments that exists within its boundary, with the environments being formed by a combination of different abiotic resource availabilities that respond as a unit in feeding back on key processes in the system (Figure 3-1). Ecosystem diversity includes several levels of hierarchical organization, because diversity must exist on the genetic and species levels in order to allow a diversity of biota in different ecosystems. In addition, within a particular type of ecosystem there may be substantial difference in the types of species present and their adaptations to local conditions.

4. Soulé (1991) noted, but did not define, *assemblage diversity.* He noted, "an arbitrary number of biotic assemblages can be defined within ecosystems." Conservation strategies can be built around these assemblages (probably a better strategy than habitat management for a single

species), which is why they are being distinguished as a separate type of biodiversity.

5. *Population diversity* or *meta-population diversity.* Population diversity may refer to the different versions of genes or *alleles*. Within a species, different populations have different versions of genes that evolve to promote survival in the particular microclimate in which the species evolves. This diversity within species has been called the "raw material for evolution" (Norse 1993), and populations with greater genetic diversity are more likely to have some individuals that can survive under stress or perturbation. This has also been suggested to be a better approach to examine diversity than using the single-species approach.

6. Addressing the functional aspects of diversity, Steele (1991) used the term *functional diversity* to refer to the variety of different responses to environmental change, especially the diverse space and time scales with which organisms react to each other and to the environment. Functional diversity, similarly, can be assessed at a number of biological scales.

While definitions of biodiversity are fairly complex and are defined in different ways by different individuals depending on their biases, conservation strategies are even more so. Once a level of diversity is selected for a conservation focus, one must define the goal of the conservation effort. For example, is the aim to prevent extinction of endangered species, to preserve the most taxonomically distinct set of species, to preserve the greatest number of species, or to preserve those species characterized to be highly appealing to humans, while exterminating the myriad viruses and bacteria considered to be threats to human health or crops, or is it to balance the costs and benefits of human activities that lead to losses of biological diversity? The management goals will have broad implications for policy, as targeting endangered species represents a different set of activities than does setting aside preserves in biologically diverse regions or making trade-offs between economically important activity and biodiversity reductions, especially given the uncertainty regarding the value of diversity loss.

In much of the management dialogue, the moral importance of preservation of biological diversity has been emphasized, while the intentional exterminations humans have caused have been overlooked. None would refute the validity of fighting hard for extermination of HIV or of malaria-carrying mosquitoes, but if this is acknowledged, the moral arguments for conserving nonhuman life on Earth can be placed in perspective. Similarly, the role of biodiversity in ecosystem function has not received much attention in the biodiversity conservation debates. However, this incorporation of ecosystem attributes may be one of the key determinants of the success of conservation strategies because the maintenance of ecosystem functions may determine how effectively a species may survive in a particular habitat. As much attention should be given to the preservation of ecological functions that support life (such as soil fertility) as to the moral and political issues that surround the allocation of conservation efforts to particular species (i.e., to endangered pandas rather than liverworts).

Examples

Conservation efforts have emphasized four areas: (1) those directly harvested (i.e., trees) or grazed in the wild; (2) those populations that have been used as a source of propagating material for planting elsewhere, and also those that have been a source of genetic variation in many agricultural breeding programs; (3) those species considered crucial or as an important link for the integrity of an ecosystem (i.e., dominant or keystone species); and (4) those species considered endangered or threatened (Frankel et al. 1995). Out of the four areas identified, the third has received the least attention due to the difficulty of identifying species that have critical roles in ecosystems.

Again, these four areas emphasize the conservation of species and do not really address the conservation of habitats or ecosystems. This is especially interesting because the Endangered Species Act does not only specify saving of endangered or threatened species but also their habitats. According to Houck (1995), habitat loss was identified as the primary cause (80% of the cases) of endangerment of species when species were listed in the *Federal Register*. Many of the single-species (e.g., the northern spotted owl, salmon, Puerto Rican parrot, etc.) strategies were based on conjectures as to what were the habitat requirements of the key species and did not consider the components and processes needed to sustain the functioning of the ecosystem as a whole.

In some cases, however, indicator organisms have provided useful information about ecosystem state. For example, biological indicators such as lichens have been used to monitor the effects of air pollution on forest ecosystems (Wittig 1993). Lichens are particularly sensitive to air pollution, and studies have demonstrated a positive correlation between lichen abundance and air quality based on various measurable parameters (total lichen community, frequency of a single species, growth rate, and chlorophyll content; Wittig 1993). Thus, gradients of ecosystem condition can be developed based on the relationship between air quality and lichen characteristics. This technique may be strengthened by transplanting specific lichen species known to be particularly sensitive into regions of primary concern, and monitoring their response over time, providing a more statistically powerful indicator of ecosystem state. However, although lichen response to pollution levels appears to be strongly correlated, this relationship does not provide information on the mechanisms behind the effects of pollution on forest condition. This model can be further strengthened by including multiple indicators that represent different potential responses of the forest ecosystem to air pollution, particularly ones that are mechanistically linked. For example, photochemical-oxidant damage of plant tissues have been associated with declines in productivity (Bormann 1985, 1990) and changes in plant respiration and carbon allocation patterns (Smith 1991), which may be monitored in conjunction with lichens in order to develop stronger multiparameter indices.

Another example in which a single indicator species has been particularly effective is the monitoring of stream state or quality based on fish characteristics. In this case the response mechanisms by the fish had been identified. For example, Bortone and Davis (1994) investigated the impact of effluent discharged from a paper mill on the stream environment using fish intersexuality as an indicator of environmental stress. Increased segmentation and elongation of the anal rays (a structural characteristic) served as a simple measure of the degree of masculination of female fish that was directly correlated to chemical inputs from the paper mill into the stream (the chemicals functioned as "endocrine disrupters," which affected the ontogenetic developmental sequences in the female fish). Thus, not only was a correlation observed between anal ray size and concentration of pollutants, but information regarding the mechanisms behind these observations has been determined, allowing for verification of the power of the indicator to predict environmental change.

Only recently has a shift occurred to considering the ecosystem as an integrated whole that must be maintained in order to sustain the various interconnected parts. As a consequence of this shift, an important focus of indicators in health assessment has been the exploration and development of multiple-indicator approaches. A good example is again provided by monitoring water quality primarily using characteristics of the fish community. In this example, an Index of Biotic Integrity was formed that was based on a combination of 12 characteristics of the fish community that related to three ecosystem qualities (species richness and composition, trophic composition, relative abundances) and that reflected five major classes of variables influenced by human activities (water quality, habitat structure, stream flow regime, energy cycling, biotic interactions; Karr and Dudley 1981, Miller et al. 1988). Since its development, this index has been modified and applied to different water bodies throughout North America, suggesting that a basic framework for identifying key parameters may be transferred among different ecosystems.

Another potentially useful approach to identifying small-scale parameters for assessing ecosystem health is the use of functional groups. This approach specifically addresses a major limitation of past approaches that neglected to consider the functional role those species have within the ecosystem. *Functional group* can be defined as a set of species that have similar effects on a specific ecosystem-level biochemical process (see Section 2.1.2). The success of the functional group concept to assessing ecosystem state is strongly dependent upon our ability to understand the significance of various functional groups in relation to the behavior of the whole system (Körner 1993). Consequently, it is important to identify those groups that help describe the major part of the flow of matter and energy in the ecosystem (Körner 1993). Relative growth rate has been proposed as a key adaptive strategy for separating plants by function, as it is known to be positively correlated with the potential of a plant to acquire resources and to take up nutrients and to be negatively correlated with plant size and storage (Chapin 1993). For example, the rate at which a transition from plants with low to high relative growth rate occurs following a disturbance may provide a useful index of ecosystem state as related to resource availability and recovery processes. The expected relationship between recovery processes following disturbance and changes in species' relative growth rate may be evaluated based on hypotheses that resource supply and time elapsed since disturbance interact to determine the physiological traits that are supported by the ecosystem (Grime 1977, Chapin 1980, Tilman 1988), thus providing a comparative baseline for observed patterns.

An important limitation to recognize in using biological systems as monitors of ecosystems such as indicator species or functional groups is that their reaction to some environmental factor depends not only on the (1) quantity or quality of that factor of interest, (2) organism-specific parameters (e.g., age, condition), and (3) environmental parameters (e.g., soil type, soil moisture, nutrient status, temperature). These factors thus limit the capacity to standardize the monitoring system using indicator species or group of species. To some extent, problems associated with the inherent variability found in biological systems can be addressed through the use of *internal standards* (see Section 4.2.2), biological indicators having greater uniformity. However, monitoring of the abiotic environment using biological indicators will continue to be a critical component in ecosystem health monitoring programs.

There exist many current uses of indicator species to monitor ecosystem integrity, and a few are listed as examples:

1. *USDI Fish and Wildlife Service*—uses Habitat Evaluation Procedure (HEP) to document the quality and quantity of available habitat for selected wildlife species. *Evaluation species* are selected using socioeconomic (public interest, economic value) and ecological (sensitive species, keystone species, species representative of a guild) criterion.

2. *USDA Forest Service*—mandated by National Forest Act of 1976 to use *management indicator species* (MIS) in forest planning. Five categories of species are considered as MIS:

   (a) Threatened or endangered species

   (b) Species sensitive to intended management practices

   (c) Game and commercial species

   (d) Nongame species of special interest

   (e) Indicators of effects of management practices on broad set of species (Salwasser et al. 1983)

3. *National Pesticide Monitoring Program*—initiated in 1967. Selected four species of birds from different trophic levels as indicator species.

4 | Detecting Resistance and Resilience of Ecosystems

Many useful tools exist that can be utilized at appropriate spatial scales to study ecosystems as described in Chapter 3. These methods, as presently utilized, have been very useful for increasing our understanding of the functions, structures, and feedbacks controlling processes at the individual plant to the ecosystem level. However, they have not been particularly effective for early detection of changes in ecosystems due to either chronic pollution or to small changes in abiotic environmental parameters (such as air temperatures). Most of our tools are very effective at detecting sudden, major changes (i.e., pulse perturbations) in the ecosystem in which a part has been dramatically manipulated by either its elimination (e.g., tree removal and herbicide applications to limit plant regrowth; Bormann and Likens 1979) or due to natural disturbances (e.g., hurricanes or landslides; Lugo and Scatena 1995). Difficulties associated with these questions and tools used to answer them include the following:

1. It is difficult to identify the present state of an ecosystem with respect to other ecosystems whose land-use and natural disturbance histories have differed.

This difficulty stems from uncertainty as to what type of difference should be expected between impacted and non-impacted sites, and a set of measures that will allow discernment of differences between these ecosystems. Historical land uses can leave a legacy on the site that may or may not be difficult to detect (Hamburg and Sanford 1986, García-Montiel and Scatena 1994). None of the criteria currently available allow us to determine where the site exists along a gradient of potential states and whether two sites with presently very different total productivities should exist in similar states or conditions. For example, one ecosystem may exist at a different state because it was influenced by human disturbances (e.g., pollution or previous land use) that overrides normal abiotic controls on plant growth (see further discussion of this in Section 4.2.1).

2. It is difficult to detect how far the current state of system is from the non–human-impacted condition.

3. It is difficult to identify how past land-use activities and other ecosystem legacies may be modifying how the current ecosystem is responding to disturbances or other manipulations.

4. It is difficult to identify the probability of adverse ecological changes, that is, current tools do not provide estimates of the risks associated with a management activity and/or natural disturbance(s) (i.e., make accurate predictions about future ecosystem state).

5. It is difficult to characterize the resilience and/or resistance of ecosystems to various disturbances.

6. It is difficult to determine if sustainable ecosystem management is being achieved in the short or long term.

Because of the limitations in the way that current data are collected or utilized to analyze ecosystems, several different approaches are being proposed that need to be further developed and utilized in more creative ways to help address the six points made earlier. This chapter will begin by briefly reintroducing the measurable signs that indicate that an ecosystem is being degraded or its state is changing. This will require an analysis of the terms that are being utilized to discuss the health or integrity of an ecosystem and a realization that they incorporate many economic, social, political, and ethical concerns (Allen and Hoekstra 1994, Kay and Schneider 1994). Measurements of *ecosystem health* and *ecosystem integrity* are presently problematic because they are not measurable characteristics of ecosystems. While the term *ecosystem health* may be useful for discussing broad scale goals of management in ecological terms, it is not something that is measured; rather, certain variables are measured and compared with either what *was expected* or what *was wanted*. The same statement could also be made for the use of the word *ecosystem integrity*, because *integrity* is not an intrinsic property of ecosystems and it incorporates ecosystem health as part of its definition (discussed in the next section). However, *ecosystem integrity* in its purest form attempts to integrate an objective and realistic assessment of an ecosystem's present and potential states with society's perceptions of ecosystem health. It is worthwhile considering the utilization of other terms (i.e., ecosystem stability, resilience, and resistance) that have been used to describe the state of ecological systems (Westman 1985); they offer the opportunity to measure the state of a system and subsequently decide whether that state is desirable or undesirable. They also offer a more trackable alternative framework for measuring ecosystem state than *health* or *integrity*.

Approaches, parameters, and frameworks are suggested and identified that have great potential for describing the state of a particular ecosystem and its resistance and resilience characteristics. Parameters are specifically discussed that can be used by themselves or in combination with one another. There exist several experimental techniques that are useful to consider for measuring the state of ecosystems and their responses to perturbations:

- energy flow (i.e., net primary production)
- individual species' adaptations to accessing a site's abiotic resources
- adaptations of species that mediate ecosystem response to perturbations
- the utilization of species with particular attributes that allow them to integrate several functions in an ecosystem (e.g., transplants of seedlings in a forest or mussels in an estuarine)
- indicator species when they reflect ecosystem function

Many of the approaches presented in this chapter are adaptations of tools and spatial scale analyses previously presented in Chapter 3. Some of the approaches presented need to be further tested and developed for increasing their utility in the field. It is extremely important to begin testing these approaches because currently methods for assessing ecosystem thresholds and characterizing the susceptibility of ecosystems to perturbations have not been effective. This section is followed by a discussion of the approaches used by ecotoxicologists to implement risk assessment and the appropriateness of borrowing their framework as a tool for estimating the ecological impacts that might result from management activities. Finally, the ecological assessment frameworks proposed or being used in several large-scale projects are examined. The ecosystem manager should be able to glean from this an introduction to the range of tools that can be used to inform management decisions. The scientist may also gain an appreciation of the potential applications of their analytical techniques.

4.1   Signs of Ecosystem Degradation and Indicators of Ecosystem State Change

Our ability to define the "multiple states" (see Figure 1-2) that an ecosystem can exist in through time and when an ecosystem is changing to a "state" that has been decided to be unacceptable are crucial and are a prerequisite for implementing ecosystem management. Because ecosystems move through several states as a normal part of stand development or as part of succession (see Section 2.1.1), there is a need to be able to assess when changes in particular variables are in the normal range or which parts of the ecosystem will move a system between these multiple states. Depending on the type and frequency of a disturbance, an ecosystem may change dramatically if thresholds are crossed and the system is reorganized into a state characterized by different energy/nutrient cycles, species composition, and population dynamics (see Figure 2-1). Because many of the changes that can be perceived in ecosystems are coupled to one another, it becomes important to identify how these changes are linked to one another and the temporal scale at which each variable shows a detectable response to a particular disturbance or combination of disturbances. For example, a disturbance that directly impacts nutrient cycles (i.e., atmospheric input of excess N) will probably first be detected as changes in plant NPP (especially at the leaf level). This will be followed by increased incidence of disease once plants allocate less carbon to secondary defensive chemicals, and finally it might be detected as loss of species diversity or increased incidence of opportunistic species.

Ultimately it is important to know when the system is functioning within a predicted range of system states that have been identified as normal and when the continued supplies of desired ecological amenities or services are in jeopardy (see Figure 1-2). As part of this, there is a need to know (1) what patterns and processes underlie these outputs and (2) whether these patterns and processes will change (or are changing) as a result of some perturbation. The present tools for assessing systems have been unable to index a site, to determine where that system should be in relation to previously identified ecosystem states, or if its rate of change between states is unusual. One factor contributing to this difficulty in defining the multiple states of an ecosystem is the limited time frame in which most ecosystems have been studied (typically 50 years). For example, old-growth Douglas fir forests in the Pacific Northwest are older than 400 years of age (Waring and Franklin 1979), and an understanding is just forming of the different developmental states that these ecosystems experience and the ecological characteristics of these different states (see Section 2.1.1). This inability to characterize ecosystem states has produced some of the controversy over how much land area remains in old-growth Douglas fir forests in the Pacific Northwest United States and has reduced the arguments to either being based on the youngest age that can be considered old growth or the presence of ecosystem characteristics that have been identified to be unique to them [i.e., coarse, woody debris, large trees—trees with diameters at breast height greater than 53 cm (21 inches), and the number of canopy layers—with old growth requiring two or more canopy layers but single canopy layers accepted when trees are greater than 53 cm in diameter] (USFS (Old Growth Definition Task Force) 1986b, (FEMAT 1993). It has been difficult to take a system that within a human time frame has appeared to be quite "static" and unchanging in structure and function, and to merely consider this as one of the many potential developmental stages of a forest. These other states are ones that will not be experienced within a human life time. Even this apparently "static" ecosystem may have responded to a changing environment by "self-organizing" so that the system appears to be functioning normally, even though particular parts of the system may already have been highly modified or eliminated (Anderson 1991).

Our ability to biologically identify when an ecosystem is experiencing shifts between the several multiple states that are within the range natural for that system is further modified by our social perception of what is acceptable for a particular ecosystem. This perception may consider that a fire- damaged forest of mostly dead trees (Agee 1993) or a forest devastated by insects, again with many dead trees (Torgersen 1993), to be unacceptable to society, even when these events may be an integral part of the functioning and maintenance of that ecosystem through time (Romme et al. 1986, Schowalter 1993).

The greater breakup of the landscape into mosaic patches of intact natural forests mixed with residential areas has contributed to society's desire to control fires to protect homes in areas originally dominated by fire-adapted or fire-dependent forest ecosystems (Arno and Ottmar 1993). Society's perception of what is considered appropriate may actually reduce the range of multiple states that a system experiences because either the system is managed to maintain it in a particular "desired" state, or disturbances (e.g., fire) are added or deleted that change successional patterns (Agee 1991). Managing an ecosystem for one desirable "state" can have the additional impact of changing disturbance cycles and accelerating the rate at which that system may shift into socially unacceptable states. Excellent examples of this can be found in the drier eastside forests of the Pacific Northwest United States, which are experiencing more severe insect outbreaks and a greater severity of disastrous fires due to fire control and selective tree cutting (Arno and Brown 1991, Arno and Ottmar 1993). A brief discussion of the role of disturbances in ecosystems is presented in Section 3.5.

The study of ecological change is challenging because in many instances the change of interest has not been specifically identified in the first place (Bartell et al. 1992), nor the endpoints and indicators describing these changes. An ability to assess the state of an ecosystem has been hampered by the use of terminology that describes neither properties of ecosystems nor measurable qualities. Therefore, several different interpretations of an ecosystem state can exist, creating gridlock in managing ecosystems for their "health." This is reflected by the fact that humans intuitively identify when an ecosystem is not healthy by focusing on data that records increases in the incidence of insects or pathogens, increased mortality of trees, or increased loss of plant or animal species. This categorization of insects, fungal pathogens, and/or fire as negative values in ecosystems, and therefore undesirable, ignores their critical roles as disturbance agents in maintaining natural ecosystem functions (Romme et al. 1986, Schowalter et al. 1986, Wargo et al. 1993). Ecosystem managers will have to realize that one parameter may have dual functions such that it may shift from being a positive to a negative feedback depending on how strongly it is expressed. Of crucial importance to managers is being able to identify when the impact of a parameter is changing. Furthermore, not all species are so integrally connected within an ecosystem that their loss from the system will change ecosystem function (see Section 2.1.2). It is important to accept the fact that a strong link between species and ecosystems may not always exist, and that a society's desire for the continued existence of a particular species should in some cases be valid enough for continuing to manage for that species in the ecosystem. Ecosystem managers will have to understand the biological and social definitions of what is acceptable in ecosystems. Because these definitions may not always be compatible, it is also critical that the trade-offs involved in accepting each definition are understood.

There have been several attempts to define ecosystem health based on idea that it is easy to identify when one system is more healthy than another, and to manage to maximize ecosystem health. Interesting analogies of ecosystem health to human health have been published (Rapport et al. 1985). Again, ecosystem health definitions incorporate the ecological attributes of a system combined with societal desires of what should be included in that ecosystem and are very subjective (Levin 1989, O'Laughlin et al. 1994). Societal desires strongly define the perception of a healthy ecosystem and are often directly incorporated into the definition of ecosystem health, while at other times their influence may be more indirect. The following examples are some of the definitions of ecosystem health that have appeared in the literature:

- "Ecosystem health is a state of ecosystem development where the geographic location, radiation inputs, available water and nutrients, and regeneration sources are optimal or at viable levels to maintain that ecosystem" (Woodley 1993). This implies that ecosystem structure and function are not being impaired by anthropogenic stresses.

- "The ecosystem is capable of supporting and maintaining a balanced, integrated, adaptive community of organisms which have a species composition and a functional organization that is comparable to the natural ecosystem within its geographic region" (Karr and Dudley 1981). This again implies that a healthy ecosystem is one that is not influenced by humans and that structure and function are not being impaired by anthropogenic stresses.

- "An ecological system is considered to be healthy and free from 'distress syndrome' if it is stable and sustainable (i.e., maintaining its organization and autonomy over time and is resilient to stress)" (Costanza 1992). The use of stability and sustainability is itself vague, so this definition is more confusing than enlightening.

- "Forest health is a condition in which ecosystems sustain their complexity while providing for human needs" (O'Laughlin et al. 1994). This definition assumes that complexity is the key to ecosystem health.

- "Ecosystem health is the degree of overlap between what is ecologically possible and what is desired by the current generation" (Bormann et al. 1993).

- "An ecosystem is healthy if it has the capacity to satisfy our values and to produce desirable commodities in a sustained fashion" (NRC 1994).

The first two definitions indirectly incorporate humans into the equation for defining ecosystem health by stating that a healthy system is one in which anthropogenic stresses are not present and therefore not modified by humans. Because there are no systems that have not been modified by humans in the past or present, these definition are difficult to implement as a working definition of ecosystem health. Even the Amazon forests are typically described as being "virgin" forests unimpacted by humans. Yet today it has been shown that these forests have been manipulated and managed by indigenous people for hundreds of years and the current structure and function is a result of their management activities (Pinedo-Vasquez 1995). In contrast to the first two definitions, the last three explicitly incorporate societal values in combination with either the products that are harvested from forests or the maintenance of ecological structure and function. These definitions acknowledge that human values and perceptions directly influence society's concept of a healthy ecosystem. These subjective factors must be combined with available ecological knowledge to derive a working definition of health. Ecosystem health cannot be effectively measured today in ecosystems because the current suite of tools are unable to index a site to determine how far a particular system is from a baseline state. This suggests that trying to develop ecosystem frameworks for assessing ecosystem health will not be productive. This could, of course, change if the approach presented in Section 4.2.1 is successful in assessing the current state of a system and indexing where that system is in relation to where it should be theoretically.

Another term that is frequently used interchangeably with ecosystem health is ecosystem *integrity*. However, for ecosystem management purposes, it is useful to define these words distinctly and not use them interchangeably. They really do have different implications. For ecosystem management, the term *ecosystem integrity* as defined by Kay and Schneider (1994) is useful to consider. Kay and Schneider (1994) defined the following three conditions as being integral to ecosystem integrity:

1. Knowing the current state of the system compared with normal environmental conditions

2. The ability to deal with stress in a changing environment (resistance to stress and the recovery rate after a stress)

3. The ability to continue to self-organize (change from within) in a changing environment.

Based on these conditions, ecosystem health would only encompass the first condition of ecosystem integrity, while ecosystem integrity requires an understanding of how ecosystems change over time under conditions of human and nonhuman intervention. Thus, health would be a very specific aspect of ecosystem integrity that defines the state of the system currently and would be estimated utilizing short-term, well-defined criteria. Because ecosystem health is the first of the three conditions identified by Kay and Schneider (1994), even ecosystem integrity would be impossible to determine if all three conditions are essential for defining this term. Until unbiased methods are developed for indexing ecosystems, it will not be possible to determine ecosystem health or ecosystem integrity. It is for this reason that there is a need to begin thinking of creative ways of indexing ecosystems (see Section 4.2). Because ecosystem health and integrity are not properties of ecosystems, it is particularly important not to pursue the development of these indices and to develop alternative assessment approaches. It is better to define the components of ecosystems that are worthy of protection (i.e., endpoints) and to measure attributes of ecosystems (i.e., resistance and resilience) that reflect ecosystem properties. Only these latter ecosystem attributes will hold great utility in assessing whether or not a system state will change, whether species loss will follow a change in ecosystem state, or how readily the system will recover when perturbed.

As noted in the previous sections, managers will have to take measurements at many different scales and use several different indicators of ecosystem change in order to determine how to maintain ecosystem functions within the range of states within which an ecosystem will naturally fluctuate and/or that are acceptable and desired by society (Figure 4-1). In addition to the approaches synthesized in Chapter 3 that have been used to provide proxy measures or indications of ecosystem function, a review follows in this chapter of some of the indicators that an ecosystem is functioning at a low level, or at least at a lower level than in the past. As mentioned earlier, it is important to recognize that many of these indicators are characteristic of ecosystems subject to natural disturbance cycles, as well as human influences, and that these indicators do not necessarily provide cause–effect information on why the system is in a particular state. Indicators of ecosystem degradation or change in the past have consisted of species level changes and/or changes in some ecosystem structural or functional aspects. The following conditions have been identified as being particularly sensitive indicators of ecosystem degradation (Odum 1969, Bormann and Likens 1979, Waring and Schlesinger 1985, Kimmins 1987, Waring 1987, Van Breemen and van Dijk 1988, Arnolds 1991, Vogt et al. 1993, Johnson 1994, Wilson and Tkacz 1993, Bloomfield et al. 1996):

1. An increase in disease, parasitism, and insect pest outbreaks due to decreases in within-plant carbon allocation to defensive secondary chemicals

2. A decrease in the presence of symbiotic microbes on plant root systems,

and an increased dominance of symbionts that are less beneficial to the plant

3. A decrease in species diversity (at the extreme end, a shift to a single-species dominance of a site, e.g., by phragmites or purple loosestrife) or a shift in species composition, along with a shift in species composition to more stress-tolerant species and r-strategists

4. A decrease in net primary production and net ecosystem production

5. An increasingly higher annual transfer of production to the decomposer system

6. An increase in plant or community respiration

7. An increase in the loss of those nutrients typically limiting the growth of plants in an ecosystem as the system is unable to utilize, conserve, or retain them

8. Bottleneck of nutrients in long-term pools (i.e., hurricanes in Puerto Rico resulted in above-ground litterfall not re-establishing to predisturbance levels, even after 5 years; Vogt et al. 1996b).

Several of the approaches that have been used to understand ecosystem function have focused on the ecological role of single species or small groups of species (Morrison 1986, Carey 1993). While a species-by-species approach alone will not provide the type nor scale of integrated information needed for ecosystem management, single-species approaches may provide valuable information. First, due to the relatively great emphasis on single-species approaches in science and management, much data exist on the individual species that together give rise to many aspects of structure. When integrated into a broader framework, these data may make valuable contributions to understanding the structure and function of an ecosystem. Second, new studies of selected species or groups of species will also be important under an ecosystem approach. For example, knowledge of the functional role of keystone species in a community, or those species that sensitively reflect change in a system, will contribute to a manager's understanding of the ecosystem. Third, an understanding of what factors cause certain species to become threatened or endangered in a region is of critical importance for understanding the effects of management practices and for designing effective management regulations. Fourth, the response of an ecosystem to perturbations depends on the responses of individuals and populations. Tolerance to stress (perturbations) and adaptive responses occurs at the level of individual organisms, so mediation of environmental change is played out at this level.

However, some species level and ecosystem functional aspects are interlinked so that a change in an ecosystem function can result in population decreases at the organismal level or vice versa. The Graveland et al. (1994) study is a good example of the interdependency and the close links that may exist between species and ecosystem processes. In their study, atmospheric deposition of pollution resulted in decreased Ca levels in the soil, which led to lower snail populations because snails are dependent on Ca to produce their shells. Ultimately, the reductions in the snail populations resulted in significant decreases in bird populations because birds are dependent on the snails for food and as a source of Ca to produce strong eggs. In this study, documenting population changes at the species level did eventually lead to identifying ecosystem changes that explained the mechanisms behind the decline at the population level. Unfortunately, the changes in population sizes occurred many years after the ecosystem had already experienced changes so that the population level parameter did not reflect early changes in ecosystem function. There may be strong links between species and ecosystems, but it is important to identify the time scale at which these links are expressed; if the time lag is too long, the species may not be useful as an indicator of whether the ecosystem is changing or not (even though they do provide a mechanistic understanding of the links).

4.2   Useful Approaches for Detecting Ecosystem Change

A number of approaches or tools exist to detect changes in ecosystem states. Some of these approaches are effective at specifically addressing temporal scale problems, while others are specific to spatial scale analyses. Only a few adequately address both scales. In the following sections, several approaches are articulated that are useful for ecosystem ecologists and ecosystem managers to consider as part of their repertoire of tools. Depending on the question or management problem, one or several of these approaches should be used.

4.2.1   Comparisons of Theoretical to Actual NPP, Decomposition Rates

One of the difficult problems in predicting how an ecosystem will be impacted by a human- induced or natural disturbance is not knowing what historical factors are determining the functioning of the current system. For example, two ecosystems may be identified that appear to be quite different from one another because estimates of NPP or decomposition rates vary significantly between them. However, the two systems may only appear to be different due to previous disturbance histories. In actuality, they may be the same ecosystem type and should be existing in similar equilibrium states and responding to disturbances in the same manner. Currently, no tools are capable of detecting how much past land use or other disturbances have modified the characteristics or responses of these sites to disturbances (i.e., their resistance and resilience characteristics).

A new approach is being proposed that combines the use of models with field measurements that would allow better detection of the historical legacies contributing to an ecosystem's current state. This technique uses models (not those parameterized or developed specifically using site-specific data) to predict the theoretical optimal ecosystem state or function and then to compare this to the current or present state or function. This approach could be used to measure very specific processes in an ecosystem, such as the amount of energy captured by plants (NPP), but should be tested for its applicability for examining how far the health of animals or other components of the ecosystem are from theoretically determined optimals (e.g., decomposition rates, plant carbon allocation to secondary chemical contents of foliage or root tissues, etc.). For example, this approach could be used to estimate what the theoretically optimal decay rates would be for a system using many of the existing decay models (Meentemeyer 1978, Carpenter 1982, Berg 1984, Berg et al. 1984, Mc-Claugherty et al. 1985, Berendse et al. 1987). These decay models use varying combinations of climatic variables and substrate quality information (i.e., lignin and N contents of tissues) to predict decay rates. Because decomposition rates do vary significantly based on the site nutrient status and there has been a direct link shown between some air pollutants (i.e., ozone) and changes in leaf litter decomposition rates (Findlay and Jones

1990), alterations in decomposition rates may be quite useful to measure as long as one can document what the typical decay rates should be for the site. Again, one would not want to use site-specific information on tissue chemistry as an input variable into the model but use generic tissue chemistries by species synthesized from many sites. Even decay rates of organic matter in the soil have been determined using models, so this approach could be tested for its utility to measure changes in this pool. For example, one model predicts decay rates in the soil based on climatic variables, tissue chemistry, and soil physical and chemical characteristics (Parton et al. 1987, 1988).

The following section will discuss how NPP could be used to measure resistance and resilience characteristics of ecosystems. As already presented in Section 3.6.1, NPP is a very useful measurement to obtain for an ecosystem and relatively reliable tools exist to estimate actual NPP in the field. However, it is important to recognize that even though above-ground NPP and coarse-root NPP estimates are acceptably good, controversy exists in the literature on the best method and how much confidence one should have for the published fine-root NPP values. Because in some ecosystems the fine-root NPP can be a significant proportion of the total (i.e., those with low nutrient availabilities, high soil aluminum levels, etc.), it cannot be ignored as part of ecosystem analyses (Vogt 1991). Despite the conflicts connected with measurements of fine-root biomass and NPP, the previously published values are surprisingly useful in giving reasonable correlations between abiotic variables and fine-root dynamics across different ecosystem types (see Section 4.2.6). Net primary production measurements, by definition, have to be collected over at least a 1-year period. These short-term measurements of productivity can be expanded to longer time scales using tree cores combined with allometric regression of tree growth. However, just comparing NPP values over longer time scales will not allow resilience and resistance characteristics of ecosystems to be determined because those data do not tell how far that system has diverged from what would be expected given abiotic (i.e., climate) and biotic (i.e., species composition) factors at that site.

It is proposed that direct measurements of NPP on a site can be combined with model estimates of the theoretical optimal NPP to investigate (1) the resilience and resistance characteristics of ecosystems and (2) to identify how far an ecosystem has been displaced from a range of states representative for that system. The theoretical optimal NPP (i.e., predicted NPP) is derived for a site using factors external to the site (e.g., temperature, precipitation, and solar radiation). The optimum NPP can then serve as a baseline for determining where the current ecosystem is functioning (i.e., actual NPP) with respect to its maximum performance capacity (see Figure 4-1). Knowledge of how far a particular system state diverges from another system identified as being representative of that state and the magnitude with which both system states vary from the theoretically calculated value can be used to assess their condition. The concept of indexing the current ecosystem to a theoretically derived value specific for a site can be used to assess changes in ecosystem state, a condition earlier described as being difficult to measure using only NPP estimates. These theoretical values provide an objective measure to which the actual system values can be compared. They are relatively easy to calculate and function as a comparative baseline between different systems. The most time-consuming part of this approach is the determination of the actual NPP values, which may require intensive field data collection for all new sites unless indirect measurements of productivity can be obtained using remote-sensing technology and physiological/ecosystem models (see Section 3.6.1).

When calculating the theoretical optimal NPP values, it is important to use a model that has a minimum requirement for site-specific data, which means that the models discussed in Section 4.2.8 are not appropriate to use. These other models require detailed information on many variables and are therefore better suited for estimating the current NPP of a specific site. These models are not as effective in predicting the carbon budget of forests over larger spatial scales because their intensive data requirements necessitate detailed site-specific data on climate, soil physical and chemical characteristics, and stand structural and morphological characteristics (Landsberg et al. 1996). It is suggested that the best model for estimating the theoretical optimal NPP is the *e model* (discussed later). Ultimately this model should be coupled with some of the more process level models.

The philosophy behind utilizing both the *predicted* and the *actual* NPP values is that each site has an actual productive capacity under typical conditions that never reaches an optimal level. Every site has some factor(s) reducing its photosynthetic capacity: Climatic factors (e.g., cloudy weather, fog, temperature, etc.) will reduce the time period in which plants

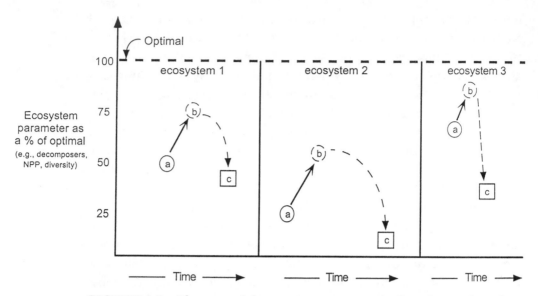

FIGURE 4-1   The potential ecosystem responses to disturbances based on how far they are from the optimal predicted state. The original ecosystem is represented by a, the short-term response of the ecosystem to a disturbance is represented by position b, and the state to which the system will recover is represented by position c.

are able to photosynthesize, low or pulsed availability of nutrients will limit plant production, and other ecosystem components (i.e., decomposer microorganisms, soil chemical matrix) may effectively compete for nutrients required by plants to maintain their growth (Larcher 1975, Waring and Franklin 1979, Gholz 1982, Boerner 1984, Bloom et al. 1985, Chapin 1991a,b, Lambers and Poorter 1992, Gower et al. 1996). Modeling efforts by Emmingham and Waring (1977) showed quite clearly how seasonal changes in potential net photosynthesis decreased in response to just one variable (i.e., summer droughts), reducing photosynthesis by 25% to 65% depending on where a system existed along a moisture gradient. How far one or a multitude of factors reduces plant photosynthetic capacities would be determined by comparing actual NPP to the theoretically derived values.

Optimal NPP values can be estimated using a simple model with a few robust parameters. A model that predicts the theoretical optimal NPP based on the amount of photosynthetically active radiation absorbed by forest canopies, called *radiation utilization efficiency*, is worthwhile pursuing (Running and Hunt 1993, Landsberg et al. 1995). This model is called the *e model* and uses an input variable that is external to the site (e.g., photosynthetically active radiation). Photosynthetically active radiation data can be obtained using several different approaches: (1) direct measurements at the ground level, (2) remote sensing of photosynthetic surfaces complemented with specific measurements of chlorophyll content, and (3) other models that indirectly predict PAR (Running and Hunt 1993, Landsberg et al. 1996).

Once the theoretical optimum NPP has been estimated using the e model, the current net primary production should be determined as discussed in Section 3.6.1. If two sites result in the same theoretical optimal NPP values but presently vary significantly from one another, one can begin to explore what current site conditions or historical land uses or natural disturbance regimes are influencing the achievable productivity of the site. To further elucidate the reasons for these differences in current production, one would have to conduct on site comparisons to identify whether (1) differences exist in the natural physical and chemical characteristics of the soil; (2) the landscape matrix in which each site exists, for example, is the site embedded within a landscape where it comprises the forested context of a block of swiss cheese or is it a "hole" surrounded by a context of grasslands or agricultural fields (with this context significantly changing the surrounding microclimatic conditions or radiation inputs; Saunders et al. 1991, Publicover and Vogt 1991); (3) is there a previous land use (e.g., agriculture, chemical dump, fire, etc.) that has left a legacy that is affecting the current site; or (4) has chronic pollution retrogressively changed plant growth on the site? The utility of this exercise is that one would be able to index and determine whether two sites should or should not be replicates of one another. If the sites are identified or predicted to have similar capacities to photosynthesize but have very different current productivities, then it is possible to explore current or historical uses of the site that may have modulated ecosystem functions.

Ecosystem resistance and resilience characteristics (discussed in Section 2.1.3) to a variety of stresses should be predictable from knowing both the theoretical and actual optimal productivities (see Figure 4-1). It is hypothesized that knowing how far the current productivity deviates from the theoretical value will identify how resistant the system will be to a disturbance and how quickly it will recover to another equilibrium state (resilience). For example, ecosystem 1 and 2 in Figure 4-1 were predicted to have similar theoretical optimal productivities, but currently ecosystem 1 has a NPP value that is 50% lower than the theoretical, while ecosystem 2 is 75% lower. It is hypothesized that ecosystem 2 will be less resistant to a disturbance because it has lost functional groups or redundancies in species or ecosystem function that would decrease its ability to resist the effects of a stress. How far the current system deviates from the theoretical value should also determine how quickly that system will recover its functions to another equilibrium state after a stress. It is hypothesized that ecosystem 2 should take a longer time to recover after a disturbance compared with ecosystem 1.

In another scenario, it is hypothesized that it is also possible that an ecosystem may achieve a current NPP that closely approaches the theoretically predicted optimum, although the system would not be considered to be functioning well in terms of providing desired goods and services (represented as ecosystem 3 in Figure 4-1). This potential to approximate the theoretical optimal NPP should not be common but may occur under several conditions:

1. Pollution inputs of inorganic N functioning as a fertilizer effect in forests (Johnson and Lindberg 1992)

2. Exotic tree species introduced into areas where they are more effective at acquiring and utilizing site water and/or nutrient resources

3. Introduction of non-native N-fixing trees or shrubs in areas with naturally low soil N levels; the increasing accumulation of Ni replaces the slow growing native species with fast growing non-natives (Vitousek and Walker 1989)

4. High N inputs into estuarine ecosystems from several different sources (i.e., waste treatment plants, fertilizers, septic systems, etc.) co-occurring with the introduction of exotic algae, resulting in large-scale algal blooms along coastal areas

In most of these examples where actual NPPs approximate the theoretically calculated values, either the availability of one limiting nutrient (e.g., N) has been increased so the site is capable of supporting a higher productive system in the short term or species have been introduced into areas where they are capable of overutilizing abiotic resources. In these cases, it is hypothesized that if productivity levels are increased to levels higher than what the site is naturally capable of supporting, the system will have little resistance to new disturbances, and its ability to recover to an equilibrium state close to the predisturbance condition is probably quite low. In contrast to ecosystems 1 and 2, ecosystem 3 should recover to a lower equilibrium state and not be as resistant to further stresses compared with its predisturbed state. The rationale for the type of response expected by ecosystem 3 can be attributed to the fact that many disturbances impact only a part of the ecosystem and do not affect the other multiple of interconnected processes that are linked to the one modified component—the impact is not occurring in a balanced manner. For example, even though the addition of high levels of atmospheric N will initially increase NPP because of the significant role of N in the photosynthetic process (see Section 3.6.3), other limitations to plant growth have not been alleviated and may even become exacerbated. The ultimate long-term effect of this will be that these limitations will feed back to reduce NPP. In this case deficiencies of other nutrients result in nutrient imbalances that cause reductions in tree growth (Ingestad and Ågren 1992). In addition, plants may be unable to produce sufficient secondary defensive chemicals due to changes in carbon allocation so that their capacity to resist pests and pathogens will be lower (Waring et al. 1985).

This latter example of increasing NPP sometimes having a destabilizing effect on an ecosystem raises an interesting point because normally it is assumed that increasing NPP would be a desirable condition for ecosystems and for management. Because timber is the product being managed for by timber companies, it is logical that management activities are geared towards increasing the rate at which biomass accumulates in a stand. Similarly, ecologists often connote positive qualities to ecosystems with higher NPP values, assuming that they are more resistant and resilient, or more buffered against disturbances. These ideas readily transfer to producing a perception that increases in NPP are positive and ecosystems should be managed to optimize productivity. It is suggested that this perception is a human value that is being imposed upon management but that does not really consider whether or not increasing NPP is a desirable attribute to strive for in ecosystems. Managing for high productivities has to be conducted in such a way that imbalances are not created in the ecosystem carbon, nutrient, and/or water cycles. When imbalances occur, the higher

NPP is being achieved at a cost because it may result in an ecosystem condition that is more easily moved beyond the thresholds of the multiple states representative of that system (see Figure 1-2) and that may have lower resiliency.

Most ecosystems have plants that are adapted to the abiotic conditions and resource availabilities inherent to that site, which in turn control the achievable productivity for that ecosystem. In some cases the introduction of non-native plant species can generate imbalances in ecosystem cycles when that species is more effective than indigenous species at acquiring limiting resources. An interesting example of this has occurred with *Eucalyptus*, which has been used as a plantation species around the world. This species has some of the deepest rooting of any plant species, with maximum rooting depths reaching 60 m and average depths consistently around 35 m (Stone and Kalisz 1991). Most other plant species are unable to acquire these deeper sources of water because they have root systems that only reach 1 to 2 m depths; even the deep-rooted tropical forests only reach 12 m deep (Stone and Kalisz 1991). Having the ability to form deep roots was originally considered a positive attribute of *Eucalyptus* because many forest plantations are located in areas that experience short- to long-term droughts. This allowed *Eucalyptus* to access deeper water sources and to grow during the drought periods, ultimately achieving higher productivities than typical for that site. However, recently this ability of *Eucalyptus* to efficiently access deeper sources of water has been recognized to be a problem because it is causing water table levels to decrease too rapidly (Nambiar Sadanandar, personal communication). This problem with *Eucalyptus* is a result of it being planted at unnaturally high densities in plantations. These negative results would probably not occur if individual deep-rooted plants were scattered throughout the ecosystem. Under natural conditions, individual plants with deep rooting have been reported to have important roles in increasing the available water to other shallow-rooted plants that grow in their vicinity (Richards and Caldwell 1987, Caldwell and Richards 1989).

4.2.2   Standard Materials or Organisms Transferred Across Sites

There are many measurements that can be used to detect how an ecosystem is responding to environmental change (see Chapter 3). However, one of the greatest difficulties with the use of ecological indicators or physiological parameters to measure ecosystem performance is the high degree of variability among different biological organisms. Even the abiotic heterogeneity within any given ecosystem further confounds the ability to interpret the responses recorded for different ecosystem parameters. Traditional ecosystem studies were designed to minimize this variability by selecting the sampling plot and experimental designs that held as many abiotic and biotic factors as constant as possible, with the exception of the one being measured (i.e., via replication). However, it is almost impossible to locate sites or to select variables that have such uniformity and constancy that the driving variables or the multitude of responses can be isolated from one another.

Another approach to assessing ecosystems is to specifically incorporate the environmental variability into the design by using specific ecosystem components or organisms with known characteristics and monitoring their response to this heterogeneity or to a disturbance. These materials or organisms, called *standards*, can be used to document how the system is responding to a change in the environment. Standards can consist of genetically similar plants (i.e., clones), litter material collected from one location and transferred to other sites to monitor their decay rates, or selected organisms (i.e., fish, mussels, corn plants to detect VA mycorrhizal inoculum potentials) that effectively integrate the response of specific parts of the ecosystem to certain disturbances or changes in ecosystem structure and/or function. The use of standards or indicator organisms for detecting changes in ecosystems has been successfully used and developed for aquatic ecosystems (Osenberg et al. 1994). Some aquatic ecosystems (i.e., estuarines, rivers) are so dynamic that it is extremely difficult to detect how that system is responding to a disturbance. In such cases, the use of a sessile organism (i.e., mussel) or fish in cages can be used as filters that integrate all the inputs into the aquatic ecosystem and retain that information as a record within their body or tissue parts.

A logical place to begin to identify appropriate standards is to focus on living organisms, because plants and animals are constantly integrating the variability they are exposed to at different spatial and temporal scales. However, perennial woody plants have a high degree of genetic variability at the species level. Therefore, the selection of trees at the species level for use as standards to be transplanted to other sites may not increase the ability of detecting change in ecosystems if the species level response is too variable. One means to address this problem is to control for genetic variability by utilizing individuals of the same genetic stock (e.g., clonal material, or sibs from the same parent tree). The genetically uniform individuals would serve as "internal standards" by providing standardized measures of the ecosystem based on selected indicator parameters. Because genetic variation has been reduced or eliminated by using these standard materials, any variation in the parameter being monitored can be attributed to differences in the environmental conditions and their effect on the performance of the organism. Internal standards should have the following characteristics if they are to be effective: (1) high signal/noise ratio (large response of interest relative to background variability); (2) rapid response to the disturbance of interest; (3) reliability/specificity of response, that is, they respond to disturbance consistently; (4) accurate reflection of the level of disturbance; (5) lower variability, that is, the same genetic stock or chemical composition; (6) survival in a range of ecosystems (degraded to pristine); (7) ease/economy of monitoring; and (8) relevance to management goals.

The utility of the internal standard technique has been demonstrated by several studies. Antonovics et al. (1987) transplanted clones of two different grass species (*Anthoxanthum odoratum* and *Danthonia spicata*) to measure the heterogeneity and the scale at which it occurred within a habitat. Using this method, Antonovics et al. (1987) were able to formulate hypotheses about how differences in the species' sensitivity to small-scale heterogeneity related to differences in their breeding systems—the more sensitive species was an obligate outbreeder and the least sensitive species was self-compatible. The application of this technique to assessing ecosystem state is demonstrated by its capacity to explicitly measure the scale of heterogeneity within the environment as perceived by two species growing in the same habitat. Thus, it may provide a powerful tool for detecting even relatively small changes in ecosystem state caused by specific factors.

A similar study based on the use of internal standards was conducted by Bell and Lechowicz (1994) using "explants," defined broadly as organisms grown in samples of the environment (in this case, soil cores) under controlled conditions (as opposed to "implants," in which the organisms are grown in the natural environment). The objective of the study was to estimate environmental variation as measured by its effect on plant performance (i.e., reproductive success) using dry-weight biomass, which is highly correlated with seed production. The study was conducted by extracting soil cores from the environment according to different spacing designs, and sowing genetically uniform material in the soil cores in a glasshouse. Despite the fact that the explant technique eliminates many sources of in situ variation, and thus provides a conservative estimate of environmental variation, the study showed a significant amount of environmental variation for characters related to fitness on scales appropriate to plant growth and seed dispersal. The major implication of this study for ecological monitoring is the high degree of sensitivity (i.e., high signal/noise ratio) that is obtainable using internal standards and the potential for obtaining more accurate estimates of ecosystem state.

Several limitations of the internal standard technique should be recognized. First, because the internal standard will also change the environment in which it is growing, there will be some degree of uncertainty or confounding influence in the measurements (Antonovics et al. 1987). Furthermore, it is only possible to transplant and measure a limited number of genotypes, which will necessarily fall short of providing an accurate representation of the population as a whole. The selection of particular genotypes known to be more common to a particular ecosystem, or more relevant to specified management objectives, may partially address this limitation.

Others have also used techniques based on the internal standard approach but did not control as extensively for genetic variability. These approaches, however, also appear to provide relatively sensitive measures of environmental conditions as compared with traditional nonmanipulative approaches. For example, Osenberg et al. (1994) conducted research to determine the impact of near-shore discharge from oil refineries on the marine environment of the Santa Barbara coast in California by identifying indicators of environmental impact (see Section 4.2.9 for greater discussion). Three kinds of parameters were used to measure ecosystem response to the discharge: (1) population level parameters (i.e., population density), (2) individual-based parameters (i.e., growth rate of individual organisms), and (3) chemical-physical parameters (i.e., element concentrations). The findings of the study showed that the transplants of biological organisms (particularly mussels) provided the greatest standardized effect size (i.e., large change in growth rate, low sample variability in background rates of growth) and required the least amount of sampling effort to detect changes in the environment. An important contribution of this study to ecosystem assessment was the approach used to assess the statistical power of the parameters measured (i.e., "standard effect size"), specifically quantifying the standard effect size of different indicators and determining the minimum sampling time plus frequency required to detect a significant change using each of the parameters.

Despite its demonstrated effectiveness, the technique of internal standards has been only marginally applied to the study of environmental impacts on forest conditions. Studies are needed that both correlate responses of specific internal standards with different ecosystem conditions, as well as on the mechanisms to explain those correlations, combined with gradient analysis using transplants and/or experimental manipulations. By transplanting internal standards shown to be sensitive to the stimulus of interest into ecosystems existing at different points along the gradient of potential states, it should be possible to develop indices of plant response that reflect changes in ecosystem-level performance. Once likely parameters are identified, experimental studies can focus on those parameters for determining underlying mechanisms.

4.2.3   Indicator Species and Functional Groups

Because indicator species and functional groups can have great utility in analyzing how an ecosystem is operating (see discussions in Chapter 2), research should emphasize determining which indicator species or combinations of functional groups can be effective surrogates for ecosystem states. When indicator species are detected, they are extremely useful because they are a substitute measure of some ecosystem attribute (De Pietri 1992). Landres et al. (1988) defined an indicator species as "...an organism whose characteristics (e.g., presence or absence, population density, dispersion, reproductive success) are used as an index of attributes too difficult to measure for other species or environmental conditions of interest."

Uses of indicators species can be divided into the following three broad categories:

1. *Indicators that reflect the presence and effects of environmental contaminants.* If an indicator species responds to a particular disturbance in a predictable manner and rate so that clear cause–effect relationships are identified, their response can be used to detect the degree of change in the environment. In effect, indicators are sensitive and reflect the effects of a perturbation. If good cause and effect relationships have been identified for a particular indicator, they can be useful in detecting contaminants in the environment. However, it may be difficult to identify one good indicator for a site because (1) ecosystem processes are complex and cause–effect relationships may not be clear, (2) the pollutants of interest may affect a wide array of species and the species known to be most sensitive (indicator) to the perturbation may be more difficult to monitor, and (3) most sensitive species may also be sensitive to and responding to other pollutants in the environment. These factors would necessitate the identification and use of multiple indicator species.

2. *Indicators that reflect changes in population trends of other species.* Indicator species have been commonly used in management to indicate when the environment might be changing sufficiently so that future population fluctuations may occur. This has potential for assessing the impact of different management activities on a larger group of species. However, using indicators to assess the condition of other populations is only useful when good correlations exist between indicators and the larger population group.

3. *Indicators that reflect habitat quality for other species or communities.* Indicators may be chosen because they have the most restrictive habitat requirements of any focus group. Again, the use of indicators to reflect changes in habitat quality has to demonstrate that clear relationships exist between the species and specific attributes of the habitat identified to be critical for maintaining system function. In this grouping, the density of indicator species is the most common measure of habitat quality.

The use of indicator species to assess the condition of an ecosystem can be useful in answering questions about perturbations and the effects of management practices, especially when time and money are limited. However, while the presence of a particular species can indicate that certain minimal ecological conditions have been met, determining the significance of the absence of a species is much more difficult and complex. A species may be absent because (1) environmental conditions are not suitable, (2) the species has not had an opportunity to establish in the area but might survive if introduced, (3) another species has assumed the functional role or niche, and (4) for a species present at low density, absence may be the result of sampling error (Cairns 1979). The use of the indicator species approach does not address how ecosystem processes have been changed due to a perturbation or the relationship between a species and ecosystem structure and/or function.

The use of indicator species can be a valuable tool for assessing and monitoring an ecosystem and its components but should not be used in isolation as the only assessment tool. This approach does have limitations that must be factored into any study utilizing this as their primary tool. The following limitations must be kept in mind when using the indicator species approach:

- Ecosystem processes are complex, and reliance on one sensitive species to indicate change may not yield a clear picture of the cause and effect relationships (Cairns 1986, Wren 1986).

- Proper identification of indicator species is problematic (Pearson and Cassola 1992) and determining the significance of the absence of a species is a complex process.

- The assumption that a correlation in population trends exists between an indicator species and a larger group of species is often not correct (Verner 1984).

These are the reasons why several researchers (Verner 1984, Cairns 1986, Wren 1986) have suggested the indicator-species approach would be more useful and widely applicable if expanded to encompass the use of multiple species (i.e., indicator guilds). Using multiple species as indicators of change in an ecosystem would provide information not usually generated by the single-species approach. It would also avoid some of the pitfalls associated with using one species, such as the proper identification of the most sensitive species. Ultimately, multiple indicators that provide information about different ecosystem components and processes can be incorporated into an integrative index of ecosystem state.

Others have suggested the development of functional groups at the species composition level to describe the way biota regulate ecosystem processes (Walker 1992, Solbrig 1993). Walker (1992) identified five criteria for defining species functional types (guilds):

1. Identify rate-limiting or otherwise important ecosystem processes

2. Develop corresponding functional classifications of biota through guild analysis

3. Further subdivide functional groups on the basis of functional attributes related to critical ecosystem process

4. Assess redundancy within functional groups (complete functional redundancy occurs when there is density compensation among group after removal of one member)

5. Assess relative importance of functional groups

This contrasts the five grouping criteria developed by Körner (1993). He based his identification of functional groups as those that at any scale have a set of structural or process features in common so they can be treated as group. Körner (1993) suggested the following five grouping criteria:

1. Degree of integration (hierarchy of complexities)

2. Quality criteria (life form, life history, size, etc.)

3. Spatial distribution of plants and plant organs within ecosystems

4. Phenological sequences

5. Distributional characteristics

The use of the functional group approach will contribute to better understanding of ecosystem function only to the extent that the significance of functional groups is understood in relationship to the behavior of the whole system. Functional groups need to be further studied, and some of the ideas just presented need to be tested for their utility and applicability for different ecosystem types.

The qualities of appropriate indicators will vary greatly between different ecosystems and management regimes. To guide selection of indicators, it is therefore important to recognize that they should primarily be chosen based on how they contribute to the overall *index* of the ecosystem state. Indicators should be selected so that they provide an index that is (1) integrative (i.e., combining different hierarchical scales and both functional and structural components) and (2) flexible (i.e., responsive to changing ecological, social, and management constraints).

Furthermore, indicator species can serve as early warning signals that a system may be shifting towards an undesired state. For example, because the discovery of the relationship between DDT in the environment and the thinning of bird's eggshells was discovered in peregrine falcons (*Falco peregrinus*) in the mid-1960s, birds have been used to successfully detect and monitor the effect of environmental contaminants (Morrison 1986). More recently, researchers have drawn attention to declines in amphibian populations on a global scale. Some researchers have suggested that their disappearance may be indicative of changes in atmospheric composition (Carey 1993), while others have postulated that these declines can be attributed to loss of habitat due to shifts in land use (i.e., agricultural lands reverting back to forests, resulting in the loss of the ponds used by frogs; Roush 1995).

In some cases, indicator species can be successfully used to monitor both habitat quality and population trends for other species. For example, Pearson and Cassola (1992) found a significant correlation between tiger beetle species numbers and butterfly and bird species numbers in North America, the Indian subcontinent, and Australia. Because habitat quality and population trends both relate to ecosystem structure, these types of relationships are important to understand and can be used effectively as part of ecosystem management.

Indicator species, as discussed earlier, refer to species native to or already existing within the ecosystem. Another use of indicator species is to transplant specific species to various ecosystem where they are monitored (see Section 4.2.2). Sensitive indicator species can act as internal standards that allow comparisons to be made between healthy productive ecosystems and degraded or declining ecosystems. They can also act as early warning signs if an ecosystem is exposed to environmental stress (i.e., canary in the cage analogy in coal mines).

4.2.4   Stable Isotope Discrimination: Element and Water Cycling

Stable isotope techniques provide a promising tool for integrating several different physiological and ecological parameters and for making linkages across scales (Ehleringer and Field 1993). The analysis of stable isotopes in ecological research is based on the characteristic discrimination that occurs between the stable isotopes of carbon, hydrogen, oxygen, and nitrogen during biochemical reactions that are preserved in plant tissues (Yakir et al. 1993). These isotopic signatures caused by natural fractionation processes provide relatively easily measurable indices of associated plant processes that are integrated over both time and space (Rundel and Sharifi 1993, Schimel 1993).

Stable Carbon Isotopes

Discrimination against the naturally occurring stable isotope $^{13}CO_2$ occurs during both the diffusion and carboxylation components of gas exchange. The overall effects are integrated over the period in which carbon is assimilated into leaf tissues, resulting in a relationship between $\delta^{13}C$ and mean intercellular CO_2 concentration (Farquhar et al. 1982). Greater discrimination indicates a higher stomatal intercellular carbon concentration. Discrimination is affected by relative changes in stomatal conductance and photosynthesis, thus providing an integrative measure of how plants balance transpiration with carbon assimilation. This may also be interpreted as a reflection of plant water-use efficiency (Ehleringer et al. 1991, 1993).

Carbon isotope discrimination has been shown to vary within a species and among different species along gradients of water availability, with lower discrimination values associated with more xeric conditions and more drought-tolerant species (Ehleringer and Cooper 1988, Ehleringer and Dawson 1992). Variation in carbon isotope discrimination has also been observed between species of different life forms, for example, higher carbon isotope discrimination values are associated with higher plant growth rates, while low values are associated with larger individuals (Ehleringer et al. 1993). A reduction in carbon isotope discrimination is also associated with increases in palisade layer thickness (Vitousek et al. 1990, Körner et al. 1991). Evergreen species have been shown to have greater water-use efficiency based on $\delta^{13}C$ values as compared with deciduous species (Valentini et al. 1992). These observations suggested that carbon discrimination patterns may provide a mechanistic link with ecosystem level patterns, particularly those related to the hydrologic balance, such as canopy moisture, which can be potentially detected using remote sensing technology (see Section 4.2.5). Furthermore, specific relationships between isotope discrimination patterns and species adaptive strategies allow for more detailed analysis about the contribution of various species guilds to large-scale patterns. Because forest productivity is strongly determined by moisture availability at larger spatial scales of analyses (Gholz et al. 1990), scaling between plant and ecosystem level water relations using carbon isotopes should provide a powerful tool for understanding how these processes are related and for making ecosystem level predictions based on more easily obtainable carbon isotope information.

Stable carbon isotope technology may also provide more sensitive and rapid measures of physiological processes occurring at ecosystem level scales compared with traditional methods. Patterson and Rundel (1993) used stable carbon isotope discrimination to identify ozone-sensitive individuals within a population of Jeffrey pine trees based on the observation that ozone-sensitive individuals operate at higher stomatal conductances than ozone-resistant individuals. Furthermore, Elsik et al. (1993) showed that stable carbon isotope analysis was more effective than instantaneous gas exchange measurements for identifying interactions between ozone exposure and water stress on loblolly and shortleaf pine. They reported that by increasing $\delta^{13}C$ values of foliage and stem tissues, ozone affected the pines by causing greater stomatal closure and also enhanced stomatal response to water deficits. These findings may have wider applications in the detection and the prediction of ecosystem-level impacts of human activities and linking them to mechanisms occurring at the leaf and individual plant level.

Scaling stable carbon isotope discrimination information from the leaf to the canopy is complicated by the fact that gradients of atmospheric ^{13}C may exist within the canopy. For example, in closed forests, respiration inputs from litter alter the $\delta^{13}C$ value of source carbon near the forest floor (Medina et al. 1986). In addition, very little is know about atmospheric gradients in source $\delta^{13}C$ values occurring on a landscape or biospheric level (Yakir et al. 1993). However, gradients in $\delta^{13}C$ values on a global scale along altitudinal and latitudinal gradients suggest that both temperature and atmospheric pressure are responsible for observations that carbon isotope discrimination decreases with altitude (Körner et al. 1991).

Global studies of carbon isotope discrimination may provide useful integrators of biospheric photosynthesis, for example, in response to changes in atmospheric CO_2 concentrations. Furthermore, increases in atmospheric O_2 concentration due to fossil fuel inputs and deforestation creates a decreasing trend in global atmospheric $\delta^{13}C$ values, which can be detected in the records of plant tissues (Yakir et al. 1993).

Carbon isotopes have also been used to assess interactions between vegetation and soil processes. Balesdent et al. (1993) related the isotopic composition of forest soil organic matter to vegetation and soil type, and found that isotope ratios of soil organic carbon were significantly related to vegetation $\delta^{13}C$. Consequently, significant changes in isotopic composition of past vegetation caused by either natural or human environmental change may be assessed by examining the isotope ratio difference between present vegetation and soil organic matter (Balesdent et al. 1993). In addition, trophic relations can be studied using carbon isotope analysis when various potential carbon sources have different isotopic ratios, allowing for the determination of the relative contribution of each resource to consumer biomass (Rounick and Winterbourn 1986).

Stable Hydrogen Isotopes

Stable hydrogen isotopes may be very useful for providing insight into the contribution of different water sources (e.g., groundwater, stream water, rainfall) to overall water uptake by plants (Dawson and Ehleringer 1991). Because no fractionation of isotopes by plant roots occurs during water uptake, the hydrogen isotope content of plant xylem water reflects the isotopic composition of the water source used by the plant (Valentini et al. 1994). Consequently, differences in the hydrogen isotope signature of varying water sources are recorded in the xylem sap and can be analyzed to determine the relative contribution of each source (Dawson and Ehleringer 1991, 1993). For example, Valentini et al. (1992) showed in a mediterranean macchia ecosystem that evergreen species obtained most of their water from precipitation compared with deciduous species, which utilized groundwater almost exclusively.

Richards and Caldwell (1987) used stable hydrogen isotope analysis to demonstrate that deep-rooted shrubs growing in desert environments were capable of absorbing water during the night due to hydraulic lift. They speculated that this additional water would be stored in the upper soil layers, where it would be available to support daytime transpiration and potentially serve as a buffer of water availability for several days. Furthermore, additional moisture in the upper soil layers provided by hydraulic lift processes may also be utilized by more shallow-rooted species, resulting in "water parasitism" (Richards and Caldwell 1987). Therefore, stable hydrogen isotope analysis provides an important tool for making linkages between individual plant water use and ecosystem functioning because of its ability to discern rooting habits and its impact on ecosystem water balance. For example, changes in the abundance of deeply rooted species as a result of disturbance or environmental stress may affect soil moisture availability on larger scales, which may have direct consequences for species composition and ecosystem developmental patterns. This would be a very useful tool to examine the problem that has been identified earlier with *Eucalyptus*, see section 4.2.1.

Stable hydrogen isotopic analysis has also been used to directly assess hydrologic processes occurring on larger scales, indicating the potential for scaling between individual plant performance and ecosystem patterns based on hydrogen isotope measures conducted concurrently at different scales. For example, long-term studies of paleoclimatic reconstruction is possible based upon the isotopic analysis of wood cellulose in tree rings, which can provide information about previous climatic patterns related to temperature, carbon dioxide concentrations, and precipitation (Dawson and Ehleringer 1993). In another study, changes in δ hydrogen values of leaf tissue in *Peperomia congesta* were found to provide a reliable indication of soil water recharge, with restoration of isotopic enrichment of leaf water occurring in response to increased water availability (Yakir et al. 1993). Thorburn et al. (1992) combined measurements of transpiration and stable isotopes of water to determine groundwater discharge from forests. This was then used to determine the potential for salination of soils at the site and to make predictions about the effects on forest health. Together these studies highlight various approaches for using hydrogen isotopes to assess water relations at multiple scales. Greater integration between these scales in the future may yield even more opportunities for enhancing our understanding about hydrogen processes in ecosystems.

Stable Nitrogen Isotopes

The cycling of key elements (i.e., N) within an ecosystem may be examined using stable isotope fractionation patterns in plants combined with analyses conducted at the soil level. Stable isotope analysis provides a means of utilizing the natural ^{15}N isotope abundance in soils to evaluate N fixation and inorganic N uptake in natural systems (Delwiche and Steyn 1970, Virginia and Delwiche 1982, Yakir et al. 1993). For example, Garten and Van Miegroet (1994) reported a strong correlation between foliar ^{15}N abundance and relative net nitrification across a wide range of forest types, soil types, and climatic environments. By integrating seasonal and spatial variations in soil N transformations, foliar ^{15}N abundance may provide a means of transferring information from the individual tree level to changes occurring at the ecosystem level. Isotopic tracers have also been used to assess nitrogen cycling in ecosystems. For example, Boyce et al. (1994) showed that spruce-fir forest canopies absorb large amounts of nitrogen from atmospheric deposition by spraying canopies with $(^{15}NH_4)_2SO_4$ or $K^{15}NO_3$ solutions and determining foliar ^{15}N levels.

In ecosystems where N-fixing plants are important contributers to the nitrogen cycle, the use of stable nitrogen isotopes provides a means for verifying the presence of symbiotic N-fixers and for quantifying the contribution of fixed N to the ecosystem level N cycles (Virginia et al. 1988). This approach can be applied at most sites where the ^{15}N abundance of soil N is distinguishable from that of atmospheric N_2 (Virginia et al. 1988). Because many leguminous plants alter their N_2 fixation rates in response to changes in internal carbon demand and to climate (Virginia et al. 1988), this technique could provide an important means of analyzing how ecosystems with high components of N-fixing plants will respond physiologically to increased CO_2 levels (Arnone and Gordon 1990).

Genetic Variation in Stable Isotope Discrimination Patterns

Genetic diversity is an important evolutionary mechanism determining ecosystem performance because it establishes the direction and capacity of ecosystems to respond to change (Gordon et al. 1992). Similar to adaptive strategies that determine the degree of phenotypic plasticity among organisms, it is important to document the range of genetic variations that exist within populations, across ecosystems and landscapes. The use of stable isotope analysis may provide a powerful technique for assessing genetic variations in resource use within and among populations. Within plant populations, a relatively high degree of variation has been reported in stable carbon isotope discrimination values. These patterns have been shown to be consistent in terms of the relative ranking of individuals over large environmental gradients of temperature and moisture availability, and thus to be strongly genetically determined (Comstock and Ehleringer 1992). Furthermore, discrimination values have been shown to be a highly heritable character in plants (Geber and Dawson 1990, Schuster et al., 1992). Johnsen and Major (1995) attributed variation in photosynthetic capacity among different families of *Picea mariana* to genetic differences in carbon isotope discrimination. Flanagan and Johnson (1995) provided information on the physiological basis for genetic variation in growth among full-sib families of *Picea mariana* by examining carbon isotopic discrimination and growth characteristics—trees with low discrimination values and high photosynthetic capacities relative to water losses had an advantage on water stressed sites. Information on shifts in the degree of variation in isotope discrimination within a population may provide an indication of changes in genetic variability with respect to key adaptive strategies for plants. However, the within plant phenotypic plasticity associated with carbon discrimination values (Rundel and Sharifi 1993) would need to be accounted for in these estimates. It is possible that the correlation between habitat heterogeneity and the range of discrimination values in a plant population (Ehleringer et al. 1993) may also be used to measure the degree of habitat heterogeneity, providing a link between genetic variation and landscape spatial characteristics.

4.2.5   Remote Sensing and Spectral Analysis

Advances in the application of remote sensing to the study of ecological systems has allowed us to literally take a broader view of ecological patterns and processes. By providing a means to detecting phenomenon at the landscape or ecosystem level, it has improved the feasibility for scaling information from the individual plant to the ecosystem level. Remote sensing is the measurement of reflected, emitted, or back-scattered electromagnetic radiation from Earth's surface using instruments stationed at a distance from the site of interest (Roughgarden et al. 1991). Satellite systems provide not only photographic representations of the Earth's surface, but also physical measurements of the absorptance, reflectance, and emittance properties of landscapes. These spectral characteristics can be used to measure biogeochemical and, in turn, ecophysiological properties at various scales (Ustin et al. 1993).

The most extensive application of remote sensing technology to ecology has been to estimate ecosystem productivity based on the relationships among total leaf areas, leaf photochemical efficiencies, and photosynthetically active radiation (APAR); the latter can be measured using remote sensing technology (Norman 1993). A widely applied but simple index is the Normalized Difference Vegetation Index (NDVI), which uses two reflectance bands (red and near-infrared) obtained from the Earth Observing System (EOS) to identify green vegetation and to estimate biomass accumulation. Because NDVI is fairly well correlated with light interception, it can be used to calculate annual light interception at the leaf scale (Ustin et al. 1991). Thus, net primary productivity and biomass accumulation can be estimated based on the absorbed photosynthetically active radiation data provided by the NDVI (Waring 1983, Field 1989).

Many opportunities exist for expanding models and indices developed from remotely sensed data to include information from other regions of the electromagnetic spectrum. For example, of the 197 reflectance bands that can be remotely sensed, only two are systematically utilized for obtaining NDVI. The other 195 spectral features may provide ecologically relevant indices associated with variables such as lignin, nitrogen, and starch, as well as the water status of canopies (Ustin et al. 1991). A key issue in the application of remote sensing to ecological models is identifying those factors that contribute most strongly to the spectral variance across scales (Ustin et al. 1993). For example, remote sensing of plant tissue chemistry (i.e., nitrogen and lignin concentrations) may be used to predict ecosystem processes such as decomposition, photosynthesis, and transpiration (Matson and Vitousek 1990).

Studies have also demonstrated the possibility of using remote sensing to investigate nitrogen stress (Wessman et al. 1988) and water stress

(Pierce et al. 1990), creating potentials for modeling interactions between ecosystem processes and their environmental controls at large scales (Chapin 1993). Furthermore, making linkages between spatial patterns in chlorophyll or canopy water and ecosystem productivity (Gao and Goetz 1990), or between canopy lignin concentrations and decomposition rates, will require tools for integrating mechanistic models with spatial patterns provided by remote sensing technology. Models predicting litter decomposition from initial lignin concentrations do exist (Berendse et al. 1987) and should be applied to remotely sensed data. However, the utility of remote sensing to detect plant physiological responses at the ecosystem scale is still in its infancy and how sensitively it can detect differences in abiotic resource utilization by plants when information is averaged across larger spatial scales has not been clearly documented.

Remote sensing may create new opportunities for identifying which parameters function as regulators of ecosystem activity at larger scales and how these variables differ across spatial and temporal scales (Roughgarden et al. 1991). For example, Rey-Benayas and Pope (1995) analyzed Landsat Thematic Mapper (TM) imagery to assess landscape diversity patterns and variability in seasonal tropical forests of Guatemala and were able to identify which variables were most important in discriminating between different landcover types. Green leaf biomass was found to be the most important variable in uplands areas, while canopy closure and degree of senescence were the most important variables in the lowland swamps (Rey-Benayas and Pope 1995). Furthermore, the dominant distinguishing properties which emerge from spectral images of different landscapes have been demonstrated to show consistent patterns, for example, in the case of foliar versus wood-litter components and soil color (Adams et al. 1990, Ustin et al. 1993). Thus, it may be possible to identify a similar suite of unifying characteristics that can be used to assess and compare patterns and processes across different spatial and temporal scales.

These observations may also have relevance for applying functional groupings to the analysis of remotely sensed data as a means of facilitating the transfer of information between different hierarchical levels. Aggregation of ecosystem components may enhance and/or facilitate the capacity for forming connections between different scales. For example, assessing the complex patterns of ecosystem response to climate change may be facilitated by grouping species into a smaller subset of functional types, each having similar response patterns (Bazzaz 1993). Dawson and Chapin (1993) proposed that form–function relationships should be used as a basis for grouping plants, particularly where the objective is to simplify landscape patterns without excessive small-scale detail. Plant forms typical to different successional stages (i.e., grass vs. shrub vs. tree), and for plants growing under different soil fertilities tended to be distinguished by common traits related to relative growth rate, longevity, and resource-use efficiency (Lambers and Poorter 1992). This broadly classifies plants according to functional attributes. Delineation of functional groups can be facilitated through the use of remote sensing technology to identify dominant growth forms of an ecosystem. For example, evergreen plants have a higher infrared albedo during the inactive season than deciduous plant, while synthetic-aperture radar (SAR) may allow for differentiation of forests from shrublands (Chapin 1993). Because of the functional relationships between growth form and ecosystem processes (Matson and Vitousek 1990), remote sensing of growth forms may allow for mapping of ecosystem functions on a regional or global basis.

Remote-sensing technology may also provide an indirect measure of ecosystem state by its ability to compare differential surface temperatures of vegetation among ecosystems. These data have indicated that the surface temperature of more developed ecosystems are colder than less-developed ecosystems. This pattern may be related to thermodynamic principles in which ecosystem development occurs in order to reduce the energy gradient impressed on the Earth by the Sun (Hollins 1993). Thus, life can be viewed as is a means of dissipating this thermodynamic energy gradient, and ecological development occurs when new pathways for degrading imposed gradients emerge (Hollins 1993). Ecosystem temperature may provide an accurate and inherently integrative indicator of ecological integrity (Hollins 1993), particularly if models or indices can be developed that describe "baseline" temperatures for different potential ecological states for a particular ecosystem, which can then serve as a reference point. However, before such a tool can be applied, there is a need to determine how sensitive ecosystem surface temperatures are to the degree of change needed to shift an ecosystem to another state or threshold, and whether or not current technology can detect these reorganizations.

4.2.6   Integrating Across Spatial and Temporal Scales

The above discussion highlighted various approaches that have been used to address questions related to ecosystem performance within the context of specific spatial and temporal scales. However, the information collected to address spatial and temporal scale issues has been quite different. For example, most spatial scale studies have concentrated on examining changes in processes occurring from the individual seedling to the level of the stand, watershed, or landscape. In contrast, temporal scale studies generally had the time scale of the study largely being determined by disturbance cycles and/or successional processes. In order to make linkages between different scales, techniques are needed for characterizing and linking spatial pattern and temporal changes, accounting for and incorporating the unique characteristics of ecosystems that are inherent at a particular spatial and temporal scale, but that often change when the scale of analysis changes. Two important techniques for integrating across scales include gradient analysis and boundary analysis, each of which are discussed below.

Gradient Analysis

A useful concept for assessing spatial pattern and temporal change is gradient analysis. Gradient analysis may be defined as an approach that relates gradients to one another on three levels: environmental factors, species populations, and community characteristics (Gosz 1992). However, this should be expanded to include a greater range of levels and to include the individual plant, the ecosystem, and landscape levels. Ecological studies have shown that plant species (Pickett and Bazzaz 1978), vegetation types (Griffin 1977, Schulze and Chapin 1987), and biomes (Gosz and Sharpe 1989) can be distinguished by analyzing changes occurring along gradients in resource availability (i.e., climate, nutrients, moisture, etc.). Likewise, an ecosystem may be considered to function somewhere along a gradient that is defined by the multiple potential states of functioning for that particular ecosystem. The gradient analysis approach can be used to identify potential indicators of ecosystem change across spatial and/or temporal scales. Ecosystem change may be indexed according to a response variable or group of variables, such as disturbance intensity, exposure to a chemical stressor, or some other factor associated with changes in ecosystem state.

The objective of applying gradient analysis to assessing spatial and temporal characteristics of ecoystems is to determine a relationship between a particular ecosystem and measurable characteristics of ecosystem components as a basis for enhancing our predictive capability for assessing ecosystem response to environmental change. An important consideration is that of scale and, even though researchers often refer to the whole ecosystem, the capacity to assess ecosystems generally relies on smaller scale parameters contained in the ecosystem. Although it might be tempting to merely develop correlations between multiple parameters with ecosystem level measures, the lack of a mechanistic basis behind these correlations will generally lead to fallacies in the overall model (Comstock and Ehleringer 1992). Thus, correlations should be used to provide a tool for selecting those parameters for which mechanistic studies can be designed to establish more powerful cause-and-effect relationships. Developing a model will involve obtaining measurements at different scales of interest and generating connections between them.

The importance of determining cause-and-effect relationships based on mechanistic studies is exemplified by looking at the forest decline research and the hypotheses that were generated on how pollution and ozone damaged forests. Although measurements indicated that cloud water near monitoring stations located in close proximity to areas with visual symptoms of forest decline (i.e., eastern United States and Europe) were significantly more contaminated with pollutants compared with unaffected or less affected areas (Bormann 1985, Saxena et al. 1989), data on cloud water chemistry alone did not provide information about how the forests were responding to the pollution. However, one can use mechanistically based information to identify which variables are worthwhile monitoring in the field to assess tree damage. For example, there exist strong mechanistic links between specific photochemical oxidants (a major class of air pollutants) and distinct symptoms of tree damage. This type of information can then be used to link growth reductions in trees with photochemical-oxidant damage in the field (Bormann 1985). Thus, a mechanistic connection of decline with air pollution was established that explained the correlations observed at larger scales between cloud water and forest decline. However, it is important to recognize that often the effects of a stress may be indirect, such as increasing the susceptibility of an ecosystem to higher levels of insect attack or disease, and thus may be more difficult to quantify mechanistically (Wargo et al. 1993).

In a spatial context, gradient analysis is a particularly useful tool to identify those areas with the greatest potential for change or that are more sensitive to change, that is, the boundary zones ("ecotones") and edges between different plant communities, habitat types, and ecosystems that tend to exhibit the greater rates of change (Gosz 1992). In fact, it may be useful to develop an analogy between ecotones in a spatial context and thresholds of ecosystem functioning in a temporal context (see discussion of thresholds in Sections 3.2 and 4.1). *Ecotones* represent transitions between different climatic conditions and abiotic constraints in space, while *thresholds* represent similar transitions in time. Consequently, by understanding patterns and processes of change occurring in boundary zones, it may be possible to elucidate similar phenomena that occur in response to changing constraints caused by disturbances but that often require longer time periods to detect. For example, along an environmental gradient where the regional climate becomes marginal for a particular species, many microhabitats are outside the species' range of tolerance and its distribution is constrained. Thus the response of a species to environmental factors is amplified or attenuated across the transition zone, and the different microsite characteristics should reflect the environmental influences having a predominant effect on the species response to change (Gosz 1992). Likewise, plant growth within the boundaries of adjacent ecosystem states will reflect those environmental factors having the greatest influence on ecosystem processes. Therefore, functions of ecosystem components in these boundary zones should be used to characterize thresholds and to identify useful indicator parameters.

In order to develop an understanding of ecosystem change or trajectories of change (i.e., its resilience and resistant characteristics), there is a need to understand both its current state and its response to changes within the environmental constraints caused by human and/or natural disturbance. Using the gradient analysis approach, ecosystems located at different points along the gradient for concurrent study can be identified, ideally using existing ecosystems already altered by natural or human disturbances, or combining the research with planned or existing management activities.

An example of the application of gradient analysis can be found in the mixed conifer forests of the Sierra Nevada Mountains, California, where changing fire regimes have created a landscape pattern that shifts from a closed forest to open woodland to a grasslands (Parsons and DeBenedetti 1979). Fire suppression has caused alterations in the age structure, size, and composition of the forests, with increases in shade-tolerant conifers and declines in black oak (Parsons and DeBenedetti 1979). In this example, the thresholds or transition zones between the states can be roughly delineated by differences in fire intervals. The transition zones are correlated with selected indicator parameters (i.e., decreases in seedbed conditions for oak germination due to increasing accumulation of duff layers due to fire suppression; Kauffman and Martin 1987) and may serve as a measurable tool to determine species distribution along this gradient. However, additional knowledge about the conditions of different forest states along this gradient would strengthen the ability to assess this forest. For example, specific changes in the location of thresholds may be measured by increases in the relaxation time from a perturbation and by increases in the variance of observed fluctuations of identified parameters (Gosz 1992). The strength of applying the gradient analysis approach as a framework for assessing ecosystem change is that it facilitates selection of effective measurable parameters for assessing changes in ecosystem state.

Boundary Analysis

Spatial pattern in landscapes can be understood based on the qualities of shape, size, distribution, juxtaposition, and configuration of different patches (Forman and Godron 1986, Milne 1992) and their associated boundaries. Although a somewhat nebulous concept, "patch" may be considered as a surface area differing from its surroundings in nature or appearance (Wiens 1976, Kotliar and Wiens 1990). Thus, *patch* implies a relatively discrete spatial pattern (Pickett and White 1985) that can vary widely in size, shape, type, heterogeneity, and boundary characteristics (Forman and Godron 1986). Patches are separated by boundaries whose delineation is scale dependent. Objective criteria for delineating landscape features at different scales are important in establishing a baseline for moving between scales. For example, patch density and structure are important constraints to consider in designing scaling experiments. If patch density is high, a relatively smaller area may be adequate for the analysis (Berg et al. 1984). In addition, functional linkages of the properties and interactions that are being studied should dictate the scale of analysis. The scale must match the processes being monitored, and it is necessary to delineate a large enough area such that the various phases of disturbances and successional development are incorporated (Berg et al. 1984).

The concept of ecosystems as hierarchically arranged entities has also been applied to the study of patch structure. Accordingly, a patch at a given scale has an internal structure that is a reflection of patchiness at finer scales, and the mosaic containing that patch has a structure that is determined by patchiness at broader scales (Kotliar and Wiens 1990). Different criteria may be needed for defining patchiness at different hierarchical levels, and it may be necessary to consider the responses of organisms to multiple scales of patchiness (Kotliar and Wiens 1990). Combining techniques for analyzing patch characteristics at different scales (i.e., within-patch experimental manipulations with landscape analysis) using satellite imagery and geographic information systems may help link landscape characteristics with patterns and processes occurring on smaller scales in the hierarchy. However, studies of gap and patch dynamics and landscape level analyses have traditionally been conducted separately. Because boundary analysis is based on utilizing different levels of "similarity" or "relatedness" among landscape components to aggregate information at different scales, it may provide a tool for integrating large- and small-scale measures and for transferring information across scales.

Boundary areas can be separated into edges (i.e., sharp discontinuities between different vegetation types) and ecotones (i.e., distinguished by the functional responses of species that create a tension zone; van der Maarel 1990). Delineation of edges and ecotones can be accomplished either by forming spatially homogenous clusters or by detecting boundaries. The former uses clustering algorithms to produce homogeneous groups, while the latter uses edge detection techniques to identify areas of change (István and Riedi 1994, Fortin and Drapeau 1995). A crucial factor in using boundary analysis to characterize landscapes is the development of criteria for delineating boundaries among different patches. Boundaries between edges and ecotones may be statistically defined as the location where the measured variables (biotic and/or abiotic) show the highest shifts or where the average rate of change is a maximum (Fortin and Drapeau 1995). Using boundary detection techniques, it is possible to rank the rates of change by their magnitude, which may provide insight into the degree of persistence or resilience of boundaries and the patches that they delineate (Fortin and Drapeau 1995). Peaks of unusually high variance may be used to indicate scales at which the between-group differences are especially large and where the scale of natural aggregation of patchiness occurs (Wiens 1989). Furthermore, boundary analysis may provide direct information about the mechanisms underlying the formation of ecotones by comparing the location of boundaries of a response variable (e.g., vegetation data) with the locations of boundaries of explanatory variables (e.g., environmental data; Fortin and Drapeau 1995). Those variables most critical in determining the formation of ecotones or edges (i.e., where rates of change are greatest) should be likely candidates to focus on in scaling because they play a dominant role in regulating ecosystem behavior where environmental change is highest. Although boundary analysis primarily focuses on spatial change, similar emergent features are likely to dominate ecosystem response to environmental change over time.

Fractal geometry provides another means of delineating ecological patterns in landscapes and associated boundary characteristics (Mandelbrot 1983, Lam 1990, István and Riedi 1994), based on a similar theoretical approach of identifying areas where rates of change are highest. The fractal dimension, D, characterizes the way in which a particular fragment differs with scale and therefore provides an index of the scale dependency associated with a particular pattern. Consequently, a change in the fractal dimension of a pattern provides an indication that different processes or constraints are dominating, and thus represents the extent of a particular scale domain and the boundary between two different domains (Wiens 1989). Fractal dimensions can also be used to develop indices for landscape diversity that take into account more components of diversity than shape alone (i.e., perimeter/area measurements), including the number and types of patches, their distribution (juxtaposition), and their shape (Olsen et al. 1993).

Fractal analysis may be particularly useful in studying boundary characteristics, as fractal indices of landscape features can be used to aggregate smaller scale data on individual plants or patch microsites, and to facilitate the transfer of information about the effects of fragmentation on ecosystem processes occurring at different hierarchical levels. Krummel et al. (1987) used fractal dimensions to delineate transition zones where landscape patterns change depending on human versus natural disturbance by (1) determining that small forest patches have smaller D values than large forest patches and (2) identifying at which patch size a distinct break region occurred with a change in D value. Mladenoff et al. (1993) used fractal dimension analysis and geographic information systems to compare the structure of second-growth and old-growth landscapes, and identified features that distinguished the two landscapes. For example, forest patches in the fragmented second-growth landscape were simpler in shape (lower fractal dimension, D), while important juxtapositions of different species (i.e., hemlock with lowland conifers) were absent (Mladenoff et al. 1993). Lower fractal dimension and lower edge/interior ratios were associated with greater fragmentation (Mladenoff et al. 1993), which provided a convenient tool for categorizing landscape spatial patterns according to structure, complexity, and heterogeneity at different scales.

Remote sensing and spectral analysis (see Section 4.2.5) may also be combined with fractal dimension analysis as a more powerful approach to classifying complex patterns occurring in a landscape. Using fractal dimensions to locate areas of highest rates of change may help elucidate those parameters most important in determining ecosystem response to change. DeCola (1989) showed that fractal analysis can be used to improve the spatial analysis, spectral classification, and functional labeling of remotely sensed data by applying fractals to Landsat Thematic Mapper images of northwest Vermont to identify and segregate different ecosystems, land-use types, and associated spatial characteristics. Based on the calculated fractal dimensions, the landscape was classified into eight landcover classes. The highest degree of complexity (i.e., high D) was associated with hardwood and brush wetland habitats, while water, cornfields, and bare fields had the lowest D; hayland, softwoods, grassland, and urban cover types had intermediate D values (DeCola 1989). The results suggested that landscape imagery combined with fractal analysis may provide a powerful tool for discerning landscape patterns without extensive field data. Furthermore, these analyses facilitate the selection of smaller scale criteria that are dominant in affecting larger scale patterns and processes, and that therefore may serve as the most useful parameters for scaling between levels and for monitoring long-term change. For example, a grid-based GIS data structure can be superimposed over the regions delineated by the fractal analysis and used to locate important features for more detailed examination, as well as to determine how spatial features are related to each other on different spatial scales (DeCola 1989).

Fractal analysis can also be applied to the study of population ranges and movement of organisms and to delineating boundaries according to differences in habitat. For example, Virkkala (1993) used a fractal dimension based on observations combined at different scales of resolution to study the distribution and range boundary of forest passerines in Finland. In addition, Virkkala (1993) characterized differences in the spatial distribution and geographic range of short- and long-distance migrants. Delineation of boundaries between different organisms is particularly important because the area in which an organism interacts with or "perceives" the environment is an important factor determining both the range of influence and the response of that particular organism (Wiens 1989).

Another important application of boundary characteristics to transferring information across scales is in the study of gradients (see Section 4.2.6). Gradient analysis with emphasis on boundary zones can provide information about ecosystem response to environmental change and the location of thresholds between different ecosystem states. Studies that have quantified changes across gradients between different patches suggest that the variation and qualities observed is dependent on specific patch characteristics (i.e., size, shape, and juxtaposition with other patches in the mosaic). For example, a study by Chen et al. (1995) investigated the changes in microclimatic variables across gradients from recent clear-cut edges into old-growth Douglas fir forests in Oregon, USA. They reported the longest and steepest gradients for air temperature, soil temperature, and relative humidity; however, the depth of edge influence into the forest varied among the different variables. These results suggested that gradient analysis along boundary zones can help elucidate which parameters may be most important in controlling vegetation response. Combining data collected for microclimatic variables with parameters for plant response or using transplants along the gradient may provide information on physiological mechanisms for ecosystem response to stress and, in turn, forest condition.

The analysis of spatial patterns in landscapes may also help elucidate the effects of changing disturbance regimes on ecosystem processes across temporal and spatial scales, and provide a tool for linking differences in disturbance effects across scales. Baker (1992) used a GIS-based spatial model and data on historical changes in fire size and interval to simulate the effects of settlement and fire suppression on landscape structure in northern Minnesota. He showed that the dynamics of landscape patches are governed by the patch birth and death rates, both of which were influenced by disturbance size and interval distributions. Baker (1992) proposed that such models can be applied to develop hypotheses as to how different human and natural disturbances alter landscape structure. Vogelman (1995) assessed spatial patterns and rates of forest fragmentation in southern New Hampshire and northeastern Massachusetts over the last several decades using digital remote sensing data. Vogelman (1995) showed a strong relationship between fragmentation and population density, and speculated as to how differences in political systems of the two states may be underlying some of the observed patterns. Understanding the relationships between various forms of disturbance and the spatial distribution of landscape features and boundaries may enhance our capability to scale the results of studies conducted within specific patches or across patch boundaries to the landscape level.

Simulation of forest fragmentation based on spatial analysis of remote sensing data also provides the capacity to assess the relationship between degree of fragmentation (based on area/perimeter ratios) and forest cover removal. For example, Vogelman (1995) reported the greatest increase in forest fragmentation as the landscape cover changed from 100% to about 80% forest area, with a threshold occurring at 60%, after which only slight increases in fragmentation occurred. In contrast, Franklin and Forman (1987) predicted that major thresholds in fragmentation would occur when about 50% of the forest was converted to nonforest as a result of clear-cutting in the Pacific Northwest. The apparent discrepancy between the two studies were attributed to differences in the fragmentation-inducing mechanisms at the two sites, leading to differences in the spatial characteristics of the forests (Vogelman 1995). These observations highlight the need for site-specific applications of landscape analysis of spatial pattern in which the underlying mechanisms behind large-scale patterns are dissected. These comparisons also warn against making direct comparisons between sites using similar spatial analysis without considering the differences in the fundamental smaller scale characteristics.

In applying spatial analysis to understanding disturbance and landscape interactions, it is important to recognize that scaling between different hierarchical levels of disturbance effects requires information at different ecosystem scales. Without concurrent more detailed measures of landscape structure to elucidate the specific characteristics behind changes observed in landscape spatial patterns, it is easy to come to erroneous conclusions about the cause-and-effect relationship between disturbance and landscape structure because factors other than the disturbance itself may be important (e.g., successional change, climate variability; Baker 1992). Thus, equally important is the development of tools for scaling between landscape patterns and the smaller scale patterns and processes at the individual patch and plant level. For example, theoretical models developed for describing population density dynamics for wildlife combined with spatial pattern analyses have facilitated the exploration of interactions between individual and population behavior related to the patch-mosaic configuration (Wiens et al. 1993). The use of experimental model systems that can be subjected to field manipulations on microlandscape scales may facilitate extrapolation across scales once parameters at the different scales are measured and linked with each other (Wiens et al. 1993), thereby predicting the response of a parameter on one scale based on the behavior of a parameter on a different scale.

Knowledge about landscape spatial patterns of key abiotic parameters (i.e., soil characteristics) may also provide useful information for linking microsite interactions with processes and patterns occurring at the ecosystem level. For example, analysis of a global forest database indicated that soil texture was strongly correlated with fine-root biomass in hardwood-dominated sites despite the fact that precipitation was not strongly correlated with soil texture (Vogt et al. 1996b, see Section 4.2.7). These observations may reflect the importance of soil water holding capacity in controlling tree growth (Vogt et al. 1996a). Interestingly, the relationship between soil texture and tree growth varied with tree species. For example, soil texture was not highly correlated with fine-root biomass for oaks, possibly due to the high degree of drought tolerance typically associated with the oaks (Abrams 1990). Thus, differences in species' adaptive strategies related to water uptake will also affect large-scale patterns.

Furthermore, some studies suggest that soil carbon accumulation is negatively correlated with soil sand content (Lugo et al. 1986), possibly indicating that in many regions water may play a more dominant factor than nutrients in regulating growth of hardwood forests. These results demonstrated the potential for making connections between landscape patterns of abiotic characteristics and emphasized the importance of focusing on those adaptive strategies most closely associated with compensating for or regulating plant responses to the dominant environmental constraints. Linkages between scales may be strengthened by conducting smaller scale and site-specific studies in boundary areas between, for example, different soil texture classes while stratifying different species according to their drought tolerance. This would identify which parameters are changing most rapidly as an indication of the driving mechanisms behind the observed large-scale patterns.

4.2.7   Analysis of Large Data Sets

Synthesis of large data sets can be very useful for identifying broad-scale patterns in ecosystems and for identifying general patterns of plant species dominance and climatic zones. Analysis of large data sets can be used to identify critical variables controlling specific components or processes in ecosystems. Observed patterns can be used as a guide to further study the mechanistic relationships in ecosystems by focusing on those variables that appear to be highly correlated. This is especially relevant when patterns emerge from data sets that were collected using very different methods and sampling protocols. More attempts should be made to synthesize the large amount of data existing in the published literature and also in the grey (i.e., referee unpublished) literature as a tool for analyzing systems across broader spatial and temporal scales. Synthesis of research conducted under many different ecosystem conditions have great utility for guiding the research focus at smaller scales and under more controlled conditions. Analyses of large data sets are also useful because they allow us to begin forming the functional groups at an ecosystem level (as discussed in Section 4.2.3) using several different variables (i.e., climatic factors, soil factors, ecophysiological responses, nutrient cycles, etc.).

The ineffectiveness of attempting to obtain one or several universal variables to predict processes occurring across all ecosystems is highlighted by looking at a large data set that has been analyzed using different group-

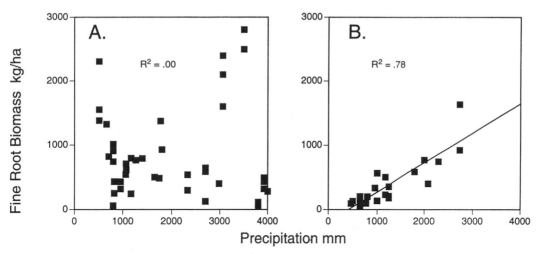

FIGURE 4-2 The relationship between fine-root biomass and precipitation for deciduous (A) and evergreen (B) forests.

ing levels of information (i.e., forest climatic groups, soil order, and species groupings). It is important to identify at what grouping levels particular parameters are useful for predicting how a system will respond to a change in the environment. In the past, there was the desire to find one model of plant growth and its controls that was applicable everywhere and at all scales. In some cases, one variable may effectively integrate the system response to particular types of change. One such example appears to be with plant leaf areas that are strongly controlled by site soil water availability: As the amount of soil water availability increases, higher leaf areas are maintained (Grier and Waring 1974). This relationship is strongly expressed along a gradient of very dry to wet forests because the change in soil water availability is significant enough that the signature at the leaf level is easily detectable (Grier and Waring 1974). This relationship also appears to scale down to the microsite scale, where plant species with varying leaf areas intermingle in the same space. In Malaysia, Palmiotto (1993) observed that wildlings regenerating in the Dipterocarp forest responded very differently to drought periods. Wildings with smaller leaf areas had less mortality than those with larger leaf areas in response to an extensive drought (Palmiotto 1993).

A recent compilation of a large data set on factors predicting fine-root

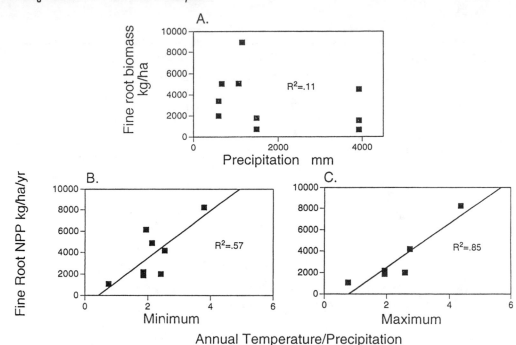

FIGURE 4-3 The relationships between fine-root biomass and precipitation (A), fine-root productivity and minimum temperatures (B), and fine-root productivity and maximum temperatures (C) for the pine forest grouping.

biomass and NPP using abiotic and soil factors (Vogt et al. 1996c) showed the utility of collating and synthesizing the existing literature. When using climatic variables to predict fine-root dynamics in forest ecosystems, no general relationship was produced that was applicable to all ecosystems. The effects of climatic variables on root growth varied by tree behavior (i.e., evergreen vs. deciduous) and by tree species (Figures 4-2 and 4-3). Also climatic variables appeared to predict the biomass of fine roots maintained in a evergreen, but not the amount of root biomass produced annually (Figures 4-2A and 4-4B). This raises exciting questions as to why climatic variables are more effective in predicting the mass of roots maintained by a tree than the amount produced annually. Interestingly, changes in root NPP were explained by nutrient variables (i.e., litterfall N or P contents), but nutrient indices were unable to explain variations in fine-root biomass (Vogt et al. 1996c). If these patterns are corroborated with further research, any disturbance that alters climatic variables would be expected to feed back to change root biomass but not annual production, while changes in ecosystem nutrient pools would be expected to affect fine-root NPP but not biomass relationships in forests. Once these patterns have been found using these large-scale data analyses, it would be important to initiate studies to understand the underlying mechanisms explaining these patterns.

Many of the other patterns identified with this large data set appear to have logical explanations and help to identify potential functional groups that should be examined for their utility. For example, broad-scale grouping of data by the deciduous and evergreen behavior of a tree species had annual precipitation explaining 78% of the variation in fine-root biomass in the evergreen forests but none of the variation in deciduous forests (see Figure 4-2). This would suggest that deciduous species may be somewhat decoupled from precipitation inputs and have more sophisticated physiological adaptations for acquiring water that compensate for the varying abilities of the ecosystem to supply water. If this pattern is maintained with further research, it would suggest that precipitation may not be as useful a parameter to measure or to model when studying global climate change in some deciduous dominated ecosystems.

Even within the evergreen-dominated forests, the ability to predict fine-root biomass from precipitation was not a correlation that transferred

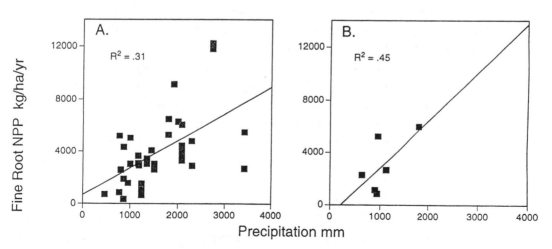

FIGURE 4-4   The relationship between fine-root productivity and precipitation in evergreen forests (A) and oak (B) forests.

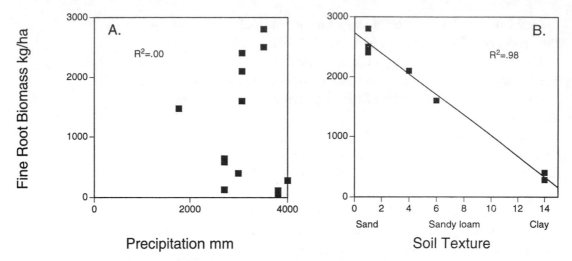

FIGURE 4-5   The relationship between fine-root biomass and precipitation (A), and fine-root biomass and soil texture (B) in tropical broadleaf forest.

generally across all evergreen forests. For example, evergreen forests dominated by coniferous species did have a strong relationship between annual precipitation and fine-root biomass, but this relationship did not exist in tropical evergreen forests dominated by broadleaf species (Figures 4-2 and 4-5). In contrast, tropical broadleaf evergreen forests had a significant correlation between soil texture and fine-root biomass, suggesting a stronger relationship between the water-holding capacity of the site and root growth (see Figure 4-5).

Again, coniferous evergreen forests had no relationship between soil texture and fine-root biomass (Figure 4-6). Regrouping information at the species level showed that even though large-scale groupings displayed a relationship between precipitation and fine-root biomass for evergreen species, these relationships were not carried over to some of the species groups (i.e., pines) (see Figures 4-3 and 4-4). This lack of a relationship with pines may be due to the problems of lumping all pine species into one group. Recent research has suggested that pines need to be examined separately by three subgroups (i.e., cold temperate, warm temperate plus subtropics, and tropics) because they respond very differently in each zone (Gower et al. 1995). Some of these differences result from the fact that pines have determinate growth in the cold temperate zone (root growth occurring from stored carbohydrates), indeterminate growth in the warm temperate and subtropics (root growth occurring at any time and from current photosynthate), while those in the tropics have a mix of both growth patterns. The pines growing in the cold temperate zone also exhibit variations in their growth patterns, similar to other conifers, by responding significantly to changes in site resource availability. In contrast, the warm temperate and subtropical pines do not respond to changing abiotic resource availability (Gower et al. 1995). Unfortunately the large database used for these analyses does not have a sufficient number of data points within each of the three zones to be able to re-examine these relationships. This highlights the importance of analyzing information at the scale that best reflects how the system is responding, making sure that the variable being used to predict activities is functional at that scale and being careful about how information collected at one scale is generalized to represent broad-scale group responses at different or multiple scales.

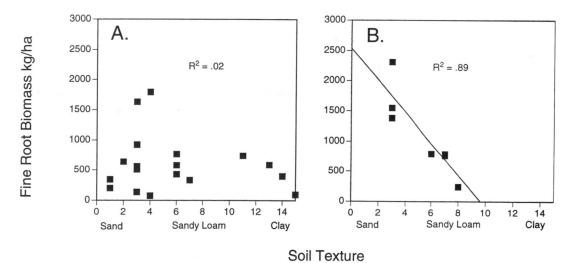

FIGURE 4-6   The relationship between fine-root biomass and soil texture for cold temperate needleleaf evergreen forest (A) and cold temperate broadleaf deciduous forest (B).

4.2.8   Ecological Risk Assessment

Attributes of Ecological Risk Assessment

Ecological risk assessment is one approach used for estimating the probability of adverse change in natural systems and may hold an important place in ecosystem management. Risk assessment has been called upon by policymakers at the federal level as the fundamental tool for changing reactive approaches to environmental policy toward more "integrative" and proactive approaches to regulatory activity (EPA 1990). At the core of this call is the desire to prioritize environmental problems and to allocate regulatory efforts in proportion to the risk posed by various hazards. Risk assessment is seen as an integral tool in the move toward an integrative policy. It could be equally important in ecosystem management, as it provides an estimate of the ecological side of the trade-offs involved in making decisions about how to balance short-term goals with long-term maintenance of ecosystems. Risk assessment provides a framework with which we can:

1. Talk about a diverse range of ecological problems impacting a management unit (from the risks of losing certain species, to the risks associated with clear-cutting, to the risks associated with air pollutants)

2. Choose the specific ecological characteristics or functions of concern, and indicators that can be measured to reflect changes in these endpoints

3. Compare the relative risk associated with each management activity and concentrate on controlling the most "risky" activities

4. Connect science with policy and management decisions by providing a clear, quantitative characterization of what are thought to be the impacts of management activities and the uncertainty surrounding those estimates.

Risk assessment could play an important role in ecosystem management because it offers a formal way to organize information related to decisionmaking. Also, it includes analysis of uncertainty, which is absent from some forms of impact assessment. Uncertainty analysis is essential for putting scientific information in the context of how much *is not known* and for elucidating the need for further research in specific areas. With respect to ecosystem management, ecological risk assessment can be seen as an effort to estimate the probability of moving from a desired ecological state (one in which desired system outputs and characteristics are present) to a less desired or unacceptable state (in which those amenities are no longer provided or are at much lower levels).

Like ecosystem management, ecological risk assessment evolved from recognition that the piecemeal approach to ecological problems is inadequate. That recognition culminated in 1987 when EPA stood back and evaluated its own performance related to environmental regulatory effort and the state of the environment. In a commendable self-evaluation, EPA noted the persistence of ecological degradation despite extensive regulatory expenditures. EPA's "Unfinished Business" report (EPA 1987) proclaimed risk assessment as the fundamental tool that will provide the basis for comparing the relative risk associated with various ecological hazards and directing regulatory efforts to the most risky ones.

Essentially, ecological risk assessment involves efforts to get a science-based, quantitative assessment of the likelihood of adverse change from a desired state to one less so as a result of human activity, and to use that information to guide management decisions. As such, ecological risk assessment is fundamentally similar to any other type of ecological impact assessment. Risk assessment is sometimes thought of as a highly technical assessment of specific impacts, which it sometimes is. However, the assessment of risks occurs in a less formal way every time the costs and benefits of our management actions are weighed. Formal risk assessment simply provides an organized framework for putting ecological knowledge together in a way that will inform management-related decisions. Risk assessment is distinguished by the *explicit recognition of uncertainty* (Bartell et al. 1992) in estimates of adverse change. Knowing about uncertainty helps in (1) guiding future research needed to revise estimates and (2) putting the scientific information in context so that managers are aware of the reliability of the information on which decisions are made.

If used carefully, risk assessment can provide scientifically defensible information to decisionmakers in a straightforward, unambiguous way (Bartell et al. 1992). This information can then be used to prioritize the portfolio of issues in environmental protection (Silbergeld 1993) or to prioritize management activities in an ecosystem impacted by a number of potential hazards (i.e., point and nonpoint sources of pollution, habitat modification, and global climate change).

Risk assessment involves looking for indications of ecological change against a background of intrinsic variability associated with ecosystem dynamics. This requires baseline monitoring and evaluation of the best potential indicators of adverse change. Risk assessment is unique in that it recognizes the need to (1) identify explicitly and carefully the processes and variables of interest to people (in effect, defining operationally what constitutes both meaningful and measurable environmental change) and (2) define a *baseline, ground state, base-state* trajectory or acceptable state against which to measure change. To do this, endpoints or ecological attributes of concern are defined at the outset of a risk assessment. Analogous to clearly stated scientific hypotheses (Suter 1988), endpoints serve to direct the risk assessment framework in a manner similar to hypotheses that control the design of the scientific experiment.

Endpoints define the scope and direction of risk assessments, because all else in the risk assessment is predicated on the defined endpoints. Endpoints for a given ecosystem will be a mix of those considered important to humans and those that are important to the maintenance of the ecosystem in a state considered acceptable by humans. Examples include maintenance of harvestable timber (socially important) and the proportion of energy cycling in the detrital food chain (important in energy and nutrient turnover, and so integral to the maintenance of the system in its current state). From the choice of *assessment endpoints* follows selection of *measurement endpoints,* or variables measured in the field, whose value reflects the current state of the system. In choosing endpoints, the choices are made obvious about what is considered to be important in the system. By selecting both socially and ecologically important endpoints, it is possible to concentrate on both the amenities people want from the system and the ecological functions that allow the system to persist over time.

Tools Used in Determining Ecological Risk

Lab Toxicity Tests with Standard Organisms

The best-known risk assessments are those developed to detect possible toxicity to humans from different chemicals being introduced into the environment. The primary concern was human health rather than ecosystem health. This was typically done by comparing observations on two groups of animals raised under precisely the same conditions: One group was divided into several smaller groups, with each being exposed to a different level of the substance being tested, while the second group was used as the control group not exposed to the substance (Barlow et al. 1992). In this way, adverse effects can be related to the dose of a chemical and a no observed effects level could be established. These results from laboratory tests with animals are then extrapolated to derive "acceptable" levels for humans after adding an arbitrarily derived "safety factor". These tests are then used to guide the regulation of chemicals with which humans potentially could come in contact.

These standard laboratory toxicity tests serve an important purpose. They provide information on *mechanisms* by which toxic chemicals exert their adverse effects. *Mechanism* refers to the processes that give rise to particular phenomenon; for instance, the changes in the functions of the body that are produced by a given toxicant and lead to its clinical effects (Barlow et al. 1992) or the specific changes in trophic relationships that result from differential susceptibility among species exposed to pollutants in the environment. This is similar to the use of seedlings to identify the possible mechanisms by which an ecosystem state is changing that was discussed earlier in this chapter. Knowledge of mechanisms is important for a number of reasons. Barlow et al. (1992), in discussing human health risks, noted that knowledge of mechanisms is important in that it may lead to early diagnosis and treatment, the possibility of in vitro testing to reduce tests with animals, and may give guidance on the extent to which the response of one organism may be extrapolated to others. With a scientific understanding of mechanisms, the basis is laid for the deployment of a variety of scientific methods to estimate risks because the modes of action are known. With knowledge of mechanisms, solutions can be tailored to the problem providing the manager the greatest "leverage" or desired results per unit of effort. Knowledge of mechanisms is similarly important in ecological risks. Such knowledge may suggest methods for early detection of risks, give guidance on the extent to which effects seen in one species can be extrapolated to other species, and point decisionmakers to the most effective risk-reduction strategies.

Standardized toxicity tests are designed to provide data about the types of adverse effects produced by chemicals as related to their dosages. These tests also shed light on the relative risk posed by various chemicals because genetically similar organisms are used to test the chemicals under standard conditions. The use of genetically similar organisms is the same idea presented in Section 4.2.2 for the use of clonal plants as a standard material to index ecological functions in ecosystems. There is a considerable degree of agreement internationally about toxicity testing methods. These methods have not changed substantially in recent years, suggesting that either (1) the methods for generating reliable data in these tests are well established or (2) political reasons facilitated the acceptance of standardized tests (Barlow et al. 1992). Barlow et al. (1992) suggested that the European Economic Community facilitated the standardization of methods as a means of reducing barriers to international trade.

Laboratory tests provide two factors that allow statistical power in the associated tests: control over confounding factors that cannot be controlled in the field and adequate sample size in the study populations for statistical power (Bartell et al. 1992, Barlow et al. 1992). Lastly, laboratory tests are necessary for assessing human health risks because it is unethical to experiment with toxic chemicals on humans or to release hazardous agents into the environment. Laboratory tests thus serve as model systems for testing effects on systems that cannot be manipulated. A similar analogy exists in the ecological literature, where studies have been conducted under very controlled conditions in the laboratory (see Chapter 3).

The use of *standard* lab organisms to evaluate acute and chronic toxicity in laboratory settings is similar to the proposed use of *internal standards* in the field (discussed in section 4.2.2). The use of standardized indicators, whether genetically similar organisms or chemically identical leaf litter bags for decomposition rate studies, can reduce some of the variability in field conditions that often clouds interpretation of observed effects. Because the responses or rates of interest are well described (such as expected rate of decomposition of leaf litter in a 50-year-old temperate hardwood stand in the White Mountains), these tests provide unambiguous information. The primary difference is that laboratory organisms address human and animal health issues in response to one or a few selected stressors, while internal standards address ecosystem condition issues by integrating complex interactions occurring in the field.

Some debate has arisen over the use of lab toxicity tests for generating estimates of risk to organisms other than those in the laboratory. Debate stems from the uncertainty regarding the extrapolation of information from the laboratory to the populations of interest. Linear extrapolation from laboratory data to field conditions will seldom be accurate (Bartell et al. 1992) because controlled laboratory conditions by definition cannot mimic ecosystem dynamics. Most tests for ecotoxicity are acute tests of the LD50 type (Barlow et al. 1992). Limits of these tests for estimating ecological risks have been well documented (Levin et al. 1986, Bartell et al. 1992, Barlow et al. 1992, etc.). Lab tests cannot account for differential susceptibility among species, population interactions among these differentially susceptible species, and adaptive or avoidance behaviors among exposed organisms, all of which greatly affect the eventual effects of chemicals on the biota. Further, laboratory settings cannot duplicate the biological, geological, and chemical cycles that dictate the effects of anthropogenic stressors and so are less appropriate for looking at nonchemical risks such as habitat modification. These types of risks require different types of analyses, including modeling and field testing and experimental manipulations in the field.

Models

Models are an integral part of ecological risk analyses. They are necessary for ecological impact assessment when attempting to predict the relationships between environmental hazards or perturbations to the system and resultant system changes. In ecosystem-based impact assessment, it is difficult to expose a whole system to potential hazards or to have unimpacted control systems because assessment at the ecosystem level is extremely resource and time intensive.

Because models are by definition *simplified* or *abstract* representations of complex systems, they force us to search for the most important components and processes that dictate ecosystem structure, function, and response to perturbations. In the model, those components are represented in the simplest way possible while modeling major components or feedbacks of the system.

Models are mathematical descriptions of defined processes. In essence, building a model makes us get down to basics. An ecological model represents what are believed to be the nuts and bolts of a system and is a useful way to organize information in a format so that a tremendous amount of detail does not confuse us. When there exists uncertainty about what are the most important structures and functions in a system, a model represents the best guess or hypotheses about the way the system is thought to function based on comparisons to similar types of systems. Models are predictive tools that allow us to explore the behavior of complicated, poorly understood systems by starting with parts of the system that are well understood and gradually building complexity. Models are tools that allow for a formalization of our knowledge so that the implications of the defined assumptions can be understood (whether correct or not). This does not mean that all the results of model output are believed, but it is a tool to help identify the implications of certain activities based on the current knowledge base.

Models allow us to think about the what if relationships between a large number of variables, more than what can be kept track of in our heads. Without models, it would be extremely difficult to isolate the possible impact of a change in one of a multitude of variables on ecosystem function. Models have become a critical tool to comprehend environmental problems (Beck et al. 1993) because the increasing complexity of the problems precludes our thinking about them all at once. Models might help us to think about the cumulative effects of a range of diverse stressors, including point and nonpoint sources of pollution, selective removal of species, habitat modification, loss of biological diversity, and stratospheric ozone depletion. Model building often entails a simplified representation of system structure (including the positive and negative feedbacks in that system) that are very useful for exploring both the direction and magnitude of change in system processes. Combining information on system structure with that on feedback cycles and their relative strengths, managers can figure out where their efforts to affect change have most potential, and where the best results come not from massive efforts but from small, well-focused actions (Senge 1990). For aquatic ecosystems, sophisticated models incorporating population and ecosystem level interactions have been developed for assessing their risk to toxic chemicals (Bartell et al. 1992). These models are innovative in that they attempt to translate existing toxicity data into estimates of risk in ecosystems by incorporating differential susceptibility among species, population dynamics, environmental factors (such as sunlight availability), sublethal effects, and uncertainty analysis.

Modeling often appears to be a highly technical and less-than-relevant exercise that does not necessarily relate well to management problems. But there are many modeling approaches and, while models can become so complex as to be impenetrable by nonmodelers, simpler models are understandable and do offer valuable insights. Models, whether conceptual or mathematical, can be extremely helpful in ecosystem management. In constructing a simple model, the manager is forced to eliminate nonessentials and to select those structural relationships that are thought to be important so that the driving and constraining variables in the ecosystem are captured and major functions are described. When attempts are being made to simultaneously assess the impacts of management activities and to make decisions in the context of changing information, this model building activity can be invaluable.

A concept captured in models that is integral to ecosystem function is *feedback*. Both positive and negative feedback occur in ecosystems. Positive feedbacks occur when an increase in some stock or flow causes a change in a connected component, which then further increases the original stock or flow. Negative feedback cycles embody the opposite effect: An increase in a component stock or flow changes other components, which then cause a decrease in the original stock or flow. An example of a positive feedback in ecosystems can be found during the early stages of high atmospheric inputs of N into a forest ecosystem. During this stage, the higher atmospheric N inputs increase the decompostion rate of litter, which increases the mineralization rate of Ni from the litter, thereby increasing ecosystem N availability. The later stages of high atmospheric N input into a forest ecosystem produces a negative feedback. For example, instead of higher N availabilities increasing NPP in this ecosystem, nitrate leaches base cations from the soil, resulting in lower NPP values because tree growth is limited by the loss of these other nutrients. Another example of positive feedback is the spread of disease in a forest. As the number of trees infected by a root rot fungus increases, the faster will be the rate of spread of that disease through the forest.

It is important that models be iteratively evaluated using field data. In this way, the knowledge of the system in question can be expanded. When the models do not explain observed behavior, there is a motivation to think through alternative reasons for the discrepancy. Field monitoring combined with mechanistic modeling provides perhaps the most direct approach to understanding ecological systems and estimating risks. It is also important to realize that many of the traditional regression-based models (e.g., allometric equations predicting the amount of tree biomass in leaves, branches, stems, and coarse roots) used by forest managers or ecologists were "derived from historical observations, without mechanistic basis, [and] are unable to predict growth responses to different climatic conditions" (Landsberg et al. 1995).

Conceptual models are very commonly used in ecology and management for making nonquantified predictions about the behavior of system components. Conceptual models guide our daily actions. They are built unconsciously from our beliefs about cause-and-effect relationships between variables in a system.

Several different types of models are potentially very useful for ecosystem analysis and management (Jørgensen et al. 1996). They can be generally grouped as being either mechanistic (based on a concept of how a system works) or empirical (based on the measurements of a defined system, e.g., allometric regression of tree diameter at breast height predicting tree biomass). A mechanistic model is process based and more predictive (but also more biased) than an empirical model, which works as long as the conditions of data collection are satisfied.

There are many different types or groups of models used by population, community, and ecosystem ecologists (see Jørgensen et al. 1996 for detailed review of many models). In general, one can group the models used in ecosystem analyses into four types based on their scale of analysis: (1) physiological; (2) population, succession; (3) ecosystem, production; and (4) regional, global. The first three model types are quite useful for testing hypotheses and organizing and focusing research prior to implementing a large project (Landsberg et al. 1995). The four model groupings focus on addressing very different types of management questions and have very different data needs. Following are summarized some of the data needs and kinds of questions being addressed with the different models (adapted from Kiester et al. 1990, Landsberg et al. 1995).

1. *Stand-level, Process-based Physiological models*

   (a) Models have been constructed to predict photosynthesis, respiration, water use, and canopy energy balance. Several of these incorporate information from the biochemical level. Some of these models are empirically based.

(b) These models are examining questions related to the physiological and abiotic (water and nutrient) controls on growth and production of trees. They have been used to examine the effects of pollutants on tree growth or the response of trees to interacting stresses (Dixon et al. 1990).

(c) Many of these models can be specific for a part of the ecosystem (i.e., decomposition, root-soil systems, etc.; Kiester et al. 1990) and do not necessarily require modeling all processes in an ecosystem.

(d) The data needs of these models will vary based on the scale of analysis, with the models examining smaller scale processes needing more input variables. Some variables are rarely included due to our inadequate understanding of the mechanisms affecting the response variable, which may vary significantly within one tree (e.g., respiration; Sprugel et al. 1994).

(e) Some of the following parameters are incorporated at different levels of complexity: estimations of photosynthetic rates and the amount of carbon annual fixed by a plant; respiration rates; allocation of carbon to leaves, stems, roots, stored carbohydrates, and secondary defensive chemicals; and nutrient uptake by roots.

(f) Some of the input variables may include:

> amount of total solar radiation and radiation absorbed photosynthetically
>
> active radiation
>
> leaf areas
>
> maximum and minimum air temperature
>
> humidity
>
> precipitation
>
> wind speed
>
> nutrient pools and fluxes (especially N, P, K, Ca)
>
> nutrient mineralization rates from litter
>
> soil morphology and topography
>
> soil water potentials, soil nutrients
>
> soil physical + chemical + biological properties

rates of photosynthesis, respiration, transpiration

carbon allocation to leaves, foliage, stems, storage carbohydrates, and secondary defense chemicals

litterfall, grazing by consumers

2. *Population, Succession models*

(a) Population-based models were constructed to examine tree growth, population dynamics, competition, and birth-growth-death cycles determined from data on stand structure, spacing, and mortality (Shugart 1984). Most of these are empirically based and consist of a series of multiple regressions.

(b) Succession models are important to understand what drives the changes in species composition or the loss of species over time in forests (Shugart 1987).

(c) These models require information by species on:

stocking or density of species

dominant species

tree diameters by species

litterfall

tree mortality

seed production

seedling establishment

other physiological characteristics, such as shade tolerance of each species

3. *Ecosystem, Production models*

(a) These models were constructed to examine the flow of carbon between plants, the soil system, and the atmosphere. Components of ecosystems are analyzed by lumping information (primary producers are divided into functional categories of above-ground biomass, leaf biomass, coarse-root biomass, fine-root biomass, etc.).

(b) These models are mechanistically based and will simulate how an ecosystem responds to abiotic driving variables.

(c) Typical information collected on this type of model is forest growth and the impact of climate change or other disturbances on their growth.

(d) These models are bounded in space and in time.

(e) These models require information on:

> total stocking or density of all species
>
> dominant species
>
> tree diameters
>
> litterfall and tree mortality
>
> daily maximum and minimum air temperatures
>
> precipitation
>
> plant nutrient pools
>
> soil solution nutrient contents
>
> soil depth
>
> soil texture
>
> soil organic matter
>
> soil water
>
> soil nutrient contents and fluxes, nutrient leaching
>
> litter decomposition rates
>
> solar radiation
>
> herbivory rates

4. *Regional, Global models*

5. These models were constructed to examine ecosystem and vegetation distribution responses to climate change. These models have been used to predict changes in LAI, NPP, site water balance and runoff, and biome boundaries due to climate change.

6. The geographically based models assume steady-state conditions, aggregate information so that individual site data are not relevant, only identify the potential vegetation because they examine vegetation at a high level of classification (i.e., biomes), and do not consider which species comprise the community to be relevant. It is the vegetative community as a whole that is analyzed. These models are

mainly driven by climatic variables such as temperature and site water balance.

7. These models use maps of global distributions of vegetative biomass, NPP, litterfall, carbon storage and sequestering in different parts of the ecosystem (i.e., soil, forest floor, or tree biomass), and tree core data (Stahle et al. 1988, Shortle et al. 1995) and relate these to climatic variables.

8. These models use very broad classification systems, do a lot of regression averaging, and have a lot of assumptions. These models are geographically based.

9. The type of data needing to be collected for these models include large-scale averages for the following at the biome level:
   temperature
   site water balance
   leaf area index
   vegetative biomass
   NPP
   soil organic matter
   forest floor accumulations

These models can be extremely useful in showing what is not known about ecosystems. Drawing boxes representing major system components and lines connecting them can be a major challenge. If it is difficult to draw a model of the major stocks and flows in a system, then the system is not well understood. Also, if the boxes and their connections cannot be described in detail (a black box model), then the mechanisms underlying ecosystem functions are not well understood either.

Thus engaging in model building can illustrate the weak spots in the knowledge of the system. Even black box models have utility because they force us to think hard about the system at hand and how it behaves; they are models by definition devoid of much detail. If they provide guidance as to how the system works, black box models may be better than more detailed ones in that they do not require as much effort to assemble.

4.2.9   Examples of Ecological Assessment Frameworks

Environmental Monitoring and Assessment Program (EMAP)

The Environmental Monitoring and Assessment Program (EMAP) of the Environmental Protection Agency (EPA) is designed to assess the nation-wide distribution of ecological resources in the United States and to assess trends in their condition (EPA 1992a). This is an integral tool in ecological risk analysis, as adverse change cannot be detected if *normal* status and trends are not known. In addition, large-scale monitoring allows for trends to be observed, such as loss of wetland or loss of arable land, in aggregate. This is a potentially powerful tool, because the incremental loss of resources such as wetlands can appear insignificant at the local level, whereas the overall trend is more easily recognized as a large problem.

The EMAP program represents widespread recognition that long-term monitoring of environmental conditions is important. This large, expensive, controversial program involves probability-based random sampling from a grid covering the entire continental United States (EPA 1992a). The choice of variables and parameters to monitor has not been completely worked out. A review of their potential sampling schemes by a National Research Council review committee noted the emphasis of EMAP on probability-based spatial sampling at the expense of probability-based temporal sampling (detecting trends; NRC 1992). The NRC (1992) report suggested that EMAP should devote at least as much effort to developing the appropriate indicators of ecological change, management of data, and financial support of the program as it has to spatial sampling questions.

EMAP has been developing four types of indicators:

1. *Response indicators* to represent the condition of ecological resources at any level of biological organization

2. *Exposure indicators* to indicate the occurrence or magnitude of contact of an ecological resource with physical, biological, or chemical stress in the system

3. *Habitat indicators* to characterize conditions necessary to support an organism, community, or ecosystem (for instance, availability of snags, vegetation type, and spatial patterns)

4. *Stressor indicators* to reflect changes in ecological resources associated with natural processes or human activities

These indicators should be developed in the context of clearly stated hypotheses or questions about anthropogenically induced changes in the status and trends of natural resources.

EPA's developing framework for ecological risk assessment begins by identifying five organizing variables that define the scope of a risk assessment (EPA 1992a):

1. Ecosystem type

2. Stress type

3. Spatial scale

4. Temporal scale

5. Level of biological organization

Many of these organizing variables have been discussed earlier in this book and are a good way to organize the information needs for assessing change in ecological systems. This organizing approach can then be used as a framework for identifying the information needs particular for each ecosystem and the stress type so that the risk assessment design can be tailored for each site. These five factors help to bound the risk assessment problem. They also facilitate people thinking about the scope of the problem and thus how management should proceed.

EPA further proposes that the traditional structural classification of ecosystems (e.g., deciduous forest, freshwater wetlands, etc.) could be supplemented with functional classifications (e.g., energy and nutrient cycling, decomposition) to facilitate the *coupling of stress modes of action to the nature of the ecological response and recovery* (EPA 1992a). This change in approach is crucial if the ecological risk assessment is to have acceptance among ecologists and scientists. Only by starting with an attempt to understand the biogeochemical processes that mediate the effects of stressors and the *mechanisms* by which measurable adverse effects arise can effective mitigation of these adverse effects occur. In addition, a process-based analysis of risks allows for a more clear recording of the full range of interactions occurring among different stressors and different systems. This type of approach gets beyond the assumption that stresses and the ecological effects do not interact and that risks, therefore, can be segregated among ecosystem types. In other words, risks to estuarine systems are not independent of the adjacent terrestrial areas.

The "Reducing Risk" report (EPA 1992b) noted several difficulties in estimating the response and recovery of ecological systems to anthropogenic disturbances:

1. *There is no single method that can be applied to all ecosystems for characterizing their response to stress or disturbance.* Different ecosystems or parts of ecosystems (e.g., primary producers, decomposers, consumers, etc.) will respond differently to a given stressor. The wide array of ecosystem types—forest, wetland, estuary, desert—will respond and recover in a multiple of pathways to a given stressor, and each will respond differently to each stressor. For example, a month with no rain may cause extensive ecological changes in a rain forest but would not be a problem in a desert.

2. *Different stressors will cause distinct effects.* The effects are further modified by the timing and magnitude of the stress and the frequency with which the stressor in question has been encountered by the exposed systems.

3. Unlike the human organism (equated as being tightly organized and highly homeostatic), *ecosystems appear to be loosely connected, with many different mechanisms for positive and negative feedbacks* to occur and relatively low levels of homeostasis (EPA 1992a). Because of the dynamic nature of ecosystems and because dynamic equilibrium states are difficult to define and measure, risk assessment involves tracking or identifying changes that are co-occurring in addition to the normal changes in the ecosystem. This is difficult to implement because in many systems the normal ranges of system components are not known and the appropriate state variables that best describe the system are still being determined (Keddy et al. 1993).

4. *There is no universal, simple measure of ecosystem health* (EPA 1992ab) *or of a good ecosystem state* that is analogous to measures of human health so that tools developed for assessing human health could be more easily applied to ecological systems. A basic problem that is central to all of this is that risk is only a risk when it is acknowledged by people to fit that category. Until people judge an ecological change to be adverse, unwanted, or risky, changes in ecosystems are ignored due

to the limited resources that can be applied to ecological problems. Lack of consensus on what constitutes adverse ecological change can stall efforts to assess ecological risk. This is the crux of the science/values mix that makes risk assessment so intriguing and at the same time somewhat problematic.

National Acid Precipitation Assessment Program (NAPAP)

One of the larger efforts to evaluate ecological impacts associated with an anthropogenic stressor was the National Acid Precipitation Assessment program. This program was established in 1980 with the National Acid Precipitation Act and was funded for 10 years. The research output from this program has been summarized in 27 State-of-Science and Technology reports. Seven of these reports specifically focused on surface water changes due to acid rain, while two specifically focused on forest health and productivity (NAPAP 1993). Most of the effort by NAPAP was placed on examining acid rain effects on human health and materials degradation, estimating emissions and air quality, establishing regional deposition models, and addressing the technology and costs for controlling emissions on a nationwide scale (NAPAP 1993). Characterizing the impact of pollutants on forests was a small part of the overall NAPAP mandate.

The 10-year effort to characterize the impacts of pollutants on forest ecosystems illustrates the difficulty of quantifying the relative impact of one stressor (air pollutants) on the overall forest conditions when these forests were being exposed to a variety of stressors (including drought). When the research was initiated in forests, much of the effort was geared to beginning to understand the mechanisms behind acid rain effects on trees. In 1980, an understanding of the major ecosystems in the United States and how they might be adversely affected by acid rain was quite poor. Therefore, it was difficult to design mechanistically based experiments during the early stages of this program.

Because of the necessity of examining so many ecosystems and the perceived need to be able to identify the specific effects of one chemical on plant growth, most of the research was conducted under very controlled conditions. It was felt that this information could be readily transferred using models to address questions on the health of mature forests throughout the different regions of the United States. This research therefore had two strong fronts: (1) detailed controlled laboratory studies to elucidate the biochemical mechanisms by which pollutants cause damage and (2) model development (NAPAP 1993). The laboratory studies emphasized the utilization of seedlings to detect species differences in susceptibility to acid rain. Some field studies (e.g., red spruce forests in the Appalachians, conifer forests in California) were established using either the gradient approach, or individual branches on large trees were used as the unit of study (NAPAP 1993). These branch level studies assumed that the branch would react and function in a manner similar to an individual tree.

The seedling studies were effective in identifying the mechanisms of pollutant action on leaf biochemical structure and function (McLaughlin et al. 1990, Eagar and Adams 1992). They also showed the wide range of sensitivity that existed between species and between strains of a given species to pollutants (Barnard et al. 1990). However, the results from these studies cannot be directly extrapolated to the field or to parameterize models because seedlings' responses to pollutants are quite different from trees growing in a forest (see Section 3.4.1). The mechanistic understanding of plant responses to a pollutant or a combination of pollutants cannot be obtained from field studies and has to be obtained from seedling studies under controlled conditions. The field is a confounding environment of many factors interacting to modify the plant response or lack of response to a pollutant. Much of the NAPAP research identified the complicating effects of several factors modifying plant responses to pollutants. Some of these interacting factors are variations in atmospheric conditions (temperature, light levels, relative humidity, CO_2 levels) and edaphic conditions (moisture and nutrient availability), as well as the episodic nature of pollutant exposures, plant growth stages, and genetic characteristics of plants.

The NAPAP studies in forests determined that there was no conclusive evidence of a general, widespread decline of forests in the United States due to acid deposition (NAPAP 1993). Acid deposition was only implicated as being one of the factors causing the mortality and decline of red spruce forests located at a high elevation in the northeastern United States (NAPAP 1993). The reasons for the recorded decline of sugar maples in northeastern United States and parts of eastern Canada could not be specifically identified. This analysis for pollutant effects on sugar maple consisted of attempting to correlate wet sulphate deposition and the location of sugar maples with unhealthy crowns, but no correlation was obtained between these two variables. Because of the way the studies were designed for the NAPAP studies, results were rather inconclusive and it was difficult to really state whether the forest health was deteriorating in the United States.

The forest field studies illustrate the difficulty of identifying pollutant *thresholds* or exposure values below which there is no expected effect. For example, under field conditions, plants do not respond to a single pollutant exposure threshold for but, rather, respond to a range of values (Krupa and Kickert 1987). Any single cutoff value for biological effects is then context dependent and will change over time so that the concept of a threshold may not make biological sense in this case. The definition of a threshold may be too rigid if it is presented as a single value for safe exposure because plant–pollutant relationships are so complex. This makes it more difficult for management because a single, scientifically defensible "safe pollutant exposure" value would be easier for regulatory and management agencies to follow.

The NAPAP program was reauthorized in 1990 as part of the Clean Air Act Amendments. The four primary mandates of the current program are to "(1) develop a comprehensive, coordinated research and monitoring program and unified budget to reduce major uncertainties and provide data and information for assessment activities, (2) evaluate the costs, benefits, status, and effectiveness of the Acid Deposition Control Program, Title IV of the Clean Air Act Amendments, (3) determine the reduction in deposition rates that must be achieved in order to prevent adverse ecological effects, and (4) report to Congress on the results of its investigations and analyses" (NAPAP 1993). As part of this new phase, NAPAP, in conjunction with the Forest Service, state forestry agencies, and the Environmental Protection Agency, have initiated a Forest Health Monitoring program encompassing 850 plots scattered throughout the United States. Special emphasis was placed on monitoring red spruce and sugar maple, the ecosystems identified earlier as having higher mortality rates than normal during the first phase of NAPAP (NAPAP 1993).

In 1992, several articles were written to evaluate the strengths and weaknesses of the original NAPAP program as a mechanism of identifying how large programs can be effectively managed over the longer time scales needed to study and assess environmental change (Levin 1992). The justification for this program was that it was "to provide critical data, analyses, and assessments to inform the debate on the reauthorization of the Clean Air Act"; however, the key bills were passed before the NAPAP reports were ever published (Levin 1992). This meant that the NAPAP reports were not used to form policy. This, of course, is historically not unusual, considering how ecology, management, and policy have not tracked one another and science has not been used to develop management guidelines (see Chapter 2). Much of the criticism of the NAPAP program was related to how the program did a poor job of interfacing between public policy and science (Levin 1992, Russell 1992) and the fact that there was too much politics involved in controlling the science being conducted (Schindler 1992).

Rangeland Health

A recent National Research Council report (NRC 1994) on rangeland management illustrates the state of scientific inquiry in ecological assessment at the regional or ecosystem level. The NRC report discussed the need for new methods for classifying, inventorying, and monitoring rangeland health. The report was motivated by debate as to whether rangelands should be managed differently and whether current practices are causing their degradation. The debate over these issues has been intense. Scientists disagree about the proper interpretation of past and current range conditions and what are the best methods for assessing their ecological status. Scientists from the three agencies involved in managing rangelands—the Soil Conservation Service, Forest Service, and Bureau of Land Management—have examined the same data and have come up with different conclusions about the state of U.S. rangelands and the value of the available data. Debate about the data was followed by disagreements about whether management practices are improving, degrading, or sustaining rangelands. In cases like this, an alternative approach to index rangelands is needed that takes the argument away from people's personal perception of the condition of the rangelands. An approach similar to what was presented in Section 4.2.1 is needed that will give an unbiased determination as to where a system exists currently.

The rangeland health report outlined a set of indicators that the committee felt had utility for assessing the state of rangeland ecosystems, with a focus on the functional properties of the systems (NRC 1994). The authors selected indicators to reflect soil stability and watershed function, nutrient cycling and energy flow, and "recovery mechanisms." The chosen indicators (shown in Tables 4-1 to 4-3) seem at first obvious, but it is revealing to look at them a little more closely to see what was identified in each category as the assessment tools. This committee had the difficult task of trying to come to grips with questions on the type of information on rangeland ecosystems which would best inform policymakers, ranchers, and interested individuals. They had to struggle to develop the assessment criteria that were finally presented because the individuals who are making these conclusions disagreed about how to make inferences. The study illustrates nicely the lack of consensus about good science related to ecological change and the escalation of debate that can ensue when the interpretation of uncertain scientific information has implications for management and those affected by management.

The rangeland health report defined three conditions and described what attributes would be found for several indicators within each condition (NRC 1994). Ecosystems were grouped into three categories of healthy, at risk, and unhealthy (Tables 4-1 to 4-3). However, there is an element of judgment involved with many of the categories that makes decisions on which group to classify a particular rangeland system quite subjective. Having three health categories only does simplify the choices that have to be made by rangeland managers, but this fails to consider that there exists a range of responses complicating the task of assigning a system under one category. Unless there is a good way of indexing a site as to what its condition should be, it will be extremely difficult to conclusively state that a system is really degrading and not responding to some longer term natural climatic cycles particular to that system. If good historical data exist for a given site, then cause–effect relationships are easier to document. Still, in most cases it is quite a challenge to distinguish degradation from natural variation.

The different qualities attributed to each indicator that would shift the system to a different status have to be mechanistically based. However, there appears to be a strong subjective element for determining into which category a system should be classified in the rangeland health report. It will be difficult to determine what the cutoff point should be for putting a system in the different categories. This leaves it up to the manager to make that decision as to where to classify a system. However, this will only propagate the existing controversy on how to classify rangelands as to their health because there will be many definitions of what each indicator grouping means. For example, examining changes in soil stability by measuring changes in the presence of a soil horizon (i.e., A horizon) may not be a good diagnostic tool because it will be difficult to state in an unbiased manner when the A horizon has finally become distributed in a fragmented manner (see Table 4-1). As part of assessing the "recovery mechanisms" in rangelands, emphasis is placed on determining age class distribution of the plants, their vigor, and if germination microsites are available (see Table 4-3). These criteria do not have a direct link to the cause and effect relationships, and some of the indicators will be difficult to determine. For example, "normal" growth form must be determined as

TABLE 4-1   Indicators used to assess the health of soils and watersheds for rangeland ecosystems[a].

Indicator	Healthy	At risk	Unhealthy
A horizon	Present, distribution unfragmented	Present but fragmented distribution developing	Absent or present only with prominant plants or other obstructions
Pedestaling	None	Pedestals on rocks, mature plants only; no roots exposed	Most plant and rocks pedestaled; roots exposed
Rills and gullies	Absent or with blunted and muted features	Small, embryonic and not connected into a dendritic pattern	Well-defined, actively expanding dendritic pattern established
Scouring and sheet erosion	No visible scouring or sheet erosion	Patches of bare soil or scours developing	Bare areas and scours well developed and contiguous
Sedimentation or dunes	No visible soil depletion	Soil accumulating around plants or small deposits or dunes	Soil accumulating in large, barren obstructions or behind large obstructions

[a] From NRC (1994), with permission.

part of plant vigor, but this is a value that must be judged by the manager because no measurable indicators have been identified for this. If a set of assessment and measurement endpoints had been derived, the assessment of these indicators would have been greatly facilitated. Judgments as to which ecosystem components are important would be made by describing the endpoints, so the assessor's job would be more straightforward—they would make measurements in the field and then compare those with the predefined endpoints.

The rangeland health assessment includes an interesting shift toward assessing productive capacity rather than current output. This is an important change because it is more effective as part of management to maintain the capacity of ecological systems to provide what is desired rather than to manage for a constant output. However, the report falls short because it does not attempt to derive a mechanistic understanding of why a given rangeland exists in a particular state. The report focuses mainly on clarifying the descriptive assessment of the state of rangelands. This is in contrast to ecological risk assessment, in which an effort is made to quantify the probability of change before it happens and to derive cause–effect re-

TABLE 4-2   Indicators used to assess the health of nutrient and energy cycles for rangeland ecosystems[a].

Indicator	Healthy	At risk	Unhealthy
Distribution of plants	Plants well distributed	Plant distribution becoming fragmented	Plants clumped, large bare areas
Litter distribution and incorporation	Uniform across site	Becoming associated with prominent plants or other obstructions	Litter largely absent
Timing of photosynthetic activity	Occurs throughout growing season	Most occurs during only a portion of growing season	Little or none during most of the growing season
Root distribution	Rooting throughout soil profile	Absence of roots from portions of available soil	Rooting in only one portion of the available soil profile

[a] From NRC (1994), with permission.

lationships between ecological stressors and eventual adverse ecological changes.

Identifying Indicators in the Marine Environment: Case Study

Osenberg et al. (1994) attempted to deal with the difficulty of detecting environmental impacts in marine environments. In an effort to derive the most statistically powerful indicators that marine environments were changing, they used data from two different types of studies—long-term baseline data from monitoring studies and "after-only" data from a study of the marine environment near a wastewater outfall. Osenberg et al. (1994) used a previously conducted long-term study to provide estimates of the natural spatial and temporal variability in environmental parameters to obtain an idea of background variability. The after-only data were compared with a similar control area to measure the magnitude of response to wastewater exposure among a variety of organisms. Together this information from the long-term studies and the after-only study was used to derive a signal-to-noise ratio for each of the variables considered as candidate indicators.

Osenberg et al. (1994) screened many potential parameters that could be measured to assess the impacts of a wastewater discharge in a marine system. They looked at

1. *Chemical-physical parameters.* These were chosen to indicate the chemical and physical changes due to wastewater discharge.

TABLE 4-3   Indicators used to assess the success of the recovery mechanisms for for rangeland ecosystems[a].

Indicator	Healthy	At risk	Unhealthy
Age class distribution	Distribution reflects all species	Seedlings and young plants missing	Primarily old or deteriorating plants present
Plant vigor	Plants display normal growth form	Plants developing abnormal growth form	Most plants in abnormal growth form
Germination microsite	Microsites present and distributed all across site	Soil movement or other factors degrading microsites	Soil movement sufficient to inhibit most germination and seedling establishment

[a] From NRC (1994), with permission.

Examples of these parameters are as follows:

(a) *Chemical parameters:* bulk element concentrations (Al, Ca, Fe, Mg, Mn, P); trace elements (Ba, Zn, Cr, Cd, Pb).

(b) *Physical parameters:* scouring of sediment substrate, altered sedimentation rates; sediment particle size distribution; water temperature change.

2. *Individual parameters.* These were characteristics of individual organisms that were measured because they indicated effects of wastewater at the individual level.

Examples of individual parameters are as follows:

(a) In sea urchins, they measured average diameter, gonad mass, somatic tissue mass, and gonadal/somatic index (these parameters reflect growth and reproductive efforts, whether the animals are stressed such that their growth slows or they stop allocating energy to reproductive organs).

(b) *Transplanted mussels* of known size and weight were placed at varying distances from the outfall and were monitored to see if proximity to the outfall affected their individual growth and condition. These transplants can be more useful than field-collected animals because their initial size and physical conditions are known and they come from similar genetic stock, factors that remove several sources of variability that are found in field-collected organisms. If the transplanted mussel growth differs with distance from the outfall, then the effect is likely to be due to the impact of the wastewater rather than to some other confounding factor that might be found in field-collected animals.

3. *Population-based parameters*

Examples of these parameters include:

(a) Densities of infaunal organisms.

(b) Densities of larger, epifaunal and demersal organisms (measured along forty 1-m transects starting at the outfall and extending out to better mixed, cleaner waters).

Interestingly, Osenberg et al. (1994) showed that chemical-physical parameters showed the least variability over time (shown from the long-term study), while population-based parameters had absolute effect sizes that were larger than either individual-based parameters or chemical-physical parameters. This could lead to conflicting results about the statistical power associated with measuring different parameter groups. The population-based (and individual-based) parameters should be most powerful due to their larger average effect sizes, while the chemical-physical parameters should be more powerful due to their smaller average background variability. But the measure that ended up being of interest was the signal to noise ratio. Signal to noise ratio is the ratio of the response or change of interest in the indicator compared with the background change in that indicator over time. In this case, individual-based parameters—those intermediate to the others with respect to background variability and effect size—were the most powerful indicators of wastewater-induced changes. In particular, changes in transplanted mussels were the most powerful indicators of change. It may be that the genetic similarity of the mussels led to a slight reduction in variability of response to wastewater exposure, just as laboratory organisms bred from the same stock respond in a similar manner to toxins. This study illustrates the importance of looking at both background variability and expected effect size in choosing indicators of the state of an ecosystem.

5 | Case Studies: Degrees of Ecosystem Management

5.1 Ecosystem Management: What Is Happening in the Field

Not all organizations will want to implement the ecosystem management approach described earlier. However, there exist many different organizations for which the ecosystem management approach holds great utility for managing their resource into the future, including many who presently are not using this approach. Some organizations (i.e., federal agencies) have to implement ecosystem management (see Chapter 2), while others may utilize some aspects of the ecosystem management approach but may not want to adopt the ecosystem management approach in its entirety on their lands (i.e., private timber companies).

Organizations managing for a specific product or products from their forest lands often state that they are not attempting to utilize the ecosystem management approach on their lands but are more concerned with system "sustainability." However, managing for a single product is similar to managing to conserve a single species in that there is a defined value that has been identified as the "product" output from the ecosystem. In both cases, it is important that the management approach be embedded in a framework that examines the ramifications or implications of the decision to maintain a species or trees. Even if private timber companies state that they are not following the ecosystem management approach, they utilize many of the same tools and approaches that are integral to this management framework to presently track their resource base. For example, timber companies have had to deal with the spatial and temporal scales of their timber resources if they are interested in maintaining the long-term timber output from their lands.

Timber companies may wish to implement some form of ecosystem management because they are having to accept the fact that the land base from which they have traditionally harvested timber is getting smaller and public opinion or perception of how timber companies are managing their lands has not been favorable. Both of these facts will require companies to adopt some form of ecosystem management because (1) the reduced land base requires the maintenance of timber output from a smaller given land base in the future (see Section 2.2.1) and (2) public opinions or values of how they perceive timber companies to be managing and harvesting from their lands may feed back to control policies regulating timber companies and also the economic returns of a company (i.e., green certification of timber products appears to be inevitable and may determine which companies will be competitive).

Ecosystem management has been debated, analyzed, and dissected largely within an academic context. It is important to determine how this approach is being perceived and implemented in non-academic settings. To answer this question, four case studies are examined that represent a gradient of management intensities, a variety of ownership schemes, and a range of management objectives. The following cases are examined: the westside old-growth forest ecosystems in the Pacific Northwest of the United States, pulp and paper companies, the Adirondack Park, and the National Forest System in the United States.

Case studies can serve as a way of obtaining a better understanding of the problems involved in designing an ecosystem management plan for forests. An analysis of the Forest Ecosystem Management Assessment

Team's (FEMAT) management plan for the westside old-growth forests in the Pacific Northwest of the United States is included because it is a good example of how federal agencies have attempted to develop an approach to ecosystem management when (1) the time period for management plan development was less than a year (see Appendix A), (2) public values were polarized into groups not interacting with one another, and (3) available scientific tools for ecosystem management were in their infancy but policy was demanding a plan that could be implemented immediately. This case also illustrates the difficulty of preparing an ecosystem management plan when the laws driving management activities are based on species protection. Even though the initial approaches and ideas incorporated into the forest management plan are still evolving, scientists in the Pacific Northwest United States are clearly at the forefront of the development of ecosystem management as a tool. The concepts and approaches being developed in the Pacific Northwest will probably be used as a reference or model system by others managing forests in other parts of the United States as well as around the world.

Another case study specifically focuses on examining how readily any federal agency (in this case, the Forest Service) can implement ecosystem management when it is being driven by public values that drive the management approach they have to implement. The usefulness of the ecosystem management concept for private timber companies who mainly manage their lands for one or two products is also examined as another case study. This case study examines the difficulties encountered in utilizing the ecosystem management approach when management is restricted to the maintenance of one product output or one desired species.

5.2   Forest Ecosystem Management Assessment Team's Management Plan for Old-Growth Ecosystems Within Range of the Northern Spotted Owl in the Pacific Northwest United States

5.2.1   Developmental History of the Management Plan

Before the arrival of settlers in the mid 19th century, a nearly continuous extent of forest cover dominated the region on the west side of the Cascade Mountains. These forests were characterized by an annual rainfall much higher than in the inland forests east of the crest of the Cascades, and ranged, in the United States, from what is now the Canadian border to northern California (Norse 1990). Within this landscape, natural disturbance and burning by Native Americans maintained a diversity of seral stages (Agee 1993), but roughly 60% to 70% (by some estimates) of the forest cover, or 7.1 million ha (17.5 million acres), consisted of mixed coniferous old-growth stands (i.e., greater than 150 years age) (Barret 1995). In areas of greater relative rainfall, such as in coastal zones and at lower and middle elevations on the western slopes of the Cascades north of central Oregon, the forests were part of a band of temperate rain forest that stretched north along the coast of British Columbia through southeast Alaska (Conservation International et al. 1995). At low and mid elevations, old-growth conifer forests were dominated by Douglas fir (*Pseudotsuga menziesii*), western hemlock (*Tsuga heterophylla*), and western red cedar (*Thuja plicata*). However, other species, such as sikta spruce (*Picea sitchensis*, on the west side of the Olympic Peninsula), coastal redwood (*Sequoia sempervirens*, in northern California), and several pine species (in the Klamath and Siskiyou mountains) were dominant, codominant, and/or abundant in specific areas.

While the precise amount of remaining westside old-growth forest is unknown, an estimate made by The Wilderness Society (1993) using satellite imagery and field surveys has been accepted by the U.S. Forest Service and the U.S. Fish and Wildlife Service as the most accurate available. This estimate places the remaining old growth at about 1.9 million ha (4.7 million acres). Most old growth outside of congressionally reserved areas (such as national parks, national wild and scenic river corridors, national monuments, national recreation areas, and designated wilderness areas on national forests) has been highly fragmented by clear-cut logging and the construction of logging roads, and is scattered within a patchwork of recent clear-cuts, younger stands, and industrial timber plantations (Franklin and Forman 1987). Current fragmentation across the landscape far exceeds historic levels (Morrison 1990, Johnson et al. 1991, Barret 1995). Although a small portion of remaining old growth is within national parks, almost three fourths is located on lands managed by the U.S. Forest Service, with most of the rest on lands managed by the Bureau of Land Management (USDA Forest Service 1992).

Concurrent with the development of the environmental movement in the 1970s (see Sections 2.1.1 and 2.2.1), the public became increasingly aware of how logging practices were changing the visual appearance of the forest landscape (a similar activity was instrumental in the formation of the Adirondack Park; see Section 5.5). Concern over the conversion of old-growth forests to short-rotation timber plantations also escalated (Dietrich 1992) as studies indicated that far more species might be associated or dependent on old-growth habitats than was previously thought (Franklin et al. 1981). Estimates for old-growth associated species, continued to grow through the next two decades, climbing from around 200 (various references in Norse 1990) to 667 (SAT 1993), and subsequently to 1374 (FEMAT 1993). In addition, studies begun in the mid-1970s suggested that populations of the northern spotted owl (*Strix occidentalis caurina*) might be threatened by the loss of old-growth habitat (Forsman et al. 1980). The northern spotted owl was listed as a "threatened" species under the Endangered Species Act in 1990.

The continuing decline and fragmentation of old-growth forests prompted many scientists, policymakers, and activists to search for a legal basis for protecting the remaining remnants of the old-growth ecosystems (Smith 1990b). Because the decline of the northern spotted owl appeared to be linked to the loss of old growth, the owl was effectively invoked as a legal tool for challenging and ultimately enjoining (in 1991) further old-growth logging on federal lands (Simberloff 1987, Feeney 1989, Bonnet and Zimmerman 1991).

Initial region-wide planning efforts were driven (1) by the U.S. Forest Service's obligation to satisfy regulations under the National Forest

Management Act (NFMA), which required the agency to maintain a viable, well-distributed populations of vertebrates on National Forest System lands; and (2) by the U.S. Fish and Wildlife Service's mandate under the Endangered Species Act (ESA) to develop a recovery plan for the northern spotted owl (USDI Fish and Wildlife Service 1992, USDA Forest Service 1992). Thus, while the primary concern of many was the decline of old-growth ecosystems in their entirety, the old-growth forest debate has been largely framed in terms of conserving a single species (and later multiple species).

In 1990 an Interagency Scientific Committee (ISC) recommended a conservation strategy for the northern spotted owl (Thomas et al. 1990) that delineated Habitat Conservation Areas (HCAs) around clusters of known owl nesting sites and set standards for maintaining owl dispersal habitat. The ISC's recommendations were largely adopted by the U.S. Forest Service in its proposed management plan for the northern spotted owl (USDA Forest Service 1992). Debate continued, however, as to whether the HCAs offered adequate protection not only to spotted owls but also to other members of the old-growth community, especially rare and locally endemic species with distributions that did not coincide with spotted owl population clusters (Watson and Muraoka 1992, FEMAT 1993).

A particular concern with the spotted owl HCA-based approach was its lack of provisions for marbled murrelets (*Brachyramphus marmoratus marmoratus*), a small seabird that nests in the canopies of coastal old-growth conifer forests. Censuses of marbled murrelets at sea had indicated a pronounced population decline (Marshall 1987, 1989, Carter et al. 1988), a trend predicted to continue by viability modeling using some life history data from surrogate species (Beissinger 1994). When the species was listed as threatened under the Endangered Species Act in 1992, the marbled murrelet gained equal status with the northern spotted owl as a preeminent focus for planning.

The HCA approach, applied as the exclusive basis for determining a desired landscape-level habitat configuration, lacked an aquatic conservation strategy. Such a strategy was becoming increasingly necessary because several commercially important salmon fisheries in Washington, Oregon, Idaho, and northern California were experiencing dramatic population crashes. The decline were attributed to several factors, including loss of spawning and rearing habitats in forest headwater areas on federal lands (Nehlsen et al. 1991, Frissell 1992, 1993, Naiman et al. 1992).

In 1991, a lawsuit brought by a coalition of conservation organizations resulted in a Federal District Court injunction against further federal timber sales in the Pacific Northwest. Federal timber sales were enjoined until the Forest Service could demonstrate compliance with the NFMA. In May 1992, the courts found inadequate the U.S. Forest Service's HCA-based management plan for the northern spotted owl. In addition to finding the plan insufficient to ensure the viability of spotted owl populations, the court ruled, in effect, that the spotted owl was an inadequate umbrella species for the full diversity of late-successional/old-growth associated species. In this court case, *Seattle Audubon Society et al. v Moseley*, Federal District Court Judge William Dwyer ruled that the spotted owl management plan would not have fulfilled the U.S. Forest Service's legal obligation under NFMA to maintain viable, well- distributed populations of all vertebrates endemic to lands managed by those agencies. In making his decision, Judge Dwyer also used a report prepared by a Congressionally appointed scientific panel (Johnson et al. 1991) that documented the social and ecological trade-offs of different management options for late-successional forests.

The ruling sent the agency back to the drawing board. This time the Forest Service was required to develop a *multispecies* plan that would maintain sufficient habitat to support all late-successional/old-growth (LS/OG) associated species (both aquatic and terrestrial) endemic to national forests in the Pacific Northwest. Thus while the integrity of old-growth ecosystems as a whole remained the underlying concern, planning continued to focus not on ecosystem management per se, but rather on management of multiple species.

In early 1993, newly elected President, Bill Clinton, sponsored the "Forest Summit" to broker a solution to the impasse occurring over the management and utilization of resources from federal forests in the Pacific Northwest. The result of this conference was a presidential directive to the Cabinet mandating the development of an interagency plan for federal forests within range of the northern spotted owl. The agencies were directed to develop a plan that would be "scientifically sound, ecologically credible, and legally responsible." To implement the President's directive, the agencies created a Forest Ecosystem Management Assessment Team

(FEMAT), which drew on the expertise of over 600 scientists and other experts. The FEMAT was instructed to present the alternatives and consequences, both biological and social, of different management options for the Pacific Northwest's old-growth forests. In developing a range of management alternatives with varying land allocation systems, the FEMAT also largely built on a system of potential LS/OG reserve areas previously mapped by the congressionally appointed scientific panel (Johnson et al. 1991). In addition, the FEMAT incorporated Johnson et al.'s recommendations for streamside buffers, consisting of wide riparian reserves, and for a system of "key watersheds," in which logging and road construction would be reduced to protect salmonids and other aquatic species.

Since the Forest Service had been directed in 1992 to manage the National Forest System under an ecosystem management concept (see Section 2.3.1), the ecosystem management approach was considered an important element of the management plan FEMAT needed to develop. Despite this endorsement of the ecosystem management approach, the policies of the agency were still being driven by the National Forest Management Act (i.e., the species viability rule) and the Endangered Species Act (i.e., the need to assist with the recovery of listed species), and any management plan had to reflect these mandates. The entire planning process was thus driven primarily by the legal mandate to provide habitat for specific species of concern, such as the northern spotted owl, the marbled murrelet, and "at-risk" anadromous fish species.

While the FEMAT report provided a basic assessment (ecological, social, and economic) and structure for federal ecosystem management in the Pacific Northwest, the management process has evolved since the FEMAT report and is now in the implementation phase. Following selection of Option 9 in the FEMAT report as the recommended alternative by President Clinton in July 1993, a Draft Supplemental Environmental Impact Statement was released. In February 1994, after closure of the public comment period, the Administration released the Final Supplemental Environmental Impact Statement (FEIS) on Management of Habitat for Late-Successional and Old-Growth Forest Related Species Within Range of the Northern Spotted Owl. The FEIS included some revisions to Option 9 that were recommended by a Species Analysis Team. The Species Analysis Team evaluated options for boosting the viability rating for 490 species that FEMAT determined to have less than an 80% probability of viability (over 100 years) under Option 9.

On April 13, 1994, the Administration issued a Record of Decision (ROD) that adopted Alternative 9 (the FEIS's version of Option 9) as the final plan. The Standards and Guidelines attached to the ROD represent the final Forest Service and BLM management requirements for what is now generally referred to as the *Northwest Forest Plan* (see Figure 5-1). The Northwest Forest Plan covers 7.9 million ha (19.4 million acres) in 19 national forests and 1.1 million ha (2.7 million acres) in 7 Bureau of Land Management Districts in western Washington, western Oregon, and northern California.

Using the FEMAT report as a case study, it is interesting to examine the elements of the document and the Northwest Forest Plan that represent early signs of the evolution of ecosystem management. However, an analysis of the FEMAT report must consider the climate under which it was developed: a short time period of only a few months to produce the report, an atmosphere of uncertainty due to the infancy of ecosystem management concepts, and a need to satisfy legal mandates associated with species conservation. As an early effort at ecosystem management, the report does not incorporate many of the variables introduced in Chapters 2 through 4, and it is unrealistic to assume that the FEMAT would have considered what the driving variables were. However, many of the concepts presented in the earlier chapters have been core research areas pursued by federal agency scientists in the Pacific Northwest for several years. An active, ongoing research program in the Pacific Northwest is addressing many of the spatial scale issues relevant to managing forests both in a fragmented landscape and where ecological and political boundaries do not coincide. In fact, many of the ideas about spatial scale analyses presented in Chapter 4 were developed by scientists who were instrumental in the formulation of the FEMAT report.

The FEMAT report serves as a pertinent illustration of the development of an ecosystem management paradigm or methodology. This section will examine several aspects of the FEMAT report: (1) the report's definition of ecosystem management, (2) the proposed ecosystem management parameters, (3) the proposed indicators of ecosystem integrity, (4) the approach taken to spatial scales, and (5) the attributes of the proposed implementation regime or management plan. Collectively, these factors represent the

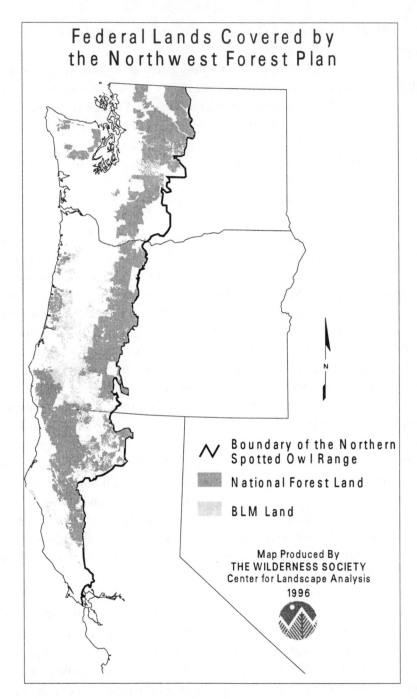

FIGURE 5-1   U.S. timberland in terms of ownership category. Data from Haynes (1990).

Northwest Forest Plan's basic approach for maintaining the long-term integrity of old-growth ecosystems within the range of the northern spotted owl in the Pacific Northwest. At the same time, they represent the plan's ability to minimize future political confrontations over timber harvesting and habitat set-asides for threatened or endangered species.

5.2.2   Review of FEMAT Methodology

Ecosystem Management Definitions and Goals

FEMAT developed and considered 10 management options of varying intensities of logging and habitat protection for forests on Forest Service and BLM lands. The management alternatives represented a range of options designed to meet FEMAT's presidential directive. In developing management alternatives, FEMAT was guided by the following basic goals (FEMAT 1993):

- "Maintenance and/or restoration of habitat for the spotted owl, well distributed through its current range, and marbled murrelet, so far as nesting habitat is concerned, to provide for the viability of each species.

- Maintenance and/or restoration of habitat to support viable populations, well distributed across their current range, of other old growth associated species.

- Maintenance and/or restoration of spawning and rearing habitat to support recovery and maintenance of viable populations of anadromous fish and other fish species and stocks considered sensitive or at risk by land management agencies or listed under ESA.

- Maintenance or creation of a connected or interactive old growth forest ecosystem in the region."

In effect, the FEMAT report and subsequently finalized Northwest Forest Plan were driven by goals related to maintenance of biodiversity at the species level, including endangered species. This focus on biodiversity specifically emphasized conservation of the species without having shown a keystone role for these species in the forests (see Section 2.1.2). This mandate to address biodiversity issues had a strong influence on how the ecosystem management concept was developed in the FEMAT report itself. The FEMAT's draft definition of ecosystem management reflected this viewpoint:

> "Ecosystem management begins with strategies that involve layering relatively independent management schemes to accommodate northern spotted owls, old-growth ecosystems, marbled murrelets and selected fish stocks. The next step is to assign multiple roles to individual land allocations in an overall conservation strategy with multiple phases to accommodate various species and ecosystems (e.g., riparian and old-growth) of concern. Including ecosystem concerns will require adaptive management action that will accelerate the transition from conservation strategies for individual species to ecosystem management." FEMAT continues, "Conservation strategies are one component of ecosystem management...A multiphase strategy ultimately could give way to a more ecosystem-oriented management."

The definitional language employed by FEMAT indicated an approach that focused on biodiversity at the species level as the primary parameter of ecosystem management. As a result, FEMAT's conception of ecosystem management stressed the viability of species rather than the maintenance of specific ecological functions and processes as described in Section 2.3. However, even though the FEMAT report did not directly deal with measurements of ecosystem processes, it was instrumental in the development of research (that is still ongoing) explicitly examining (1) the spatial scale attributes of ecosystems, (2) how the spatial distribution of different land uses affect processes occurring in adjacent landscape units, (3) how legacies (e.g., snags and coarse wood) influence ecosystem processes, and (4) potential institutional development for managing ecosystems that cross jurisdictional boundaries (see Section 2.1.1).

FEMAT acknowledged that an accepted paradigm for managing other ecosystem components is not currently available, and the need for ultimately addressing broader ecological processes other than species by offering the following:

> "Our understanding of the underpinnings (supporting science, ecological constructs, legal interpretations, and societal accep-

tance) of natural resource management is in rapid flux and deals with imprecise concepts such as "ecosystem management" itself."

"We recognize that ecosystem management as a term and as an agency or interagency agenda may be ephemeral, as similar terms and management initiatives have been in the past."

"Ecosystem management as a guiding principle, focuses attention on ecosystem well-being in senses consistent with Congressional expressions of social values. Furthermore, the concept directs attention of land managers and others to understanding ecosystems and developing appropriate site-specific management."

"Conservation strategies emphasize single-species and maximum care of the best remaining habitat. Ecosystem management, on the other hand, works with present conditions and an understanding of natural ecosystem patterns and disturbance regimes to direct ecosystems to a potentially different future."

Even though the scientific understanding of ecosystem management was in its infancy at the time the FEMAT report was developed, the Northwest Forest Plan represents a significant departure from earlier Forest Service management plans and traditional management's focus on individual species. This departure is manifest in the Northwest Forest Plan's integrated planning for all constituent species within an ecosystem. The previous regional plan delineated Habitat Conservation Areas (HCA) based exclusively on the habitat needs of northern spotted owls (USDA Forest Service 1992), whereas Land and Resource Management Plans for individual National Forests required habitat protection for specific species of concern, such as the pine marten (*Martes americana*) and pileated woodpeckers (*Dryocopus pileatus*). FEMAT's Option 9 was designed to ensure the viability of all 1374 species considered to be "old-growth associated." The Northwest Forest Plan delineated Late-Successional Reserves (see Figure 5-2) to provide habitat for all old-growth associated species based on a classification system developed by a congressionally appointed panel of scientists. This classification system (Johnson et al. 1991) categorized late-successional/old-growth forests as either most ecologically significant (LS/OG 1), ecologically significant (LS/OG 2), or other (LS/OG 3) (Johnson et al. 1991). Late-Successional Reserves under Option 9 were designed to include as much of the LS/OG 1 and LS/OG 2 as possible, as well as much of the critical habitat designated by the U.S. Fish and Wildlife Service for the northern spotted owl and the marbled murrelet. Thus, while Option 9 incorporated many of the recommendations made by the Interagency Scientific Committee to Address Conservation of the Northern Spotted Owl (Thomas et al. 1990), it represented a broader approach that accounted for all old-growth associated species as well as the "ecological significance" of forest stands identified using structural/compositional characteristics as indicators.

The FEMAT was required to assess the persistence of viable populations of all old-growth associated species, even though scientific data did not exist for most of these species. Therefore, the species viability assessments made by FEMAT were quantitatively comprehensive only for well-researched species such as the northern spotted owl. For other species, viability assessments were based on estimated population distributions as represented solely by habitat availability. Population viability analysis based on accurate distributional data and simulations of landscape level population interactions could not be conducted for most species because the data do not exist. Thus the scientific basis of FEMAT's viability assessments for most species was poor, and the short time in which this plan could be formulated did not allow for better analysis and compilation of data for poorly studied species. Nevertheless, the Northwest Forest Plan represents the first attempt at a regional-scale analysis to address an entire ecosystem's constituent biodiversity, as well as other habitat qualities, conditions, and characteristics needed to sustain that biodiversity. As such, it serves as a model that should be expanded and improved upon. This again highlights the importance of adaptive management as a flexible system to change management strategies or plans as more information becomes available on poorly understood species or ecosystem attributes.

Ecosystem Management Parameters

The FEMAT report's primary ecosystem management parameter was biological diversity, addressed primarily at the species and population levels. Its management approach was thus based on an objective of maintaining sufficient habitat to support viable, well-distributed populations of all native flora and fauna associated with old-growth ecosystems. The FEMAT

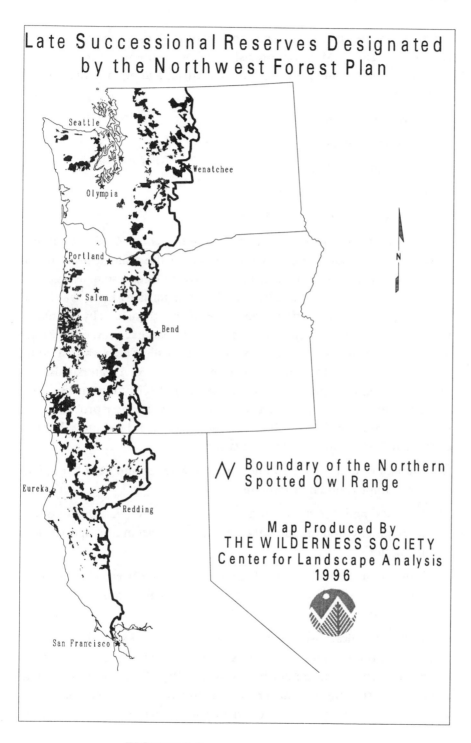

Late Successional Reserves Designated
by the Northwest Forest Plan

Boundary of the Northern
Spotted Owl Range

Map Produced By
THE WILDERNESS SOCIETY
Center for Landscape Analysis
1996

FIGURE 5-2

did, however, recognize the importance of other ecosystem management parameters. The FEMAT report discussed or mentioned the following variables as being important for both terrestrial and aquatic old-growth ecosystems:

1. Ecosystem persistence

2. Ecological processes

3. Ecosystem functions

4. Structure and composition

5. Abundance and ecological diversity

In addition to assessments of socioeconomic considerations, the FEMAT report is divided into two assessments of the biological/ecological dimensions of ecosystem management: a terrestrial ecosystem assessment and an aquatic ecosystem assessment. The aquatic assessment was based on the objective of supporting populations of aquatic species by maintaining the physical characteristics and integrity of riparian zones, establishing a level of protection for "key watersheds" and roadless areas, maintaining and restoring spatial and temporal connectivity between watersheds, and other management approaches. In the terrestrial ecosystem assessment, the management options were evaluated relative to their probability of sustaining "functional interacting late-successional and old-growth forest ecosystems" based on the following criteria:

1. Abundance and ecological diversity: the acreage and variety of plant communities and environments

2. Processes and functions: the ecological actions that lead to the development and maintenance of the ecosystem for species and populations

3. Connectivity: the extent to which the landscape pattern of the ecosystem provides for biological flows that sustain animal and plant populations

In the ecological assessments, the FEMAT report explicitly identified ecosystem structure and function as underlying determinate of species persistence. The management plan proposed by FEMAT thus sought to maintain key structural characteristics and functional processes that create and maintain habitats for old-growth associated species. For example, the proposed management plan was based on a system of "late-successional reserves" designed to protect core LS/OG habitats. Drawing on studies of the structural complexity of westside ecosystems (USFS (Old-Growth Definition Task Force) 1986b, Franklin and Forman 1987, Swanson and Franklin 1992, Franklin 1993), the management plan prescribed retention standards for late-successional forest patches, snags, green trees, and coarse wood debris. These retention standards, together with a system of riparian reserves, were prescribed for "matrix" (or intensively managed) areas to maintain a degree of connectivity between reserves (for some species) and to provide habitats at smaller scales on the managed landscape. This would maintain the landscape level structure necessary for population persistence. The FEMAT report also addressed key functional processes for both aquatic and terrestrial systems, such as the hydrologic and sedimentation regimes necessary for maintaining fish habitat and the role of natural disturbance in maintaining dynamic terrestrial and aquatic habitats.

FEMAT Indicators of Ecosystem Health

The FEMAT (1993) defined ecological health as *"the state of and ecosystem in which processes and functions are adequate to maintain diversity of biotic communities commensurate with those initially found there."* (See Section 4.1 discussion on ecosystem health.) Using this definition, FEMAT assessed ecological health (in very broad terms) under each management option. In its biological assessment of management options, however, FEMAT used the persistence of 1374 old-growth associated species as its primary criteria for rating management options. In taking an approach that focused on both functional and structural characteristics (such as species composition), FEMAT's assessment could also be viewed as an assessment at the ecosystem level. In other words, FEMAT assessed all the ecosystem elements, such as processes and species, that collectively represent an intact ecosystem.

The FEMAT report proposed indicators of ecosystem health for both aquatic and terrestrial ecosystems. FEMAT used these indicators to assess and predict the degree of ecosystem integrity that would be conferred by each management alternative considered. Examination of FEMAT's assessment methodology indicated that a combination of approaches was used that could be described as coarse filter (e.g., viability ratings based on generalized assessments of habitat availability) and fine filter (quantitative Population Viability Analysis for the northern spotted owl).

In assessing aquatic ecosystem integrity, each management alternative was assessed primarily relative to its capacity to provide the *structural* ecosystem characteristics that would "provide the necessary range of ecological functions and processes that create and maintain good fish habitat" (FEMAT 1993). A primary indicator was the structure of riparian zones (e.g., vegetational composition and vertical complexity) and extent (or width). From this structure, the corresponding ability of riparian zones to provide shade and coarse, woody debris to in-stream habitats, and to buffer erosion, was assessed. Other structural indicators included the physical integrity of channels, headwalls, unstable slopes, and logging roads, and water quality and quantity (e.g., as related to hydrologic and sedimentation regimes). The FEMAT recommended cumulative effects analysis as a central element of implementation to assess the cumulative effects of past, present, and future management activities on all of these indicators.

In addition to the assessment of structural ecosystem characteristics, the aquatic assessment included a species-specific dimension. Data on the distribution of "at-risk" as well as healthy fish stocks was used to identify a system of key watersheds. FEMAT also used viability ratings for individual species based on habitat availability as an indicator. However, these ratings were developed only for anadromous fish species and not for other aquatic species.

The primary indicator of integrity employed by FEMAT in its terrestrial ecosystem assessment was the probability of persistence of viable, well-distributed populations of individual species. This focus on individual species stemmed from the agencies' legal mandates under the NFMA and ESA (discussed in Section 5.2.1). The FEMAT relied on expert panels for individual taxonomic groups (e.g., fungi or lichens) to assess the probability of the persistence of viable, well-distributed populations for old-growth associated species under each management alternative. Each alternative was assigned a rating for each species with respect to the alternative's capacity to provide sufficient habitat and connectivity to support viable populations of that species.

The FEMAT's generalized assessment approach, based on habitat availability rather than quantitative population viability analysis, was necessary because population and life history data did not exist for most of the species assessed. Viability estimates based on accurate population data could only be made for well-researched species such as the northern spotted owl. Thus, though the FEMAT attempted to incorporate and assess the viability of all old-growth associated species, the northern spotted owl remained the primary indicator of terrestrial ecosystem health because the greatest amount of scientific information existed for this species.

Although the likelihood of persistence of the spotted owl represented FEMAT's single greatest indicator for determining the utility of management alternatives, the report correctly avoided using the spotted owl as an indicator of "ecosystem health" per se. Using spotted owls as an indication of ecosystem health would have belied the uncertainty regarding the species' functional role in old-growth forests (see discussion of species and ecosystems in Section 2.1.2). If there was a strong link between a species and the maintenance of ecosystem function, focusing on one species would greatly facilitate attempts to measure the present condition of an ecosystem. However, even if such a link could be shown, the keystone species approach does not measure ecosystem health as discussed in Section 4.1. Our ability to analyze health and integrity characteristics of ecosystems are presently highly problematic. By definition, *health* incorporates ecosystem attributes considered desirable by society that are not necessarily properties of systems, such that measurements become highly subjective (see Section 4.1). To avoid this problem, resilience and resistance characteristics of ecosystems have been proposed (see discussion in Section 4.1) as better measures of changing ecosystem states and dynamics. They have also been proposed as useful indicators for determining whether land-use activities would allow old-growth ecosystems to persist over time.

Spatial Scales

Much of the debate surrounding ecosystem management centers on the need for a spatially defined management unit. This unit, "composed of physical-chemical-biological processes active over both space and time (Tansley 1935, Lindeman 1942)," must have physical but not functional boundaries such that the unit is "nested" within the unbounded ecosystem. The FEMAT report recognized this need and provided a fairly comprehensive discussion of its treatment of spatial scales. The report suggested that ecosystem planning needs to be conducted at four spatial scales: regional, physiographic province, watershed, and site.

The regional level of scale covered by the plan is based on the range of a single indicator species, namely, the northern spotted owl. The "distributional range" approach seemed to be appropriate in this case because the spotted owl's current range correlates well with the existing distribution of westside (i.e., west of the crest of the Cascades) old-growth forest ecosystems. On the other hand, part of the northern spotted owl's range includes mixed conifer forests at middle and lower elevations on the eastern slopes of the Cascades, particularly in northern Washington state. The spotted owl's range east of the crest of the Cascades was included under the Northwest Forest Plan. This makes perfect sense as a recovery plan for spotted owls, and to some degree as a conservation plan for the small minority of westside species (such as grizzly bears and gray wolves) with ranges that extend east of the Cascade crest. However, it makes little sense in most other respects because eastside forests are considerably different ecologically (Johnson et al. 1991) and are considered to be within a very dissimilar ecoregion (Omerik and Gallant 1986). In fact, the eastside forests covered by the Northwest Forest Plan also have been included under the Interior Columbia Basin Ecosystem Management Project, a planning effort for the drier ecosystems of the inland Northwest. The eastside spotted owl forests will thus be covered by two different ecosystem management plans oriented towards very dissimilar ecoregions. This overlap is an example of the difficulty involved in defining ecosystem boundaries and ecosystem level planning units. In some cases, species' ranges do not coincide with ecological boundaries defined using other determinants such as vegetation or precipitation zones. This may be particularly problematic where management plans are intended as both species-specific management plans and ecosystem level management plans.

However, while the boundary delineation for the ecosystem of concern may be largely appropriate, the boundary delineations with regard to actual management requirements is limited by the plan's lack of jurisdiction over lands other than national forests and BLM districts. Most of the remaining late-successional/old-growth forests left in the Pacific Northwest are, in fact, on federal lands (though a small amount remains on state and tribal lands). However, private inholdings and checkerboard ownerships, especially within BLM lands, are prevalent throughout the region. Consequently, the comprehensiveness and effectiveness of management is highly fragmented in some areas. As an example, some areas key watersheds and adaptive management areas overlay large areas of private and state lands. In certain instances, private land owners in these areas have been unwilling to cooperate with federal agencies in watershed analysis and, in particular, in Adaptive Management Area planning (e.g., for the Little River AMA, Oregon).

FEMAT used physiographic provinces in a number of ways as part of its terrestrial ecosystem assessment. Distinguishing ecological characteristics, such as elevation, precipitation, and vegetation, of individual provinces were described. In addition, estimates of LS/OG acreages for individual land allocations were made for each province under each management option. The FEMAT attempted to ensure that a significant percentage of LS/OG in each physiographic province was protected in late-successional reserves, such that ecological diversity at the province level would be maintained. Potential timber production was also calculated on a per province basis for each management option. The only management prescription tiered to the physiographic provinces, however, was the establishment of province teams to provide the agencies with policy recommendations. Ecological processes, such as fire regimes, that might have discernible characteristics at the province level were not described, identified, or addressed by management prescriptions.

The watershed has become the scale at which perhaps the most active management is occurring under the Northwest Forest Plan. Because salmon fisheries are a central feature of the Northwest's economy and culture, and because anadromous fish populations and habitats have been severely degraded throughout much of the region, it is not surprising that a primary goal of the Northwest Forest Plan is the recovery and maintenance of anadromous fish populations. Thus, a large part of the Northwest Forest Plan (i.e., the Aquatic Conservation Strategy) is devoted to one particular ecosystem type (i.e., the aquatic/riparian habitat) and much of the management direction mandated by the Northwest Forest Plan concerns watershed restoration and management to maintain aquatic and riparian ecosystems, water quality, etc. Watershed analysis is required prior to any management activities in the 164 key watersheds delineated by the plan, as well as in riparian reserves and inventoried (RARE II) roadless areas.

Consequently, much of the initial implementation effort is concentrating on watershed analysis for key watersheds and non–key watersheds where high priority projects have been proposed. In addition, the Forest Service and BLM have completed 15 plot watershed analyses that developed and tested methodologies for watershed analysis.

The FEMAT identified the *site* the appropriate level of scale for project implementation. Standards and guidelines for determining the appropriateness of site level management practices were also prescribed. However, as a regional plan, FEMAT deferred most site-specific assessment and planning to procedures for smaller spatial scales, such as watershed analysis, management assessments for late-successional reserves, and project level planning.

A potential weakness in FEMAT's treatment of spatial scales is its major reliance on levels of scale that are immediately identifiable, such as *watersheds* or *river basins* (FEMAT 1993). These do not inherently represent scales at which ecological processes other than the maintenance of aquatic ecosystem characteristics occur (see discussions of scale in Sections 2.1.1 and 3.2). For instance, processes such as terrestrial natural disturbance regimes are often independent of watershed boundaries. In addition, "regional" and "site" levels of scale are designated arbitrarily with respect to ecological processes. Many authors have suggested that a primary consideration of ecosystem management should be the identification of scales at which manageable ecological processes occur (Körner 1993, Levin 1993, Reynolds et al. 1993). As Gordon (1992) pointed out, because ecosystem management must recognize both the existence and the transcendence of "real boundaries," location-specific management is required to address identifiable "mechanisms" of ecosystem function. While, as previously noted, units of ecosystem scale do not have functional boundaries, physical boundaries can be delineated based on management objectives. If these objectives included the persistence or sustainability of specific ecological processes or functions, the physical boundaries required for management to maintain these processes at a desired level can be delineated. However, given the brief time allotted for preparation of the FEMAT report, as well as limitations in our current understanding of ecological processes in the westside old-growth ecosystem, it is understandable that the FEMAT report does not incorporate this approach for designating spatial scales of management. This has been a significant research area by agency and university scientists to develop an understanding of how processes change with scale and along gradients in the Pacific Northwest.

Because much of the remaining old-growth forests on national forest and BLM land is severely fragmented, a central element of the Northwest Forest Plan is planning at multiple scales to restore and maintain connectivity across the landscape. Morrison (1988) found that approximately 25,110 ha (62,000 acres) of the estimated 120,285 ha (297,000 acres) of remaining old-growth forest in the Mt. Baker-Snoqualmie National Forest have been surrounded and "ecologically altered" by clear-cuts, young forest plantations, and nonforested areas. More than 27% of the remaining 42,930 ha (106,000 acres) of old growth in the Olympic National Forest, 55% of the 48,195 ha (119,000 acres) of old growth in the Gifford Pinchot National Forest, 37% of 72,090 ha (178,000 acres) in the Mt. Hood National Forest, 44% of 121,095 ha (299,000 acres) in the Willamette National Forest, and 46% of 57,105 ha (141,000 acres) in the Siskiyou National Forest are exposed to fragmentation or are in highly fragmented landscapes (Morrison 1988). Due to this high degree of fragmentation, both watershed and site level planning is supposed to assess connectivity potential. For instance, the Northwest Forest Plan requires connectivity to be maintained by determining where linkages remain intact between watersheds, and by assessing the ability of riparian reserves to provide connectivity. This is to be assessed through the watershed analyses process. Fragmentation of the forest landscape has also become a core research area for scientists in the Pacific Northwest. Research is exploring how fragmentation affects ecological processes (see Section 3.2).

Implementation Regime

About 30% of federal lands within the range of the northern spotted owl are already protected in congressionally reserved areas, such as national parks, national monuments, and designated wilderness areas. An additional 6% is protected in administratively withdrawn areas on national forests and BLM lands. The Northwest Forest plan divides all unprotected areas on national forests and BLM lands within the range of the northern spotted owl into five additional management categories. The management categories are as follows:

- Late successional reserves, covering about 30% of federal lands

- Riparian reserves, 11% of federal lands
- Adaptive management areas, 6% of federal lands
- Managed late successional areas, 1% of federal lands
- Matrix, 16% of federal lands

Each category has different management prescriptions and varying intensities and types of permissible logging, road construction, and recreational management. Management prescriptions amend U.S. Forest Service Regional Guides for Regions 6 and 5, amend existing forest plans for individual national forests, and will guide the development of forest plans for those national forests that do not currently have management plans. Approved and draft BLM District Resource Management Plans will be similarly amended.

Where logging and road construction are allowed, management prescriptions set restrictions that must be followed in developing logging and road construction projects. Other prescriptions mandate specific activities, such as road decommissioning, that the Forest Service and BLM must undertake to restore and maintain terrestrial and aquatic habitat. Considerable discretion, however, is given to Forest Service and BLM personnel using "professional judgement" to interpret and modify restrictions in the field.

The Northwest Forest Plan delineates 164 key watersheds that were identified as important refugia for anadromous fish and for producing clean water. These delineated watersheds overlay all other land allocations and ownerships. The Forest Plan requires a process called *watershed analysis*, or ecosystem assessment at the watershed scale, before management activities can proceed in all key watersheds, riparian reserves, and inventoried roadless areas. Watershed analysis provides a baseline characterization of a watershed to guide subsequent management activities.

Salvage logging and thinning are allowed in all management categories, including late successional reserves (LSRs) and riparian reserves. Thinning in LSRs is intended to speed the development of old-growth structural characteristics, whereas thinning in riparian reserves is permitted where it will promote structural characteristics needed to provide in-stream habitat and dispersal or riparian habitat for terrestrial species. Salvage logging is allowed where it will not "negatively" affect late-successional habitat, a considerably undefined standard. The Northwest

Forest Plan requires management assessments for LSRs before salvage or thinning can proceed. Experimental silvicultural techniques, longer rotation periods, and innovative logging methods are encouraged, but not mandated, for adaptive management areas. Intensive logging and silvicultural management under relatively short rotation periods (50 to 60 years on average) is allowed to continue in matrix areas. Logging is not permitted where watershed analyses have identified on-site sensitivity (such as unstable soil and risk of slope failure) or off-site concerns (such as the potential to contribute unacceptable levels of sediment to streams). Identified sensitive areas are to be included in riparian reserves. All critical habitat on federal lands identified to date by the U.S. Fish and Wildlife Service is included in LSRs (though a small proportion of this is to be managed using an adaptive management approach). Most of the critical habitat identified by the Northern Spotted Owl Recovery Team is also protected in late successional reserves and riparian reserves.

The FEMAT included a spatial analysis team of over 80 specialists. The team compiled a Geographic Information Systems database that it used to delineate land allocations, such as key watersheds and late-successional reserves. Much of the GIS team's initial work involved assembling a database because no comprehensive database had been compiled for the region previously. However, due to time limitations, it was not possible to conduct original or new data collection and mapping. For this reason, the GIS team gathered and integrated individual data layers produced by previous mapping efforts. For instance, the team utilized maps of key watersheds for anadromous fish that had been produced for the Scientific Panel on Late-Successional Forest Ecosystems (Johnson et al. 1991). It used data for LS/OG distribution and locations of spotted owl nesting sites also produced by previous mapping efforts. Other data layers, such roadless area boundaries, had to be digitized from hard-copy maps. The GIS team produced overlays of these and other data layers that allowed the FEMAT to determine optimal land allocation configurations. However, the short time frame precluded the team from conducting more than only limited spatial analysis of the potential impacts of management options, such as impacts on species viability. The FEMAT thus did not conduct the spatial scale analyses (discussed in Section 2.1.2) needed to address the mechanistic basis for relationships between species and their habitats.

Interagency or Interorganizational Coordination of Ecosystem Management

The Northwest Forest Plan provides an interesting case study of the creation of new administrative structures to coordinate interagency cooperation in implementing an ecosystem management plan. This development supports the idea that past administrative structures are not well formulated or flexible enough to implement ecosystem management (see discussion in Section 5.3). Under the plan, a hierarchy of five administrative levels has been established. This includes, in descending order of authority (1) *The Interagency Steering Committee,* consisting of Washington DC–based representatives from the Interior, Agriculture and Commerce Departments, the Environmental Protection Agency, and the White House Office on Environmental Policy; (2) *the Regional Interagency Executive Committee (REIC);* (3) a *Regional Ecosystem Office (REO),* based in Portland, Oregon; (4) *12 Province Teams* (each province covers one of the physiographic provinces identified and evaluated by the FEMAT report); and (5) *local teams:* subgroups created to work on specific watersheds in some areas. In addition to these new administrative entities, an *Office for Forestry and Economic Development,* located in Portland, has been established to serve as a liaison between the White House and the agencies and as a general troubleshooter for the Administration's Northwest Forest Plan.

While the REIC and REO have already become fully functional, the timber industry successfully challenged the legality of the province teams under the Federal Advisory Committee Act (FACA). FACA prohibits federal agencies from utilizing advisory committees to obtain advice and recommendations unless those committees are officially chartered and adhere to various requirements intended to ensure public access to the committee's work. The agencies complied, and have only recently completed the official chartering of each province team. The charter for each province team specifies the number of team positions awarded to each agency and interest sector (e.g., Forest Service; BLM; Park Service; Fish and Wildlife Service; National Marine Fisheries Service; Environmental Protection Agency; Washington, Oregon, or California State agencies; Native American Tribes; timber industry; tourism industry; fisheries industry; environmental groups; etc.).

REIC and REO have been moderately successful at establishing authority over certain decisions that would previously have been made by individual agencies. For instance, all management assessments, fire management plans, and salvage sales in late-successional reserves have to be reviewed and approved by REO, a process that is now underway. REO also coordinated the Pilot Watershed Analyses through the publication of a federal guidance manual, and recently released a manual for all other federal watershed analyses as well. However, most management activities are still planned at the individual national forest or BLM district level, leaving most implementation decisions up to individual administrative units without interagency or interunit coordination. The effectiveness of the province teams approach is still unclear. It has been suggested that disagreements between interest groups represented on the province teams may paralyze the ability of these teams to effectively coordinate watershed restoration projects and other activities within each physiographic province. Moreover, it remains unclear how province teams will function within the bounds of the existing project approval processes that have in the past varied considerably between individual national forests and between individual BLM districts. The Northwest Forest Plan does not specify precisely what the role or operating procedures for the province teams will be. Thus, it remains unclear how effective province teams will prove as facilitators of intersectoral coordination.

Socioeconomic Management Parameters

Although multiple-use requirements under NFMA and FLPMA are legally superseded by the "viability" rules, the agencies have a mandate to provide multiple uses and extractable resources to support local and regional communities and economies. Timber, fisheries, and recreation considerations were therefore central to the FEMAT assessment.

The socioeconomic parameters were not specifically identified by FEMAT. However, they can be distilled from the general objectives identified by the FEMAT report (FEMAT 1993). The socioeconomic objectives could be interpreted as including the following:

1. Provide multiple uses to accommodate all user groups:

   - Timber interests

   - Nontimber forest products (e.g., mushrooms, berries, boughs, etc.)

   - Fish interests: commercial and sport fishing companies

- Recreational interests: outfitters, guide services, equipment retailers
- Environmental interests

2. Maintain support to local communities through profit sharing in lieu of taxes (i.e., PILT payments), providing revenue-generating resources, and assisting with economic diversification

3. Reduce interagency conflicts and improve coordination between agencies

The FEMAT report included an Economic Evaluation of Options and a Social Assessment of Options that examined the economic and social ramifications of each management option presented. These assessments predicted the social and economic impacts to timber dependent communities as well as to the states of Washington, Oregon, and California. In particular, the associated changes in state and local revenues from timber sales on federal lands and jobs related to the forest products industry were predicted. However, the social and economic assessments did not consider direct economic revenues and benefits provided by tourism and recreation on federal lands, as well as indirect economic impacts of logging, such as reduction in water quality, air quality, and amenity values associated with public lands.

As a management plan for National Forests and Bureau of Land Management Districts within range of the northern spotted owl (USDA 1994a,b), the Northwest Forest Plan does not include economic or other community assistance programs. However, the plan is accompanied by an economic package called the Northwest Economic Adjustment Initiative (NEAI) that will provide $1.2 billion ($270 million per year for 5 years) to assist affected timber-dependent communities in making a transition to a more diversified economy. The money provided in the NEAI is divided into four categories: one being ecosystem investment, which provides funding for watershed analysis and restoration to create "jobs in the woods" for dislocated timber workers. In addition, public participation in the implementation of the plan is promoted through province and local teams (the planning process for adaptive management areas), and to some extent through the watershed analysis process. Furthermore, the processes mandated by the National Environmental Policy Act for public notification and commentary regarding federal land management decisions still apply to most management activities and projects that will be implemented under the Northwest Forest Plan, although these were suspended through 1996 by Section 2001 of Public Law 194-19, also known as the salvage or timber rider.

Adaptive Management

The development of an adaptive management framework was identified in the FEMAT report to be important for implementing ecosystem management in the Pacific Northwest. The tools for conducting adaptive management were to be developed in areas set aside specifically for this purpose. Much research and development of the concepts of adaptive management has occurred since the publication of the FEMAT report that will not be articulated here (Bormann et al. 1993). The important point is that adaptive management was clearly identified as an important research area within the FEMAT report. Three primary mechanisms were identified in the FEMAT for developing adaptive management in the forests in the Pacific Northwest:

1. *Adaptive Management Areas (AMA).* The Northwest Forest Plan delineated 10 Adaptive management areas in which forestry research was to be conducted. An example of the type of experiment envisioned by FEMAT is the Demonstration of Ecosystem Management Options (DEMO) project. This was a study of population responses of multiple species to variable harvesting intensities and retention configurations in late-successional stands. Some of the DEMO study sites are located in the Little River AMA on the Umpqua National Forest in Oregon and in the Cispus AMA on the Gifford Pinchot National Forest in Washington, although other study sites are in matrix areas and on state lands.

   Under Option 9, AMA management plans are to be developed that are based largely on local public input and participation. An example of such public involvement is the Quincy Library Group in northern California, which brought together representatives from timber companies, town chambers of commerce, and environmental groups to hammer out a mutually agreeable plan for timber harvesting that was consistent with environmental protection. The Quincy Library Group's plan was adopted verbatim by the agencies in northern California.

2. *Iterative planning.* Watershed analyses and management assessments for late-successional reserves are to be conducted iteratively as new information becomes available and as new projects (e.g., timber sales or restoration projects) are developed and prioritized.

3. *Monitoring.* FEMAT and the Northwest Forest Plan stressed interagency monitoring and the development of feedback mechanisms to guide planning and management. The Northwest Forest Plan called for (1) implementation monitoring (determine if the plan is fully implemented as intended), (2) effectiveness monitoring (determine if it is having the desired effect for all interests), and (3) validation monitoring (determine if there is a cause–effect relationship between management activities and indicators of resource conditions). It also called for the creation of an interagency monitoring network. The plan encouraged research on all land allocations, although highly manipulative research was only allowed in matrix and adaptive management areas.

5.2.3   Conclusion

The FEMAT plan provided an useful early model for the development of the ecosystem management approach in managing forest ecosystems. This management approach was strongly controlled by social values driving what the management plan was to conserve—in this case, conservation of old-growth associated species was mandated. The FEMAT report was interesting in that the social and biological constraints both had to be incorporated into a plan that addressed the trade-offs and consequences of different management options. This was one of the early reports to acknowledge the importance of both the natural and social sciences in influencing the effectiveness of management. Attempts were made to integrate ecosystem management into the management plan, even though a primary focus was species decline. The plan developed in the FEMAT report has evolved since 1993, and its future was uncertain when this book went to press, but an examination of the original report serves a very useful function to demonstrate the early development of the ecosystem management system and what kind of system can be developed under conditions of high uncertainty.

The FEMAT report demonstrated the evolution of the social and natural science management systems within the U.S. Forest Service and BLM. These agencies concentrated their implementation efforts on four main areas initially presented in the FEMAT report: (1) formation and chartering of new administrative entities, (2) initiation of the planning process for the 10 adaptive management areas, (3) initiation and development of watershed analyses for key watersheds and for proposed projects in riparian reserves and inventoried (RARE II) roadless areas, and (4) initiation and development of management assessments for late-successional reserves. Of these areas, the watershed analyses (as the analytical backbone of the implementation process) have received a considerable share of the financial resources, staff time, and general attention by the agencies. As part of management in the Pacific Northwest, the spatial scales of analyses should be expanded to incorporate appropriate new spatial tools discussed in Chapters 3 and 4, with explicit attention being made to identify the mechanisms explaining the patterns of vegetation distribution within the landscape. The resistance and resilience characteristics of the fragmented old-growth ecosystems also need to be examined to determine the ability of management to maintain the integrity of all seral stages of these ecosystems and associated species into the future (using the tools presented in Chapter 4).

5.3   National Forest Ecosystem Management

5.3.1   Overview

During the past three decades, Congress has adopted a diverse body of law that relies heavily on the concept of *sustainability* or continuous production of the diverse resources commonly associated with national forests. There is a natural tension built into the Forest Service's management mission: They must decide the most appropriate mixture of resources to be produced and harvested from the forests. If there has been one discernible trend in congressional intent, it has been to elevate the importance of noncommodity resources in the service's land management decisions. The dominant elements of congressional strategy have included: (1) improvements in the scientific basis of sustainable management; (2) extensive, long-term, aggregated planning at the forest, regional, and national levels; and (3) considerable public participation. While this change in purpose by Congress seems to reflect increasing diversity of resources that the American public values in national forests, the primary effects have been on the analyses that precede decisionmaking rather than on the actual decisions themselves that govern the mixture of resources produced on national forests (Wilkinson and Anderson 1987).

The effect of these laws on how research and planning is occurring in the Forest Service has been extraordinary. For example, these laws have resulted in enormous volumes of data being generated on resources in national forests, new analytical methods have been designed, elaborate management plans have been constructed, and perhaps most importantly, there is a heightened sensitivity among service employees to noncommodity values and their management. Significant methodological advances have also been made on evaluating other resources in national forests, such as recreation and wildlife.

Despite these innovations, there appears to have been little significant shift in the mix of resources produced or managed for on national forests. For example, the traditional resources (i.e., timber, minerals, and other commodity resources) continue to be the dominant products being harvested from national forests, while other resource uses or products (i.e., recreation, wilderness, fish, wildlife, soil, and water) are receiving little attention. Evidence for this dominance comes from a diversity of sources:

1. The Forest Service Chief in 1987 presented the 1985 Recommended Renewable Resources Program (USFS 1986a) before a House Subcommittee and was forced to justify support for the "low bound" alternative, which emphasized a high level of timber production in concert with a low level of investment in reforestation, recreation, timber stand improvement, land acquisition, and research. Robertson (1987) stated:

   "The high bound represents what we think are wise investments, a good way to manage the National Forests, but the Forest Service is part of the Federal Government and the Federal Government has other priorities, and a Federal deficit problem. So when you...are trying to eliminate a $200 billion deficit, it gets you down to looking at the low bound, which really is fairly realistic because our

1988 budget is actually at or near the low bound...It is kind of like a savings account. You invest in a savings account and you live off the interest. The high bound would be living off a high level of interest and in the low bound we would be spending some of the savings account."

2. The Congressional Research Service found that "high bound" production targets for timber and other commodities were largely exceeded by the Forest Service, and that "high bound" production goals for recreation, wilderness, wildlife habitat, land acquisition, and human resources were consistently unmet (Gorte 1986). Short-term investments necessary to ensure long-term resource sustainability are deferred (Lyons and Knowles 1988).

3. A more recent comparison between Forest Service funding targets and actual appropriations was conducted by the Conservation Foundation (Sample 1990). They found that between 1981 and 1986, actual appropriations for timber management fell short of RPA high bound targets by 5% to 15%. By contrast, appropriations for the soil and water program, fish and wildlife program, and recreation program fell short of high bound targets by 40% to 50%.

4. Since 1960, annual appropriations for soil and water, fish and wildlife, and recreation have together ranged between 15% and 20% of all Forest Service resource management funding, while allocations for timber have ranged between 65% and 80% of the same total. In 1989 and 1990, actual funding of the timber program fell short of planned funding by approximately 10%, while the shortfall was approximately 35% for recreation (Sample 1990).

5. In the early 1970s the Forest Service admitted that only one third of the cutover land within Rocky Mountain National Forests was successfully regenerating (USFS 1974).

6. In response to a question posed by Senator Hatfield, the Forest Service admitted that it is "obvious that the high harvest levels of the past 3 years in the Pacific Northwest, cannot be continued indefinitely." The service argued that these high levels of harvest still constitute "sustained yield management" if averaged over long periods of time during which harvests are substantially reduced (USFS 1990).

7. Financial sustainability is threatened by recent disclosures that timber sales on 53% of the national forests return less than what it costs the government to grow and sell the timber. The service has reduced the appearance of the costs by amortizing them over extraordinary periods of time. For example, in 1988 the service spent $10.7 million to harvest timber on the Nez Perce National Forest and received $4.5 million in revenue for a net loss of $6.2 million. However, by the service's accounting, the program showed a profit of $300,000, which was accomplished by amortizing costs over 302 years (Rice 1989).

8. Twight and Lyden (1989) analyzed policy and value statements made by 400 District Rangers (the field-level line managers of the Forest Service). They compared these responses to those made by two other groups (environmentalists and a "utilizer" constituency) and found that the ranger group was "quite close to the utilizer constituency and relatively far from its environmental constituency." The authors suggested that the agency's "socialization process" had changed little from the findings of Kaufman's landmark study conducted in 1960.

9. During the summer of 1989, an newsletter was initiated by Forest Service staff that is highly critical of Forest Service policy. In an open letter to then-Chief Dale Robertson, the editor stated (DeBonis 1989):

"We are over-cutting our National Forests at the expense of other resource values...Examples include moving spotted owl habitat area boundaries to accommodate timber sales; exceeding recommended cover/forage ratios on big game winter range; ignoring non-game wildlife prescriptions; exceeding watershed/sediment threshold values of concern in areas with obvious cumulative damage. We rarely, if ever, exceed our objectives in non-timber resources, even though these objectives are set at the absolute minimum we can legally get away with. These practices are so common place, they are the standard operating procedure. We have taken the politically mandated timber harvest level and manipulated the Forest Plan data to support the cut level, rather than build the plan up from the bottom, letting the harvest level be determined by sound biological and ecological considerations mandated by our resource protection laws."

The evidence is patchy yet there are clearly enough pieces to suggest that the call by Congress for improvements in science and planning have not contributed to the sustained production of diverse resources associated with the forest commons in the United States. (It is important to be cognizant that some members of the U.S. Congress are mandating the continued extraction of certain resources that is, timber, mining, etc., from National Forests.) In addition to this dominance of which resources are harvested from the national forests, it also appears that there has been a failure to promote sustained production of timber on these lands. This case study examines three hypotheses that might help to explain these patterns of nonsustainable resource management occurring on national forest lands:

- *The first hypothesis is that nonsustainable management is caused in part by the failure of science to adequately analyze the conditions of the resources.* For example, the boundaries of analysis have often been drawn too tightly in space and time, and have probably not examined the issues using appropriate scales. Historical data and analytical methods have been inadequate to identify thresholds for maintaining ecosystem production into the future. Sources and magnitudes of uncertainty in the data have not been identified with much precision. Under these conditions, forecasting the behavior of an ecosystem far into the future is probably not going to produce reliable estimates of ecosystem state or responses to stress or disturbances.

- *The second hypothesis is that nonsustainable management is being driven by the utilization of methods for evaluating resources that are predominantly based on economic evaluations of commodities.* Reliance on quantitative measures of commodity values necessarily favors the production of commodities such as timber, animal unit months, and minerals, while other resources (i.e., wilderness, recreation, and wildlife) remain more difficult to quantify or to place a commodity value. Analytical methods to estimate the current "value" of diverse forest-based resources offer little confidence that, first, these values are representative of those held by the public, and second, the values will persist into the future.

- *The third hypothesis suggests that nonsustainable management is caused in part by congressional failure to alter traditional patterns of control over decisions to pursue specific resource production targets.*

The idea of sustainability underlies most renewable natural resource law. Yet sustainability is an ambiguous concept, the usefulness of which is dependent upon understanding at least three characteristics of a social-ecological system:

1. What scientific knowledge is required to secure critical elements within or surrounding the ecosystem of interest?

2. What do people want to sustain?

3. How should people manage themselves—through law and implementation—to achieve the desired end state defined by those values chosen by people to be sustained?

These questions demand an exploration of the science, values, and politics associated with sustainable resource management of national forests.

5.3.2   Science of Sustaining Multiple Resources

Sustainable resource management requires a basic understanding of the structure, function, and dynamics of ecological systems. What are the appropriate boundaries surrounding the ecosystem? What are the fundamental survival requirements of species and communities? What conditions are necessary to maximize production? What types of human stress can be sustained by ecosystems without the loss of important species, communities, or system functions?

These three questions seem fundamentally important regardless of the species or community of concern, for example, an endangered species or a commercially valuable species. Nearly a century of silvicultural research has provided an impressive scientific basis for individual species management. By contrast, the scientific basis for community management, particularly when defined to include species that are not of commercial interest, is often more primitive. The evolution of the biological knowledge of the California condor is an excellent example. Debate over the species' endangered state spanned nearly half a century before a consensus was reached that the species was on the brink of extinction. Our lack of understanding of the condor's habitat requirements, mating behavior, and reproductive biology all contributed to a *do nothing* managerial response while the population levels of this species continued to decline. Debate over spotted owl population sizes and habitat requirements within the LS/OG forests of the Pacific Northwest has followed a similar pattern (see Section 5.2).

Sustainability and Thresholds

Sustainable resource management requires direct consideration of system thresholds. What level of stress can be sustained without loss of critical system conditions, for example, loss of individual health, loss of a minimum viable population size, loss of community diversity, etc.? The concept of a threshold is an important basis for managing human behavior or stress to ensure that a system is not pushed beyond its ability to recover to a pre-stressed condition (see Section 4.2.8). It is the foundation of the science of toxicology that attempted to differentiate between doses of a chemical that resulted in an adverse health effect and a safe level of exposure for humans. The threshold effect is defined by using carefully controlled laboratory settings, often using test animals, and the results become a basis for estimating safe levels of human exposure to the substance in question. This type of dose–response assessment is precisely what is necessary as a foundation for sustainable resource management, however, ecosystem thresholds cannot be determined from these type of laboratory studies (see Sections 3.4.1 and 4.2.8) and will depend on the development of new methods, as suggested in Section 4.2.1. The complexity of human and nonhuman stress types, their combinations, and varied durations make the identification of thresholds for individual species and communities an enormously complex problem. This complexity and inability to detect thresholds cause one to question whether thresholds have been identified by scientific hypothesis testing or by processes of social valuation (Burch 1984).

Sustainability and Uncertainty

Often laboratory-based science has been used to predict how the ecosystem would respond in the field. In the past, there were few long-term analyses of complex ecosystems that could be used as a basis for the development of large-scale resource management decisions. A couple of clear exceptions to this statement were the research watersheds established by the Forest Service in North Carolina and New Hampshire, where data collection spans some 40 years (see Section 2.1.1). In contrast to the past, today there are many long-term ecosystem studies scattered throughout the world that can be used to formulate effective resource management decisions (see Section 2.1.1).

The National Environmental Policy Act of 1969 (NEPA) requires federal agencies to forecast the environmental effects of their actions and to fully disclose their findings to the public (P.L. No. 91-190, 83 Stat. 852 [1970]; codified as amended at 42 USC 4321-4370 [1982]). One of the more interesting effects of NEPA has been to motivate the Forest Service to prepare detailed natural resource inventories for each national forest. Soils, hydrology, timber, slope, wildlife, fisheries, and other resources have been mapped in an effort to provide a scientific foundation for the decisions that require determining the trade-offs among conflicting land uses, that is, timber harvesting and endangered species protection (Wilkinson and Anderson 1987).

Many of the earliest environmental impact statements were simply inventories of the biophysical environment with a poor mechanistic understanding of ecosystem functions that would support future forecasts of changes in the system. Similarly, agencies maintained control over proposed alternatives that were often designed to make their preferred alternative appear to be optimal. Several decades of litigation brought on by environmental interests have improved the quality of environmental forecasts and have caused abandonment of those projects that could not financially sustain the delay associated with judicial review.

If there is one consistent lesson from the discipline of ecology, it is to explore probable causal chains and to realize that the impacts of a particular activity in the ecosystem may not be expressed until at a much later time (therefore, the time horizon of concern should be lengthened rather than shortened). Lawmakers often demand that scientists should establish clear ecological and human health thresholds of sustainability. Because scientists have not been able to establish these thresholds (but see Section 4.2.1), it has resulted in lawmakers using their own values or desired outcome to determine at what level to define the existence of these thresholds. Because of the inability of defining these thresholds and the difficulty of finding evidence for direct cause–effect relationships that define thresholds, ecology, like medicine, has had to rely on identifying subtle interconnections to identify where these thresholds might potentially exist. This has been further complicated by the fact that seemingly insignificant stresses in one part of a complex system at one point in time may affect distant components of that system perhaps far in the future. This highlights the importance of identifying the temporal scales of feedbacks in ecosystems.

How should resource managers cope with this combination of limited scientific evidence and the statutory obligation to forecast effects far into the future? One answer lies in the formal characterization of scientific uncertainty: Because confidence in forecasts normally diminishes as time horizons are pushed further into the future, it becomes extremely important to identify the limitations of our current knowledge base. Assumptions and uncertainties are rarely explicitly incorporated into most large computer models used to estimate ecological responses or changes due to perturbations.

Brewer (1986) suggested that analyzing complex problems should be approached using simulation and gaming models. He presented an overview of those approaches that have their historical roots in the design of military strategies for contexts that are complex, uncertain, and dangerous. Playing "what if"' games in a sequential process allows one to see more clearly the implications of behavioral responses to alternative system "conditions" (Brewer and DeLeon 1983). Others have adopted this approach to model uncertainty associated with human health risk estimates used to regulate agricultural chemicals (Wargo 1996). The complexity of the forest planning process seems ideally suited for the design of simulation and gaming models to estimate alternative futures, while poorly suited to the type of linearized optimization modeling process that will naturally favor market-based methods of resource valuation and therefore commodity production.

Sustaining Production

A body of U.S. resource management law has evolved over the past three decades that demands a scientific rationale for chosen production levels. The Multiple Use and Sustained Yield Act of 1960 (MUSYA) promoted forest management practices that ensured a sustainable yield of timber and that did not diminish noncommodity resources. This law required the Forest Service to afford "due consideration" to other forest-based resources, such as recreation, water, wildlife, range, and fisheries in its management efforts. The phrase "due consideration" is ambiguous and has been implemented by the Forest Service primarily through more intensive planning.

The accurate estimate of what levels of cutting will ensure a sustainable yield of timber is also required by MUSYA. The service, by their own admission, has had some difficulty in achieving this goal, evidenced by their finding in 1974 that only one third of cutover land in Rocky Mountain National Forests was successfully regenerating (Wilkinson and Anderson 1987). Similarly, back in 1988, the Chief of the Service admitted that the 1988 budget called for the harvesting of more timber than was justifiable to ensure sustainable yields, given their low level of investment in reforestation and timber stand improvement (Robertson 1987). The reasons seem less based upon the failures of scientific prediction than upon the politics of the budgetary process in concert with interest group pressure to gain access to commercially valuable timber stands.

Sustaining Diversity

Amid growing controversy over Forest Service timber-cutting practices (particularly the reasonableness of clear-cutting), Congress in 1976 passed the National Forest Management Act (NFMA), which specifically specified ecological suitability criteria for timber harvesting (16 USC Sec. 528-531 [1982]). A second relevant provision of NFMA requires the Forest Service to "provide for diversity of plant and animal communities" (16 USC Sec. 1604 (g) (3) (B) [1982]). Ostensibly, this requirement is designed to prevent the national forests from becoming single-species plantations, while encouraging natural levels of ecological diversity. Finally, NFMA requires the development of 10-year interdisciplinary forest plans for each of the 123 national forest administrative units. The planning process is forcing the integration of disciplinary knowledge from specialized fields such as economics, wildlife biology, hydrology, soil science, and forestry.

There is a natural tension between achieving the goals of single species and multiple resource production (see Section 5.2). Optimizing production of multiple resources within the same forest will rarely be possible, again, highlighting the importance of utilizing the ecosystem management approach to identify the trade-offs of management decisions. Examples of the tensions that will exist with multiple resource management include the following: the dilemma of harvesting old growth timber in the Pacific Northwest while at the same time decreasing habitat for endangered species, cutting roads within roadless areas to remove timber that destroys wilderness character, and reforestation using trees selected in genetic tree improvement programs that may have lower genetic diversity at the species level, therefore diminishing naturally occurring diversity. Sustainable, multiple resource management requires a scientific understanding of how the single-species production systems differ from that of the multiple resource production systems. The information that has already been collected on the single- species production system will not transfer directly to the multiple production system.

Boundaries: Units of Analysis

Sustainability is often considered only within carefully defined spatial and temporal boundaries. Shifting the boundaries of these units of analysis can have an enormous effect upon conclusions regarding what constitutes sustainable patterns of human use. Consider the effects of herbicide application on national forests to suppress undesirable growth. If the spatial unit of analysis is defined only as national forest land and the temporal unit of analysis is defined as the period within which the herbicide may be detected in the soil, one may easily conclude that the environmental effects are minimal and the action is therefore "sustainable." By contrast, consider the possibility that the herbicide is a likely human carcinogen and that spray drifts not only onto adjacent private property but also follows streams to drinking water supplies. Because oncogenic effects in humans commonly take several decades to appear, searching for soil residues several years after the application of the herbicide is unable to address any of the human effects. In this case, the original conceptual unit of analysis was defined too narrowly and therefore is unable to detect effects in the subject group of interest.

Summary

In summary, the laws just described have had an enormously positive effect on the quality of information available to Forest Service administrators responsible for planning future mixes of forest resources. This improvement of the planning process has also forced advances in data collection protocols, data management techniques, and methods of analysis.

The simple collection of data, however, does not by itself ensure improvement in the accuracy of forecasts of sustainable resource production or reasonable ecosystem management. Similarly, the availability of resource inventories, forecasts of production, and projections of costs and benefits do not imply that an obvious course of action will be supported by all interests. Instead, the quality of data and its interpretation commonly become a focal point for controversy among competing interests groups over appropriate management strategies.

Each of the laws just described has had the effect of expanding the data available to the Forest Service to make resource management decisions. This should not necessarily be considered a broader scientific foundation for management, because there seems to be a very real difference between data and an understanding of causal patterns within complex systems. Fundamental resource inventories are crucial to estimates of sustainability required by the MUSY Act, estimates of environmental effects required by NEPA, estimates of ecologically appropriate levels of harvest and levels of ecological diversity required by NFMA, and planning efforts required by RPA. Yet inventories alone are insufficient to meet the predictive requirements of laws that demand forecasts of ecological change in response to alternative levels of management, or hypothetical patterns of human behavior. It is this necessity for prediction under conditions of extraordinary uncertainty that places the greatest strain upon limited scientific evidence, often forcing technical experts to make uncomfortable judgments (see Section 5.2).

The increasingly technical character of decisionmaking also has had an important influence on patterns of participation in decisionmaking. Few individuals have the expertise or resources to learn and interpret complex computer programs such as FORPLAN. Technical complexity therefore seems inversely related to the democratic character of the decisionmaking process, and antithetical to the congressional objectives of improved public participation stressed in the National Environmental Policy Act, the National Forest Management Act, and the Resources Planning Act. Technical complexity is too often used as an excuse to safeguard the concentration of knowledge and power among those who traditionally have controlled decisionmaking.

5.3.3 The Pursuit of Value or Who Cares?

In the previous section, the scientific basis for ecosystem management was explored, asking the question, *Does sufficient knowledge exist to sustain the resources of interest?* The central question to be discussed in this section is instead, *What resources do humans care about sustaining?* The answer demands inquiry into human values, and any response will likely vary among and within individuals over time, depending upon the context within which it is asked (Rokeach 1979).

What Does the Public Want to Sustain?

How should publicly held values be discerned? Democracy leaves much to be desired for several important reasons. Law passed by elected officials often reflects values of the politically influential at one point in time. While public values shift or become more refined and the scientific knowledge base constantly changes, law remains far more stable. Statutory ambiguity normally gives decisionmakers considerable discretion. It is this discretion, then, that becomes a focal point for the resolution of conflict among interests promoting diverse values.

The democratic character of agency policy is therefore dependent upon the representativeness of groups that attempt to influence agency policy. Interest group reflection of publicly held values is difficult to gauge because interest groups actively shape those values. They shape these values through marketing and advertising in the private sector and by defining what is a *problem* and escalating it on the public sector agenda. While examples of private sector value-shaping efforts are obvious in any edition of any magazine or newspaper, nonprofit sector efforts are often more subtle.

The statutes described in the previous section reflected a shift in the values that Congress associated with national forests, elevating the importance of sustaining biological diversity, recreation, wilderness, and wildlife relative to the production of commodities such as timber, cattle, and minerals. The frustration lies less with Congressional intent than with Forest

Service administration, which determines the particular mix of resources to be produced within specific national forests. There is clearly a scientific dimension to the debate, described in the previous section; however, there is also a value-based or an "evaluative" dimension. How does the Forest Service value diverse mixtures of natural resources that might be produced?

If one demands a rational foundation for any chosen mixture, the problem requires methods of evaluation that permit comparison among diverse resource values. There are diverse methods for valuing board feet of timber, barrels of oil, animal unit months, acre feet of water, recreation visitor days, wilderness acres, and endangered species. The current market value of any of these resources is only one method of estimating its value, usually an incomplete one.

Valuing the more subjective characteristics of life, including aesthetics, health, biological diversity, and leisure, is more difficult because the units of measure are neither standardized nor agreed upon. This does not mean that they are unimportant. In fact, these are the concerns that appear to most shape policies and human responses to them. Although difficult to quantify, these values tend to define the quality of our lives and environment.

For example, an individual's "marginal willingness to pay" for a day of backpacking in the wilderness has been used as a reasonable reflection of the value of wilderness (to that individual for that period of time) so that recreational value may be compared with other commodity-based uses of forest land. Yet the value of camping beside a wilderness lake in the northern Sierras far exceeds the $76 spent per day on gas, food, equipment, maps, and insect spray to keep mosquitoes away. Burch (1987) suggested that the most enduring value of outdoor recreation may lie in the formation and solidification of social bonds. Similarly, West (1987) urged us to consider the psychological, family cohesion, and status values of outdoor recreation—all difficult to quantify but essential components of most people's idea of a quality experience.

The Forest Service's Renewable Resources Programs propose two alternative production levels for each resource—a high bound and a low bound level (USFS 1986a). Binkley and Hagenstein (1987) analyzed the efficiency of reallocating the service's budget among resources by moving from the low bound to the high bound estimates of output for each resource: timber, range, minerals, wilderness, recreation, water, and wildlife. They asked the question, *Is it possible to achieve a higher output of benefits by shifting investments among programs without increasing the overall budget?* They concluded that budget reallocations should be made away from the inefficient range program and the marginal timber program and shift toward the minerals and noncommodity programs, such as recreation, wilderness, water, and wildlife.

There seem to be several flaws with this method. First, the outcome is completely dependent upon the method chosen for valuing each resource. Binkley and Hagenstein (1987) conceded that their analysis was conducted using the Forest Service's own estimates of relative resource values. Second, there are no assumptions that investments in one resource area will increase or reduce investments in another. Investments in wilderness, for example, will often produce clear benefits for soil and water quality, nongame wildlife, and recreation. Clear-cutting, by contrast, may reduce these resource values. Modeling these feedback and substitution effects is an analytical nightmare, yet Congress and an increasingly sophisticated public have demanded more logic to the mixture proposed yearly by the Forest Service.

Congress requires the service to develop integrated resource management plans, predicting the effects of hypothetical mixes of resources aggregated within forests, regions, and the nation. One response to the congressional planning directive included the design of a forest planning model known as FORPLAN. This program helps national forest planners forecast the effects of shifting resource output targets on both spatial and temporal scales. The designers' original vision anticipated the ability to simulate the national forest system's response to alternative management strategies. Criticisms of FORPLAN include: (1) the ability to hide assumptions; (2) the level of sophistication required to comprehend why a specific management action produced results that the computer found to be "optimal"; (3) many of the problems of managing dynamic ecosystems may not be linear in nature, making a linear modeling technique inappropriate (Iverson and Alston 1986); (4) because many resources are not commonly traded on commodity markets, there are no reliable and commonly accepted measures of their value, which frustrates any attempt to study substitution effects; (5) the quality of ecological forecasts is highly dependent upon the quality of the science that produced the data employed;

(6) there is currently no simple method to incorporate and track scientific uncertainty associated with numerous data and assumptions; and (7) most importantly, there may be a great disparity between the coefficients that modelers assign to value resources in computer programs and the values held by concerned and influencial members of the public (Barber and Rodman 1990).

Yet there are clearly some tangible benefits to formal mathematical modeling of complex systems (see Section 4.2.8). The primary value lies in forcing the modeler to contemplate the causal links, temporal lags, and feedback relationships among important variables. Simulations can be run and rerun based upon a variety of assumptions to forecast system responses (Meadows and Robinson 1985).

Planning has sharpened our perception of the limitations associated with traditional methods of resource valuation. While some progress has been made by social scientists in developing innovative methods of resource valuation, significant problems of value additivity, value commensurability, and value dynamics remain. Quantitative approaches to valuation have traditionally favored commodity production, because market-derived units of measure have been thought to be most reliable. If a computer program is chosen as our "value accounting system," the disadvantage of noncommodity interests is reinforced. The dilemma as follows: *Increasing rationalization of the analytical process seems to diminish the importance of resources that are difficult to quantify.*

The Forest Service might be thought of as a value broker, capable of entering into a variety of possible exchange relationships among special interests. It is naive to assume that the service is a neutral broker that will quickly respond to value shifts within the public, even those expressed by diverse interest groups (Pfeffer and Salancik 1978). A more accurate image might be one of the service attempting to manipulate its environment through analysis, with a success that has been limited by the public participation requirements of key statutes, litigation, and a forest plan appeals process. In a sense, democracy is confronting "rational" and centralized planning and analysis, infusing it with diversified values.

At the outset of this section the following question was posed: *What values do the public care about sustaining?* The answer will vary among interests, and within interests over time. Perhaps the more difficult question pertains to value convergence between an agency and its public, or the interest groups purporting to represent the public. When the distance between these different groups is wide, decisionmaking is conflict ridden, as was observed in the plan developed for the westside old-growth forests in the Pacific Northwest (see Section 5.2). Those who perceive their interests to be at risk will fight against protective policy (constricting spotted owl habitat boundaries to maintain logging jobs or reducing logging areas to maintain the spotted owl habitat). Those who perceive opportunity will pursue legislation that offers exclusive protection for their valued resource (wilderness designation).

5.3.4 Pursuit of Control

Imagine a state of perfect knowledge of (1) biological resilience and (2) publicly held values. Would this type of knowledge clearly demonstrate the management approach necessary to sustain biological resources and diverse human values? Probably not, because the core of the management problem lies in the governance of competition for access to the forest commons among divergent interests. Perfect knowledge would do little to reduce this conflict. Approaches to managing the desire for control are considered next.

Subtle distinctions among the terms *control, power, authority*, and *property* seem central to an understanding of the strengths and weaknesses of alternative conflict management strategies. The following definitions are offered merely to avoid semantic confusion. *Control* is the effective regulation of behavior. *Power*, by contrast, is the ability to control behavior that may or may not be exercised. *Authority* is the right to control behavior; it may be legitimated by law, tradition, and/or charisma; and commonly has associated obligations. Finally, *property* is a type of authority that may be infinitely fragmented and distributed among individuals and groups. In this sense, property is not a thing but a set of social relationships that establish rights and obligations among competing interests with respect to valued and scarce resources.

Using these definitions, the Forest Service might be thought of as a property regime that holds the authority to assign rights and obligations among interests competing for access to scarce forest resources. The Forest Service's multiple-use planning efforts constitute a type of zoning of public land resources. This planning process defines and distributes property rights to national forest resources among competing private interests. While fee simple rights are rarely privatized, private interests have succeeded in securing less-than-fee interests from the federal government, primarily in the form of lease agreements allowing specified levels of resource use and extraction. Managers of these resource agencies might be thought of as common property right brokers, planning to allocate rights among competing private interests.

At any point in time, a specific *pattern* of property right distribution among public and private interests may be identified. The multiple-use planning process is the dominant mechanism that Congress and the Forest Service have chosen to establish the pattern or mix of resources produced within the forest commons. The pattern's stability over time seems to be related to the magnitude of benefit shifts that would be affected by redistribution efforts (i.e., the larger the shift in benefits, the more difficult it will be to effect the change). Property right stability also seems to be related to the political influence of those holding the rights at any point in time: It is much easier to disenfranchise those with no influence.

Unfortunately, planning is a process that may be controlled by interests currently in power. Control may be maintained or gained in numerous ways: by the design of data collection protocols; by choice of methods of data analysis; by selection of possible alternatives; by choice of methods of valuing alternative futures; and perhaps most critically, by gaining access to those making the choices. Shifting the balance from commodity resource production to noncommodity resource preservation will require insulation of the planning process from traditional, dominant commodity interests, both within and outside of the Forest Service.

Synoptic Versus Strategic Planning

The failure of science-based planning to effect a significant redistribution of control over forest resources may be explained, in part, by the natural tension between two approaches to resource allocation: synoptic or comprehensive planning versus strategic decisionmaking. This is a tension identified by Lindblom (1977) as a fundamental distinguishing feature between planned and market-based economies.

The synoptic approach requires an idealists' complement of knowledge, methods of comparative valuation, and uncanny forecasting ability to carefully plan the allocation of scarce resources far into the future. It relies heavily upon computer-based modeling to keep track of production volumes and schedules, and hopes to achieve a level of coordination among producers and consumers that achieves the most socially desirable mix of benefits, thereby maximizing social welfare. It sounds a lot like the RPA process, as ideally envisioned by its initial drafters.

In response to the vulnerabilities of comprehensive planning, especially as associated with limited resources, Lindblom (1977) argued that there is a tendency to be forced to be more strategic. A strategy of making modest changes in behavior to achieve short-term increments in welfare is employed in a world of limited knowledge, unexpressed or highly dynamic values, inadequate or nonexistent analytical methods, and diverse conceptions of social welfare. The strategic approach relieves an enormous analytical burden associated with synopsism while setting the stage for later adjustments of policy following recognition of the failures of the initial choice.

The historical behavior of the Forest Service is clearly more strategic than synoptic. In response to more than a decade of intensive planning, the upper administration of the Forest Service has not significantly shifted mixes of resource production. Their short-term strategic decisionmaking has favored commodity production, while synoptic planning would have encouraged the production of noncommodity resources and investments in long-term maintenance of biological resources. Control over the production of diverse forest resources remains uninfluenced by sophisticated internal synoptic planning efforts. Strategic or incremental decisionmaking is forced by external control over the budgetary process.

Summary

In the previous sections it was argued that neither an expansion of the knowledge base nor evolving methods of outcome valuation hold much promise for shifting patterns of resource outputs from national forests. Differing conceptions of sustainability seem to lie at the core of this failure. To professionals within the forest service, sustainability is defined by units of natural resource production and in terms of organizational stability and job security. To the various commodity-based interest groups competing for access to forest resources, sustainability is defined primarily in terms of financial security. To the Office of Management and Budget, sustainability is defined by the immediate quantitative relationship between total federal revenues and expenditures, or in the case of specific agency programs, the existence of a positive marginal benefit/marginal cost ratio. The shift in control over production targets from professional resource managers within the Forest Service to federal budget examiners has clearly overshadowed the substantial biophysical and social science advances of the Forest Service within the past decade (Sample 1989).

5.3.5   Conclusion: Equity and Resource Allocation

In an arena where land-use behavior is governed less by science and more by the values and power of traditional controlling interests, the potential abuse of science and knowledge by the powerful is of great concern. Comprehension of the sources and magnitudes of scientific uncertainty embedded within estimates of alternative futures often translate directly into power. While inventories of national forest resources have improved dramatically over the past two decades, our analytic methods for interpreting data to understand ecosystem function and resilience often remain primitive (see Sections 4.1 and 4.2). Also the concentration of knowledge and analytical expertise within the Forest Service merely increases the potential for this to be overshadowed by commodity interests.

Equity in value distribution has long been thought of as a fundamental goal of political organization, particularly the pursuit of Bentham's and Pinchot's concept of the *greatest good*. There are at least two problems in the pursuit of the greatest good. First is the difficulty in coming to any consensus on how it should be defined. Because *good* is clearly a value-based term, there is a tendency to fight over appropriate definitions of value and methods to measure it. Second, to achieve equity in value distribution, it will be important to move beyond questions of valuation and to focus directly on special interest group power. Increasing the power of noncommodity interest groups will not be a simple prospect, particularly because it often implies decreasing the power of commodity interest groups. Equity, in this sense, requires a deliberate attempt to balance power relationships among diverse interests so that a sort of *mutual control* is achieved over the value distribution process (Dahl 1982).

Finally, equity of control over forest resources will never be achieved. Equity in control is really a measure of the democratic character of the process, requiring not just equal participation but equal power to influence resource management decisions. It is also a measure of the level of accountability of decisionmakers to the public they purportedly represent. There will never be a national referenda to determine the appropriate uses of specific national forests. What can be achieved is a higher degree of equity in the ability to influence resource allocation decisions. Public participation in plan development and in the selection among alternative futures is an important mechanism to ensure higher levels of decision-maker accountability. As Burch (1976) reminded us, the environmental movement might be interpreted as an attempt to effect a fundamental shift in participation rights in natural resource allocation decisions. The primary danger is that the science-based planning process will be little more than an exercise in rhetoric that rationalizes the choices of a powerful minority.

5.4   Developing Sustainable Management: Cases from the Pulp and Paper Companies

5.4.1   Overview

To date, the majority of information concerning ecosystem management has been largely generated by academia and the governmental sector. As a result, a perception has been created that suggests ecosystem management concepts and tools are being developed and utilized primarily by these two groups. Most pulp and paper companies do not state that they are using the ecosystem management approach but discuss a management approach that has a goal of being sustainable and follows the *principles of good stewardship* (Champion International Corporation 1995). What approaches are followed by these companies and how closely they mimic ecosystem management is difficult to state. The specific approaches utilized by forest product companies are not readily available to the public because individual companies have no wish to inform their competitors of the practices they are or are not using as part of their management.

Several companies (i.e., Weyerhaeuser, Mead Corporation, Champion International, and Westvaco, to name a few) are publicly stating that they are in the process of shifting their management paradigms from maximum sustained yield and multiple use to a broader, more holistic approaches. These approaches are designed to maintain the health, integrity, and productivity of other noncommodity resources found within these forested ecosystems and at the same time to maintain the *best management practices* and *forest stewardship* on timber company lands (Champion International Corporation 1995, Georgia-Pacific 1995, International Paper 1994, Weyerhaeuser 1995). Unfortunately, in most cases, the general public awareness of management practices occurring on lands belonging to the forest products companies coincides with the companies being legally pursued because of the concern for a threatened or endangered species living on their lands (i.e., Georgia Pacific Corporation owns large sections of the southern forests inhabited by the red-cockaded woodpecker, which is listed under the Endangered Species Act of 1973; Simberloff 1993).

Despite the fact that many of the issues and concerns related to ecosystem management are not new to the forest products companies, changing public values have created a new climate that has caused timber companies to take a harder look at how the ecological approach fits into their management schemes. There are strong incentives to consider and respond to public opinions on how they think forest product companies are managing their lands because it can directly affect their economic returns. There is a strong desire by the general public to have forest product companies consider all the values that are contained in their lands (some of these being threatened or endangered species), and the repercussions of not heeding the public values can result in the public refusing to buy products from that company (i.e., "green certification" is one approach being suggested as a vehicle to identify which companies are using ecological management of timber on their lands). The perception among many timber companies is that green certification is something that the companies will have to incorporate into their management regimes in the future because it will directly affect the productivity of their business. This means that some companies may consider implementing some *form of ecosystem management* because they will have to balance maintaining the product that is their business (i.e., timber) with the other nontimber forest products and amenities that the public is demanding from their lands.

In addition to the public opinion driving how forest management is conducted on forest timber company lands, the reduced ability to harvest timber from federal lands has resulted in companies becoming more dependent on their own land base to supply their timber needs. A reduced timber harvesting land base means that management will have to consider the implications of becoming more dependent on a smaller land base and the importance of long-term maintenance of the site to supply timber at current or higher rates. This will again motivate companies to consider a form of ecosystem management because management will have to consider how the manner and extent of timber harvesting feeds back on processes occurring in the whole ecosystem.

Shifting Paradigms

Although ecosystem management has become a slogan of the 1990s, the concept encompassed by these words is not new to forest managers, who have been concerned with secondary forest functions, such as watersheds, wildlife, and recreation for the last 40 years (Leopold 1949, Aplet et al. 1993). Many of the basic tenets of ecosystem management can be found in Leopold's *Sand County Almanac*, in which he describes his vision of a land ethic. In this collection of essays, published in 1949, Leopold alluded to the bias of management, "I always cut the birch to favor the pine. Why?" The idea of ecosystem health: "Health is the capacity of the land for self-renewal," the complexity and interdependence of species, and that management should be concerned with both commodity and non-commodity resources. Although these issues were described over 45 years ago, they are still hotly debated today.

Industrial managers are quick to report that they have been thinking in terms of landscapes and watersheds long before they ever heard the term *ecosystem management*. Despite this claim, industrial forestry practices and forestry practices, in general, are undergoing the most rapid and intense changes since the establishment of the field over a century ago (Sample et al. 1993). Gordon (1994) referred to this deep-seated change as a "paradigm shift," in which the former management paradigm (comprised of land allocation measures, maximum sustained yield principles, and multiple use objectives) is being replaced by an approach that focuses on a *sustainable forests* rather than *sustainable forestry*. The forest products industry has experienced a paradigm shift for primarily three reasons: changing public views and values, the recognition that federal lands alone are inadequate for conserving biodiversity, and a general desire to avoid the type of political and legal stalemate that has plagued the timber industry in the Pacific Northwest United States (see Section 5.2).

Public Views and Values

Recently the views and values of the American public have changed their emphasis from single or multiple resource management objectives to a more holistic approach that incorporates both commodity and noncommodity resources (see Section 2.2.1). On both public and private forest lands, the public's concern that forest lands be managed for biodiversity, hydrological, recreational, and climatic values has never been greater (Sample et al. 1993). This heightened concern is evidenced by the consumers' willingness to accept higher timber product prices in exchange for habitat protection and other environmental values (Roper Organization Inc. 1992). Concerning industrial forest lands, timber products corporations have begun to address noncommodity issues in addition to concerns of fiber and profit, largely due to the environmental concerns of forest product consumers, corporate shareholders, and forest industry employees (Owen and Sweeney 1995).

Conservation of Biodiversity: Federal Lands Are Not Enough

In 1949 Leopold recognized that the national parks were not large enough to sustain the larger carnivorous species, such as wolves and grizzlies, and recommended that "the most feasible way to enlarge the area available for wilderness fauna is for the wilder parts of the National Forests, which usually surround the Parks, to function as parks in respect of threatened species." More recently, ecologists have concluded that conservation of biodiversity will not be effective if carried out on federal lands alone (Franklin 1989). Instead, collaboration is necessary across public and private ownership boundaries.

Because the timber companies own significantly large tracts of forest lands in the United States, they can potentially contribute to advancing many of the goals of ecosystem management and help in conserving biodiversity on their lands. For example, of the 198 million hectares (490 million acres) of forest lands in the United States, forest companies manage approximately 15% of the total, which is almost equivalent to the 18% of total forest land base in national forests (Haynes and Brooks 1991, Powell et al. 1992; Figure 5-3). If the goal is to maintain biodiversity in the United States, it will be important to form some type of partnership with owners of industrial forest lands because half of all the federally listed species and subspecies are not found on federal lands (Simberloff 1993). In fact, in some parts of the country most of the forest lands are privately owned, so conservation efforts will not succeed without the participation of private land owners. For example, in the southern United States, where the red-cockaded woodpecker is endangered, its prime habitats are the longleaf pine forests, which are 90% privately owned (Simberloff 1993). This again highlights the importance of landscape level management of habitats and the utilization of an ecosystem approach to sustain the ecological characteristics needed by this species.

Furthermore, the nation's timber companies are a significant economic player by generating revenues exceeding $200 billion annually. Moreover, nearly every sizable forest product company in the United States has begun to broaden its management objectives from growing more trees in a shorter amount of time to including management of noncommodity resources through ecological research programs (Schneider 1993). Several forest product companies have begun to conduct pilot projects in order to test implementation of *sustainable* and *good stewardship approaches* to man-

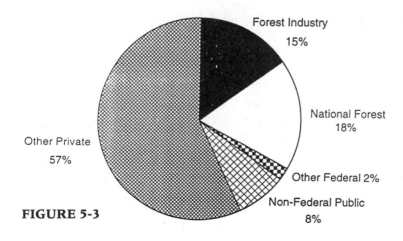

FIGURE 5-3

agement. Given both their geographic and economic importance, forest product companies have an opportunity to contribute to the advancement of the ecosystem approach or some variation of this approach. Because the industry does not publish information on the obstacles they have encountered or the successes they have achieved, it is difficult to present a solid case study for them. However, it is interesting to still examine, albeit with a small database, how they have modified their management activities to respond to the public concern about "sound" ecological sustainability and conservation of species.

Industry Adopting Sustainable Forestry Principles

In October 1994, the American Forest and Paper Association (whose members comprise many of America's forest products producers) adopted a set of sustainable forestry principles. These principles call for members to

- Practice a land stewardship ethic that integrates the reforestation, management, growing, nurturing, and harvesting of trees for useful products with the conservation of soil, air, water quality, wildlife and fish habitat, and aesthetics

- Use sustainable forestry practices that are economically and environmentally responsible

- Protect forests and improve long-term forest health and productivity

- Manage unique qualities (protect unique sites)

- Continuously improve forest management practices

How these principles will significantly affect day-to-day management decisions remains to be seen. However, their adoption represents a significant shift from former management objectives. This is just one example of how industry is changing from a maximum sustained yield paradigm to one that embraces more than commodity resources. The next sections examine the approaches and initiatives being used by several individual timber companies to develop sustainability and good stewardship approaches to managing their lands.

5.4.2 Mead Corporation's Approach to Assessing Sustainability

To date, the majority of land managers who use an ecosystem management approach have been government agencies that were often given a federal directive to do so. There are, however, a number of forest products companies that have chosen to implement this management philosophy but are calling it *sustainable management, best management practices*, and *good stewardship*. Mead Corporation represents one such company. To understand how the Mead Corporation envisions this management approach and how this paradigm shift occurred, a discussion will follow of their management philosophy and how this perspective is reflected in the company's daily management regimes.

Background

The Mead Corporation owns over 0.6 million hectares (1.5 million acres) of United States domestic timberlands, of which 283,500 hectares (700,000 acres) are in Michigan. The Mead Corporation is Michigan's largest landowner, comprising 1.9% of the state's land base. Over 243,000 hectares (600,000 acres) of its Michigan landholdings are open for hunting, fishing, and other recreational activities for the general public.

In Michigan, the forest industry has contributed over $6 billion to the state's economy each year, and Mead alone employs over 1300 people in the Upper Peninsula region. The Mead Corporation is a leading manufacturer of coated textbook paper in North America and is one of the top suppliers of paper for periodicals, catalogs, and promotional literature. The division's Escanaba Mill produces more than half a million tons of coated paper per year from approximately 350,000 tons of bleached softwood and hardwood kraft, and 60,000 tons of refiner mechanical pulp.

A New Management Paradigm

During the late 1970s to early 1980s, as the spotted owl conflict developed in the Pacific Northwest United States, the Mead Corporation became sensitive to escalating environmental concerns surrounding the harvesting practices used in harvesting timber. Starting in the late 1980s, the company embarked upon a proactive, comprehensive approach to deal with environmental concerns related to forest management on their lands. The primary components of this approach included the development of a management and ownership philosophy that outlined Mead's responsibilities as a land steward. A major change in their management plans was to proactively identify environmental concerns and to manage their lands to prevent potential problems from escalating into significant management concerns. As part of this philosophy, Mead decided to identify and map those sections of their land that could be characterized to be more sensitive to human activities or contained species or habitats that were threatened or endangered.

Mead spent 2 years developing their Forest Resource Ownership and Management Philosophy, which was completed in 1989 after much discussion and debate. The philosophy is broken into two sections: (1) Description of the Resource and (2) Guiding Principles. The Description of the Resource stated that certain basic resource attributes (i.e., variety of forest tree species, forest soils, etc.) will determine future productivity on their lands and ultimately what will be the levels of sustainable supply of wood fiber.

While satisfying economic returns is the *primary* reason for Mead's ownership of forestland, Mead acknowledges that their forests can potentially support a broad variety of other important human values or nontimber forest products (i.e., helping to maintain global biodiversity, hydrologic cycles, salmon, etc.). As an owner and user of timberlands, Mead claimed that it has a "stewardship responsibility to care for these resources in ways that maintain the legitimate value of these other, non-economic dimensions." These are goods and services that do not directly have a market value for the company but do strongly contribute to positive public opinions, and therefore indirectly affect their economic conditions. This acknowledgment of other values existing in their forest lands is a movement away from the multiple-use management approach in which a manager focused only on a few specific forest attributes (i.e., timber, game management, and recreation).

Mead's stewardship philosophy comprises seven guiding principles that were developed "to guide the practice of forest management on both its own lands and lands of others where Mead is in a position to influence the selection of forest treatment alternatives." These seven guiding principles are published in an October 1989 brochure entitled, "Mead's Forest

Resource Ownership and Management Philosophy" and are synthesized below:

1. *"For areas where intensive management is suitable we rely on both long- and short-term economic criteria in selecting specific forest management alternatives and assessing their impact on recreation, the environment, and aesthetic considerations.*

2. *Mead will care for and manage forest sites in ways which will enhance, not detract from, future productivity.*

3. *Mead will manage its forest lands in ways consistent with a code of "best management practices" (BMP's) which when applied, will minimize any undesirable environmental consequences of timber harvest and transportation. BMP's will address road construction, stream crossings, wetlands operations, construction and use of skid trails and loading ramps, and other practices peculiar to each operating region.*

4. *In areas of forest ownership not economically suitable for intensive management or single species management, we will encourage the viability of naturally occurring plant and animal communities. In practice, this involves resisting the temptation to "bring every acre under management." We recognize that there are some stands that are mediocre from a productivity perspective but are valuable for ecological reasons.*

5. *Mead will identify unique or rare ecosystems, historical sites and natural phenomena which occur on land that it owns. We will explore ways to exclude these areas from normal forest management by set-aside, sale, gift, or exchange of these areas to organizations that are in a position to care for, use, and preserve them over the long run.*

6. *Many plant and animal species require large areas of unbroken, relatively homogenous forest cover ("core") for survival. Other species are best adapted to small areas and the transition between different forest conditions ("edge"). Mead strives to achieve balance by accommodating both large and small harvest and treatment areas in forest management planning. This recognizes that smaller is not always better and that larger management areas on industry land can complement and balance the trend toward increasingly smaller treatment sizes on non-industry private and public land.*

7. *To obtain the information necessary to intelligently and sensitively manage this multiplicity of forest influences, Mead will seek ways to efficiently measure both*

the current status of the non-economic dimensions of its forest ownership and the rate of change in these characteristics over time. An initial assessment of these aspects of the forest will be developed by a qualified outside expert for update on five year intervals thereafter."

Summarizing these principles, Mead plans to

- Rely on both long- and short-term economic criteria as a basis for decision making
- Protect future productivity
- Resist the temptation to bring every acre under intensive timber management
- Identify and manage for unique and rare ecosystems
- Manage properties for the maintenance of biodiversity
- Strive to obtain the best possible information for decisionmaking

The process of drafting, debating, and articulating these principles is significant for several reasons. It is during this time that the company spent considerable energy and resources in identifying their own land steward-ship philosophy. The company not only deliberated over the philosophy but confronted how it would affect daily operations.

Although Mead's primary objective is fiber production, management admits that to maintain this product they must begin to consider this objective in an ecological context, accounting for the health and functioning of other biological attributes in order to ensure future productivity. Sustainable management is perceived as a more holistic form of science-based management that directs attention at the relationships and interactions between various resources (both commodity and noncommodity) and requires planning on larger spatial and temporal scales, and across jurisdictional boundaries. These guiding principles are being tested in Mead's management of the Mulligan Creek Riparian Area.

The Mulligan Creek Riparian Management Area Project (MCRMA)

One of the most scenic river gorges in the Upper Peninsula of Michigan is the Mulligan Creek area, located in Marquette County. In addition to its outstanding scenic quality, this area is ecologically significant for several reasons: It serves as a corridor between the McCormick Tract of the Ottawa National Forest and forested regions northwest of Marquette, it contains rare plant and animal habitats, and it provides a large expanse of unfragmented landscape utilized by many migrating and nesting bird species. The MCRMA project was created by a landscape management coalition that began with the cooperative efforts between industrial forest managers at Champion International Corporation, Mead Corporation, and Longyear Realty and then later expanded to include the Michigan Department of Natural Resources. The project area comprises a total of 10,125 hectares (25,000 acres) that are under several ownerships: Mead Corporation (3240 ha or 8000 acres), Longyear Realty (2430 ha or 6000 acres), and Champion International (1013 ha or 2500 acres).

This project developed in 1990 when Mead embarked on identifying unique features within its properties. The Mead Publishing Paper Division hired the services of an environmental consulting firm (White Water Associates) to assist them in accomplishing this goal. After a year of research, the Mulligan Creek Area was selected as an area requiring special management attention. In 1992, Mead prepared a landscape level plan for the Mulligan Creek watershed and shared the results of this effort with neighboring land owners and other major players, including Michigan's Department of Natural Resources, the U.S. Forest Service, Champion International Corporation, and Longyear Realty. Shortly after this meeting, Mead, Champion, and Longyear agreed to work cooperatively across ownership boundaries (Premo and Schwandt 1993). The primary objective of this planning process was to coordinate management across ownership boundaries. For example, the plan provided a mechanism for activities, such as coordinating harvesting activities and road building and closures.

The management of the MCRMA included several different components: research designed to examine the impact of management activities on species and water quality, education and training of their personnel to deal with sustainable management issues, and the use of technology to facilitate data acquisition and analysis of their lands. These activities are discussed in more detail later.

Research

In order to better understand the impact of management practices on their lands, Mead initiated a series of research projects with the assistance of White Water Associates. Research studies were specifically established to determine the effects of different management activities on water quality and on bird and mammal populations. Both research topics were designed to help address specific problems that could occur on these lands due to *defined management activities*; however, they were *not designed to examine the ecosystem level impacts of management.*

When conducting the different research projects, Mead used biodiversity indicators as a means of assessing or indicating the health of their forests and to assess how management is affecting the sustainability of the forests. The rational behind this decision was that biodiversity is both a measurable and practical index of ecosystem health, making it a logical integrator of ecology and sustainability-based management (Premo 1993). In addition, issues concerning biodiversity are a major concern to the general public, and Mead felt that it would be remiss not to address these concerns. A combination of species and water quality indices were used to assess the extent of riparian zone buffer needed when harvesting trees to maintain the health of the system. Avian and mammal tracking studies were used to indicate if riparian areas are important mammal migratory corridors and songbird habitats. Another study examined the effect of tree harvesting at different distances [buffer zones varying from 15.2 m (50 feet), which is Michigan state's present requirement, to 45.6 m (150 feet), to 76 m (250 feet)] from stream edges on water quality. Water samples were collected prior to and after harvest, and were analyzed for total suspended solids, total organic carbon, and total phosphorus. They concluded from several of their studies that the 15.2 m buffer zone was sufficient to maintain water quality but inadequate for mammal migratory corridors and songbird habitat. In response to these results, Mead expanded the extent of their riparian corridors beyond those required by the state of Michigan (Schneider 1993).

Education and Training

Because forest management is largely conducted by foresters working at the ground level, foresters have to be updated as to what factors needed to be considered when implementing sustainability-based management and to monitor the health of their sites. With this in mind, Mead initiated a program to educate and train their forestry personnel. According to John Johnson, Technical Services Manager from Mead's Woodlands Department in Michigan, "It is no longer sufficient to have a generalist background to do forestry. This is what most foresters receive from a four year degree. To address the problems of today, foresters need expertise in ornithology, soil microbiology, and a host of other specialties." One educational program called TEMS (Total Ecosystem Management Strategies) was designed to help foresters understand rare and endangered species, biodiversity, forest fragmentation, and nongame wildlife issues. Mead hired the White Water Associates to teach and train their foresters about the principles of sustainable forest management. This was achieved through workshops, newsletters, and study guides.

Technology

Technology, in this case geographic information systems, has allowed the Mead Corporation to begin assessing their resource base using larger temporal and spatial scales—the landscape level. For example, in the Eastern Upper Peninsula of Michigan, Mead has been trying to coordinate its management activities with the other major private land owners, and federal and state agencies in the region. For instance, as part of the planning process, a cover map of this region was created using Forest Service Ecological units. These GIS maps were then used in discussions with other land owners to facilitate the discussion of resources management issues that cross boundaries. How successful these types of cross-boundary management scenarios will be is still not clear, but they are an important mechanism that needs to be pursued because resources are not contained within political or ownership boundaries. This type of approach has also to move beyond a simplified mapping operation to address how the configuration of different land units within a landscape affects the quality of the system within and between different landscape units. This has to consider the spatial scale factors discussed in Sections 3.3 and 4.2.6, and to develop a mechanistic understanding of why the quality of the environment in different landscape units change in response to landscape patterns.

Mead has used GIS to obtain a better understanding of the distribution of their resources on their lands and with respect to adjacent land owners. For example, GIS has been used to analyze road densities on an area basis as opposed to a traditional ownership boundary approach in which land managers only examined properties under their own management. On Mead's properties, the GIS has been used to identify potential features of concern (i.e., endangered species habitat), and once these have been identified, to clearly demarcate these areas for different management options.

Mead has also found that GIS is particularly useful in estimating timber fiber potentials on a landscape basis and uses it to graphically envision future desired conditions of its properties. This technology has given Mead the ability to model potential outcomes under varying management scenarios, thereby increasing its ability to manage for both commodity and noncommodity resources.

Summary

To summarize, the Mulligan Creek project has served many purposes for Mead. It has

- Developed practical ways in which the sustainable management focus can be tested and implemented into daily forestry operations
- Provided continuing education and training to its forestry staff
- Served as a multi-ownership management demonstration project
- Enabled resource managers coming from a range of backgrounds to share and blend management perspectives
- Fostered the formation of additional landscape coalitions

Since 1993 and using the Mulligan Creek project as a model, Mead has identified 10 other landscapes for which it has created comprehensive management plans. According to Mead, implementation of this ever-evolving approach has been successful largely due to the following components:

- Training and continuing education of forestry staff
- Intensified research on issues associated with impacts and resource interactions
- Increased cooperation and collaboration with other landowners and resource managers
- Acquisition and use of a GIS database

The Mead Corporation is not conducting ecosystem management as defined in Section 2.3 of this book; however, their goal has never been stated to be the management of their lands using the ecosystem approach. Instead, the approach being used by Mead is defined as *sustainable management* and *good stewardship*, both terms that incorporate aspects of ecosystem management. They are definitely conducting spatial scale analyses of their systems, and it will be worthwhile, as the technology develops, for them to incorporate more explicit spatial scale analyses that are better at defining changes in the quality of the spatial units within the landscape (see Section 4.2.6).

Because that is their business, a private timber company will always have to focus on improving the quality and maximizing the output of the specific timber products harvested from their lands. However, if a company becomes more restricted to maintaining the timber outputs from fixed land bases, they will need to begin managing their lands in a manner that considers the impact of management activities at a mechanistic level on their resource base. A mechanistic understanding of the feedbacks that occur in ecosystems will allow them to begin detecting if there is a change occurring in the ability of a given land base to produce products to be harvested in the quantity desired at a later time scales (using tools articulated in Chapter 4). This long time interval between the regeneration phase and the final harvesting phase contributes to the importance of understanding ecosystem level processes in these forests. It will be important to incorporate an understanding of how human disturbances or human modifications of the natural disturbance cycles modify the resistance and resilience characteristics of forest ecosystems. A lack of understanding of these processes can feed back negatively on a company's ability to have sustainable management at later time scales.

5.4.3 Westvaco and the ACE Basin Coalition

Because cooperation between land use managers and collaboration across property boundaries are integral to ecosystem management, it is appropriate to examine the Ace Basin project. This project is an example of collaborative efforts between government agencies, conservation groups, and private landowners. This 141,750 ha (350,000 acre) project is located 64 kilometers southwest of Charleston, North Carolina, USA and is perhaps one of the most expansive and biologically diverse ecosystems along the Atlantic Coast (Muckenfuss 1994).

The ACE Basin is biologically significant for several reasons. This area contains the Ashepoo, Combahee, and Edisto rivers. This basin is home to more than 500 plant and animal species (including 17 endangered species) and is one of the most productive estuarine areas on the Atlantic Coast of the United States (Campbell 1989). The basin is considered to be the most significant eagle-nesting region in the United States and provides staging and wintering grounds for 14% of the migratory waterfowl that use the Atlantic flyway (Campbell 1989).

It was apparent that the publicly managed lands within the basin would be important in maintaining habitat diversity, recreational access, and educational opportunities. However, there was concern that without collaboration and cooperation from the private landowners, the goals on public lands could not be achieved. In 1988, a coalition was formed to protect and enhance the traditional land uses of the ACE Basin. This coalition included private landowners, the South Carolina Wildlife and Marine Resources Department, the United States Department of Fish and Wildlife Service, The Nature Conservancy, and Ducks Unlimited. Within the Basin, 22,275 ha (55,000 acres) are under protected status, and the remaining acreage is privately held. Management of the ACE Basin was founded on five principles (Muckenfuss 1994):

- To maintain the natural character of the area

- To maintain traditional natural resource use (i.e., hunting, fishing, forest management, and farming)

- To acquire additional conservation land (without land condemnation)

- To maintain and improve public access

- To improve wildlife habitat

Westvaco owns nearly 6,885 ha (17,000 acres) of forest lands in the basin that it manages for fiber production. For Westvaco, the ACE Basin project represented only one of many its efforts to implement an ecosystem approach to management. Although Westvaco's primary objective is the production of "a continuous sustainable supply of fiber for its mill in North Charleston," the company manages all of its properties on "an ecosystem, multiple-use basis to provide habitat diversity and protect environmental quality" (Muckenfuss 1994). On the ground, Westvaco's ecosystem, multiple-use approach is implemented through stratifying land to dominant uses: (1) areas to receive increasingly intensive management for fiber and timber production when they have been classified as being highly productive for trees and (2) those areas to be managed for nontimber values that are not highly productive for trees. In South Carolina this means that two thirds of the company's 202,500 ha (500,000 acres) are managed using high technology silvicultural systems that are compatible with the state and federal *best management practices*, while the remaining one third is managed for ecosystem values other than timber production (Muckenfuss 1994). Westvaco's management plans are based on a watershed approach whereby they identify special areas for enhancing water quality and habitat diversity that are then delineated for protection. In addition, because the public is often concerned with the visual impact of forest operations, Westvaco has tried to improve aesthetics by creating roadside corridors along well-traveled highways (Muckenfuss 1994).

This approach of separating land areas into intensive use and non-use forestry areas can be successful if a large land base exists so that the type of edge environment created within the landscape is not dramatically modified. But this type of approach does not address the landscape level distribution of these habitats and whether the different landscape units are modifying the quality of the environment in other adjacent or close proximity units. It becomes important to establish that the non-use areas have ecological processes that continue to function in the normal range for that system depending on how the system is situated with respect to other land management units. As already mentioned in the Mead Corporation example, it is important to have a mechanistic understanding of how the changing landscape mosaics affect the maintenance of species distributional patterns and ecosystem functions. Just doing spatial scale analyses without having an understanding of other processes and feedbacks occurring in the system means that this is not ecosystem management. One attribute of ecosystem management (i.e., spatial scale) is being addressed but the other attributes of ecosystems that need to be examined as part of ecosystem management have not been addressed (see Section 2.1.3).

5.4.4 Endangered Species and Industry

Timber companies have had to increasingly deal with the presence of endangered species and the legal ramifications associated with having these species on their lands. The Endangered Species Act does not require private landowners to "restore" or "recover" endangered species. It does, however, prohibit private landowners from "taking" any listed species. The issue of "taking" has become an increasingly controversial subject. For example, the U.S. Supreme Court recently upheld that timber companies must assist in habitat protection. It is likely that the forest industry will become increasingly sensitive to issues pertaining to the management of endangered species, especially in light of this recent case.

Private landowners have become increasingly wary of sharing natural resource information with state or federal agencies, conservation non-profits, and the general public, especially when it concerns endangered species. This is no wonder because there have been instances when timber forest products companies have been forthright in sharing information with interested parties only to have their own information used against them, often ending in court battles. In an attempt to assuage these fears, Champion International Corporation and the United States Department of Interior's National Biological Service (NBS) have signed a Memorandum of Understanding (MOU) to develop a model process for biological inventory, survey, research, and monitoring on private forest lands (Owen and Sweeney 1995). Because federal agencies such as the NBS are subject to the Freedom of Information Act, this requires them to comply with requests for information from the general public. Therefore, in the past, private landowners, such as timber products companies, have been apprehensive about interacting with these agencies.

On September 7, 1994, Champion and NBS signed a Memorandum of Understanding in which Champion agreed to make the Gulf Tract [a 2,714 ha (6,700 acre) property in Tennessee] available as a research and demonstration site: "This watershed will be operated as a working industrial forest while NBS and Champion field-test survey procedures, habitat relationship models, and long-term monitoring efforts that would be compatible with private land-owner goals and objectives" (Owen and Sweeney 1995). As part of the MOU, Champion and NBS have agreed to initiate discussions on procedures for "handling, disposition of, and access to the data collected." This process could then be used as a model for cooperation with other private forest landowners.

Another example is the management of the red cockaded woodpecker on a Champion International property. In early 1995, Champion International, United States Department of Fish and Wildlife, USDA Forest Service, and the Texas Parks and Wildlife Department signed a Memorandum of Agreement in which Champion committed 810 ha (2000 acres) in Texas to foster the recovery of an endangered species (i.e., the red cockaded woodpecker). Although Champion will continue to harvest timber from this area, it will also foster endangered species management by increasing roosting and nesting sites through the addition of artificial cavities, by prescribing fire burns to prevent midstory encroachment by hardwoods, and by using silvicultural prescriptions that favor longleaf pine (*Pinus palustris*), and by extending rotations in order to improve the woodpecker's habitat. It is hoped that the management area will be able to support up to 20 breeding groups, thereby creating a strong population of red cockaded woodpeckers in the area, which will satisfy the Forest Service's endangered species restoration goals.

5.4.5   Lessons Learned and Conclusions to be Drawn from this Case Study

Mandates Versus Internal Motivations

Federal agencies are often required by the organic acts that created them to implement management objectives that are conflicting in nature. For example, the Organic Act of 1916, which created the National Park Service, mandated that the Park Service conserve the national parks for their scenery, natural and historic objects, and wildlife; provide for pubic enjoyment of these areas; and preserve them for the benefit of future generations (16 U.S.C. §1). From the beginning, the Park Service has had to contend with the dual objectives of preservation versus public access. More recently, the 1978 amendments to the Organic Act, as well as a series of lawsuits, have clarified that the preservation takes priority over access in the National Park system. The Forest Service, on the other hand, has operated under a multiple-use management philosophy, according to which it must balance competing interests for the maximum public benefit. Since its inception, this philosophy has evolved as a series of laws have been enacted through time (see Appendix A). For example, the Forest Service's Organic Administration Act of 1897 mandated the management of forests for the objectives of timber supply and watershed protection (16 U.S.C. §475). In 1960 this philosophy changed with the passage of the Multiple Use-Sustained Yield Act, which extended the Forest Service's mandate to include "outdoor recreation, range, timber, watershed, and wildlife and fish purposes" (16 U.S.C. §528), and then again under the National Forest Management Act of 1976, which stressed a "systematic interdisciplinary approach to achieve integrated consideration of physical, biological, economic, and other sciences" (16 U.S.C. §1604).

Given this history of federal directives, it is no surprise that ecosystem management has become yet another federal mandate. On June 4, 1992 Forest Service Chief Dale Robertson informed his staff that ecosystem management is now the "Forest Service way" (Gerlach and Bengston 1994). Less than a year later (February 1993), Secretary of Interior Bruce Babbitt announced that an ecosystem approach would be used in environmental and endangered species protection (Gerlach and Bengston 1994). In effect, the Department of Interior and the Forest Service have been directed to manage over 280 million ha (700 million acres) of land using this new management paradigm.

Recent governmental mandates have raised concerns among industry that it will soon be mandated to "do ecosystem management." The current atmosphere created by the concerns of forest product consumers, corporate shareholders, and forest industry employees has forced industry to be more creative in addressing environmental concerns. The question remains whether mandates or publicly driven motivations will foster better management.

It has been proposed that the former management paradigm, consisting of land allocation for species protection, sustained wood yield, and multiple-use management is no longer appropriate today because of the multitude of uncoordinated environmental laws and regulations (Gordon 1994). There have been attempts to create statewide ecosystem management laws, such as in the State of California. This raises the question, *Would environmental concerns be better addressed in a single all encompassing ecosystem management law, or should regulation continue under several pieces of environmental legislation comprising single resource management goals?*

Information and Uncertainty

How do government agencies and corporations operate under uncertain circumstances? How do they make their decisions? In the Pacific Northwest, natural resource management regimes have gone into a noticeable holding pattern, because in many instances management of federal lands has been a result of the processes of litigation and appeal—a costly and ineffective use of resources (Lucier 1994). Industry, on the other hand, has no such public mandate, is not confronted with the staggering public review process, and does not have the ability to wait until the best knowledge is available. In order for a company to survive it must keep producing, and it does this by using the best information available at the time.

The financial success of a forest product company relies heavily its ability to anticipate potential problems. As mentioned earlier, forestry professionals are quick to relate that large- scale planning in terms of geographic scope and time is not new to the industry. However, the tools that further enhance this approach have only recently been developed. The industry consistently cites remote sensing and GIS technologies as having profound effects upon its ability to manage and proactively plan for potential obstacles (Lucier 1994). *Again, it is important to consider that spatial scale analyses are insufficient in themselves to do ecosystem management.*

Boundary Delineation and Interagency Collaboration

Boundary delineation has been cited as one of the key components to implementing ecosystem management. Instead of managing ecosystems based on jurisdictional boundaries, for example, proponents recommend that management boundaries should reflect the needs of the ecosystem of interest. Timber forest product companies consistently reported the use of geographic scales based on watersheds or a combination of landscape ecology and watershed analysis. As mentioned in the FEMAT case study (see Section 5.2), these landscape and watershed scales are not always appropriate for every management consideration and should be used with some precaution. However, industrial forest management dictates the use of larger spatial scales, and it appears that in terms of industrial forest management, landscape or watershed scales are often likely to be the most conducive.

For federal agencies it appears that boundary delineation has posed considerable obstacles. Take, for example, the case of the Greater Yellowstone Ecosystem (GYE). Clark and Minta (1994) cited that disagreement among the major actors over the boundaries of the GYE has been the primary obstacle to implementing ecosystem management in this area.

Disagreement stems largely from the lack of agreement over definitions, size, and locations of GYE. A compromise between ecological and administrative boundaries has yet to emerge.

Timber forest products companies have been attempting to implement boundary based analyses on their lands. The Mead case involving the Mulligan Creek project is an excellent example of boundary delineation that was not based on ownership patterns. The Mulligan Creek planning effort was based on a combination of watershed management and landscape ecology. Once Mead was confident with the implications of this analysis, it approached adjacent landowners, Champion International and Longyear Realty, to encourage their support and participation. More than likely, timber products companies such as Mead may find it easier to succeed in the boundary delineation process because they operate in less politicized atmosphere then, say, a federal agency. Federal agencies are required by law to undergo intensive public review processes that may make boundary delineation all the more difficult.

Mead's Mulligan Creek case demonstrated that not only was the company able to define ecosystem boundaries on its properties, but it was highly successful in cooperating and collaborating with adjacent landowners or other major players such as the Michigan Department of Natural Resources. Some conservation organizations viewed this collaboration in a positive fashion, while others remained skeptical. For example, Michael McCloskey, Chairman of the Sierra Club (Washington DC) called this effort "leading edge" and said that it "could turn out to be a more constructive, mature relationship between the environmental community and the timber industry" (Schneider 1993). On the other hand, local environmental groups, such as the Superior Wilderness Action Network, saw this as just an effort to create the "right image" (Schneider 1993). This perceived lack of credibility among local environmental groups led to considerable delays in the implementation of this management area. A lesson learned from this situation is that despite collaboration and cooperation adjacent landowners, if other stakeholders are not confident in the process, obstacles will arise. This scenario highlights the need for considerable communication efforts among all stakeholders.

5.5 Adirondack Park Case Study

Adirondack Park is located in upper New York State. Today, nearly 2.4 million ha (6 million acres) of land are located within the park's boundaries, including roughly 1 million ha (2.5 million acres) of state forest preserve lands interspersed with 1.4 million ha (3.5 million acres) of private lands. This mixture of private with public lands has created a unique conservation and sustainable resource management problem. This park represents an interesting mixture of ownership groups and different legitimate activities that can occur within the park boundaries, not that different from the scenario existing in the coniferous forest landscape of the Pacific Northwest (see Section 5.2).

Though ecosystem management is presently not being implemented in the park, the Adirondack Park is a good case example in which to demonstrate the utility of the ecosystem management approach. Park management will have to deal with the mosaic spatial patterns of land utilizations and the ecological consequences of these land uses on adjacent landscape units where human activities have been restricted. When the park was formed, people maintained ownership of small to large tracts of land within the boundaries of the park, resulting in multiple use being an integral part of park management. This has resulted in long-term conflicts between people's desires for managing their own land (i.e., private property rights issues) and also satisfying the needs of park management to maintain a "sustainable" park for usage by people living outside the park boundaries. A more detailed discussion of the historical and planning process utilized in the development of the Adirondack Park follows. The type of scientific information that was used to assess the intensity of development allowable in the Adirondack Park is also presented, focusing on one specific example of development in a pond watershed. The Adirondack Park is an excellent example illustrating the long-term conflicts that can occur in natural resource and environmental management, and the different roles of science and human values in contributing to management decisions.

The approach used to manage the Adirondack Park in the 1970s appeared to be quite successful because it identified several different types of sensitive areas for protection and allowed for a balance to coexist between conservation and development in the newly forming park. Furthermore, based on a 1970s understanding of ecosystem function within a landscape framework, the approach used was the best available at that time to deal with the low availability of data, uncertainty in our understanding of ecosystems, and the short time scale in which to implement the rules for the development of a park (a very similar scenario to what was happening in the westside old-growth forests in the Pacific Northwest when the FEMAT report was formulated; see Section 5.2). In the 1970s, the decisionmaking framework for determining the map boundaries within the Adirondack Park lacked a scientific basis to this process. If this planning process was occurring today and was based on an ecosystem framework, many of the approaches presented in Chapter 4 could be utilized to assess how the boundaries should have been demarcated between different developmental zones.

5.5.1   History of the Formation of Adirondack Park

At the time of the American Revolution, the State of New York owned approximately 4.4 million ha (11 million acres) north of the Mohawk River valley, known as the Adirondacks. During the early 19th century, the state permitted the sale of timber and also lands to private interests, so that by 1883 public ownership of land had been reduced to only 324,000 ha (800,000 acres). At the same time as the amount of public land was decreasing, the public's attention was directed towards many examples of "land degradation" resulting from extensive overharvesting of forests within the Adirondack area. The public linked poor forest harvesting practices to the cause of extensive forest fires, soil loss, and the cycles of floods and low water outputs from the watersheds draining the park (water sources for communities outside of the park). Even commercial traffic on the Erie Canal was perceived to be threatened by the mismanagement of forest lands in the park.

In 1885, the state legislature responded to these environmental problems by attempting to restrict any additional transfers of public lands to private ownership by creating the Adirondack Forest Preserve. Despite this, an additional 40,500 ha (100,000 acres) of public land was acquired by private hands through title disputes. By 1893, timber interests had also persuaded the state legislature to allow them harvest mature timber from the preserve. The people of New York were upset by these continued abuses of the public domain and, in the following year, adopted a constitutional amendment by public vote that prohibited public land sales or exchanges, and prevented the sale, removal, or destruction of timber without a constitutional amendment (Pomeroy 1961). The protected status of the Adirondacks was complemented by a land acquisition policy that also allocated funds to buy land.

Significant increases in recreational use of the Adirondack Park occurred in the late 1950s and early 1960s. During this time, the development of the highway system in New York, combined with increased interest in vacation homes, created considerable developmental pressures within the park's 107 towns and villages. In addition, hundreds of thousands of tourists began to visit the park during the summer time, many of them taking advantage of the campgrounds developed by the state. Thus the increased ease of access and utilization changed the way that New Yorkers used the Adirondacks, and at the same time contributed to rapidly expanding the political support for conservation of the natural resources contained in the park by the public.

Few Adirondack communities had zoning regulations linked to goals of resource conservation or environmental quality during the 1960s. The most intensive commercial and residential developmental pressures focused on shorelines of the hundreds of lakes and ponds, and lands adjacent to thousands of miles of Adirondack rivers (Liroff and Davis 1982). On lakeshores, not only was development visible, but it often threatened water quality in the same lakes that were used by these municipalities as their water supply.

Growth pressures clashed with the rise of environmentalism during the 1960s. For example, the core of the Adirondacks was once even proposed to be included in the U.S. National Park system in an attempt to control the escalating development that was occurring in the park. However, private, local, and state interests joined in vehement opposition to this federal plan. As an alternative, the state proposed a highly restrictive plan for state control of private lands, along with strict guidelines for the management of public forest preserve lands. The result was the passage of state laws in 1971 and 1973 that created a highly controversial set of land-use regulations to be implemented by the Adirondack Park Agency (created in 1971).

The Adirondack Park Agency was granted broad authority to control private land development within the park (State of New York 1991). The primary purpose of the agency was to insure that the park's environmental quality and open space characters were maintained (see below):

"The basic purpose of this article is to insure optimum overall conservation, protection, preservation, development and use of the unique scenic, aesthetic, wildlife, recreational, open space, historic, ecological, and natural resources of the Adirondack park" (Adirondack Park Agency Act).

Any threats to maintaining the "quality" environment of the park were to be addressed "within a land use control framework which recognizes not only matters of local concern but also those of regional and state concern" (State of New York 1991). Thus, the purpose of the law was to achieve a balance between local and state concerns and interests in the management of the park (i.e., management across political boundaries).

The statute was conceived to balance "conservation, protection and preservation" with "development and use" (State of New York 1991). Although the environmentalists who drafted the law never intended this utilitarian purpose, concessions in the law's mission statement were necessary to appease vocal opponents who lived within the park boundaries and who often made their livelihood from either resource extraction, tourism, or land development.

The 1971 statute required the newly formed Adirondack Park Agency to prepare a land-use plan, including a map, which specified the types and intensities of allowable land uses throughout the park. The plan and map were designed over an 18-month period following a quick appraisal of the vegetative communities existing within the park boundaries. At the same time, those areas that needed special protection because they were considered to be ecologically "vulnerable" back in 1971 were identified and recorded on the maps.

Since 1970, two dominant legal strategies evolved to attempt to balance development while at the same time maintaining the open space character of the park. The first strategy was the design of a comprehensive land-use plan that specified allowable land and resource uses on all of the private lands throughout the park. The second strategy consisted of having the Adirondack Park Agency and the State of New York review all development or resource extraction projects. Prior to any construction or resource extraction in the park, a state permit was required, which had to be approved by the Adirondack Park Agency staff scientists as well as the politically appointed commissioners. The first strategy defined the spatial utilization of the landscape of the park, which was not based on any mechanistic understanding of ecosystem processes occurring at larger scales or how different landscape units interacted with one another. The second approach had the potential to introduce science into the decision framework but again has at very site-specific scales and was not ecosystem based.

5.5.2 Creating the Land-Use and Development Plan

In 1969 Ian McHarg published *Design with Nature*, a book that had significant influence on the future of the Adirondack Park. McHarg was a landscape architect at the University of Pennsylvania who argued persuasively that humans normally overwhelmed ecological processes as they developed the landscape. Using examples, including the New Jersey shoreline and the Brandywine River valley in Pennsylvania, he demonstrated a form of landscape planning that was governed by an understanding of the inherent sensitivity of ecological systems to disturbances, such as the fragility of coastal barrier dunes or the hazards of developing in floodplains. As part of his landscape planning, McHarg (1969) developed an approach of using several maps charted on acetate overlays. These maps, when superimposed on top of one another, could be used as a tool to highlight locations within the landscape with pre-identified characteristics that would make them appropriate or unsuitable for development (Table 5-1).

Separate maps were developed on those characteristics considered important to understand to maintain the supply of resources from these lands: hydrology, slope, soils, wetland areas, floodplains, wildlife habitats, along with cultural features (i.e., scenic vistas, historic sites, and parklands). This mapping approach was a simple but potentially powerful tool that was readily adopted by environmentalists who were searching for a "scientifically legitimate" tool to assess environmental impacts. The ap-

TABLE 5-1   Physical, biological, and public resource values used to determine the impact of development in the Adirondack Park

Resource category and specific considerations	Source and nature of inventory data	Resource evaluation and percent screening
Physical resources		
Soils: depth, bearing capacity, erosion potential, permeability, drainage, depth to water table, sewage effluent capacity, floodplain soils	Portions of park covered by SCS soil surveys but with variable resolution; remainder mapped in 1972 by SCS with aerial photos, with no field checking	Evaluation based on use limitations in Cornell/SCS (1972); 0–80% screening
Slope: erosion potential, sewage effluent capacity, visibility	Estimated from topographic maps	Slope categories and evaluation based on earlier work in VT (McHarg 1972) and Finger Lakes, NY (Bailey 1972); 16–25% slopes screened at 20%; > 25% slopes screened at 40%
Elevation > 2500 ft: low temperature, high precipitation, high water storage capacity Other considerations (soil, slope, vegetation) recognized in other categories	Topographic maps	Vogelman et al. (1969) as scientific justification; VT regulations (Act 250) as legal precedent; 40% screening
Unique physical features: waterfalls, unusual geologic formations	APA inventory; survey of geology and recreation experts and knowledgeable local residents (G. Davis, personal communication); No objective criteria for inclusion	These features given point locations instead of screening Presence of a unique feature tended to "tip the scales" towards more restrictive development categories (G. Davis, personal communication)

TABLE 5-1 (*Cont.*)

Resource category and specific considerations	Source and nature of inventory data	Resource evaluation and percent screening
Biological resources		
Fragile ecosystems: bogs, alpine and subalpine communities	Cornell land use/natural resource (LUNAR) mapping, based on interpretation of aerial photos (G. Dupree, personal communication)	All fragile ecosystems at 60%
Ecotones (characterized by high levels of biological diversity): juxtaposition of two or more vegetative communities with abrupt boundaries/transitions	Based on LUNAR mapping, with ecotone identification based on APA staff judgment of boundary sharpness (G. Dupree, personal communication)	All ecotones screened at 40%
Vegetation: old-growth forest, rare plant habitat	APA identification of old growth and consultation with NY State Museum re. rare plants	Old-growth screened 60%; rare plant habitat screened 40% or 80%, depending on level of rarity
Wildlife: habitat of rare and endangered species; key habitat for deer and native trout		Habitat for species rare within park and for deer and trout screened 40%; habitat for species rare nationwide screened 80%
Public Resources		
Scenic vistas: viewsheds accessible to the public	APA inventory of all significant scenic views from roads accessible to the public; inventory based on staff judgment with no objective criteria (G. Dupree, personal communication)	Viewsheds screened at 60% in foreground ($< \frac{1}{4}$ mi.), 40% in midground ($\frac{1}{4}$–5 mi.), 20% in background (> 5 mi.)
Travel corridors: areas visible from roads	APA inventory of areas visible from undeveloped roads accessible to the public (G. Dupree, personal communication)	Visible areas within $\frac{1}{2}$ mi. of roads screened at 40%; visible areas beyond $\frac{1}{2}$ mi. screened at 20%

TABLE 5-1 (*Cont.*)

Resource category and specific considerations	Source and nature of inventory data	Resource evaluation and percent screening
Proximity to public lands: areas within sight or sound (usu. $< \frac{1}{2}$ mi.) of popular sections of wilderness, primitive, or canoe areas	APA inventory; no objective criteria	All mapped areas screened at 40%
Proximity to wild, scenic and recreational rivers	Designated wild and scenic rivers system, with APA inventory of areas within sight/sound of wild and scenic rivers (up to $\frac{1}{2}$ mi.) and within 500 ft of recreational rivers	Evaluation based on interpretation of legislature's stated concern for the different categories; area around wild and scenic rivers screened at 80%; area around recreational rivers screened at 20%
Historic sites	Inventory by State Historic Trust and Blue Mountain Lake Museum	All sites at 80%

proach introduced by McHarg (1969) was interesting in that it attempted not only to include natural science information as part of the maps, but it also included cultural elements, such as existing development, scenic attributes, and historic sites in resource categories. Today, it is apparent that this approach is useful as a first overview of a site (this is in actuality an early GIS approach), but it is not very effective in being able to really state whether a site is resistant or resilient to the proposed developmental activities (see Sections 3.2 and 3.3).

Those charged with the responsibility of preparing a land-use and development plan for the Adirondacks turned to McHarg's methods for keeping track of ecological and social information. Overlays were combined to provide an image of where development was appropriate to occur within the landscape. Several areas were immediately identified as being unsuitable for development (i.e., floodplains, wetlands, rare species habitats, locations characterized by steep mountainous terrains, and shallow soils). Areas considered suitable locations for future growth and development were those characterized by deeper soils, which were typically found in valleys; these lands were especially desirable when they had reasonably well-drained soils and existed near village centers. If these desirable physical features were located far from existing villages, they were deemed unsuitable for development because it was felt that delivery of public services at these distances would become a burden on local taxpayers. Lands lying adjacent to state forest preserve lands were normally judged less appropriate for development than lands with similar resource constraints located far from state boundaries.

The planners also recognized the possibility of overcoming natural limitations through design and technology. Their primary goal was to identify areas "that, from a natural resource standpoint, are best suited for development," but also to identify areas "where the physical characteristics of the land will require that certain standards be imposed if development is to provide positive values..." and "where the potential costs of development to the developer, the community and the prospective home owner are so great that serious consideration should be given to other kinds of uses" (Temporary Study Commission Adirondacks 1970). The process thus identifies three fundamental categories of land: areas that are suitable for development, areas that are suitable with strict limitations and/or standards, and areas that are unsuitable. The categories that were eventually adopted are simple refinements of these three broader classifications. By recognizing the potential of design and technical standards to mitigate development impacts, the planners believed the process they followed was consistent with the authorizing legislation, which required balance between resource protection and development.

The information used to determine resource capability of withstanding development was categorized as (1) physical resources, including soils, slope, and unique geological features; (2) biological resources, including fragile ecosystems, ecotones, vegetation, and wildlife; and (3) public resources, including scenic vistas, travel (road) corridors; proximity to public land, designated wild, scenic, and recreational rivers; and historic sites (Table 5-1). All of these resources were inventoried by Adirondack Park Agency staff (either through original research or compilation of existing data) and were evaluated for their sensitivity to development, sometimes based on published sources but more commonly on the judgment of the staff or outside advisors. The values that were examined, the sources and quality of data used, and how resource sensitivity was evaluated are outlined in more detail in Table 5-1.

Once summary overlays showing physical, biological, and public re-

TABLE 5-2   Five categories groupings used to define type of developmental activity possible on lands in Adirondack Park

Category 1
0% screening: No special resource limitations for development and no significant environmental impact expected
Category 2
20% screening: Minor resource limitations for development, which need to be taken into account to avoid adverse environmental impact
Category 3
40% screening: Moderate resource limitations for development, which requires special considerations to avoid adverse environmental impact
Category 4
60% screening: Severe resource limitations for development, which would make it very difficult to avoid adverse environmental impact
Category 5
80% screening: Overriding resource limitations, which make development without unacceptable environmental impact unlikely

sources were completed, they were combined to form a map of overall landscape suitability for development. The Adirondack landscape was separated into five category groups, reflecting different levels of land sensitivity to development that were delineated according to the percent screenings shown in Table 5-2. The five categories were then converted to three more general development suitability types: categories 1 and 2 were combined to show areas suitable for development, category 3 was depicted as suitable for development with standards imposed to overcome limitations, and categories 4 and 5 were combined to show areas where development was generally believed to be inappropriate. Once the resource capability map had been completed, Adirondack Park Agency staff inventoried a variety of cultural features [i.e., including existing areas of development, existing or proposed infrastructure (public sewers and water) and existing land use (developed areas, extensive forestry, and agriculture)]. Where existing infrastructures were capable of overcoming physical or biological constraints (for instance, where public sewers eliminated the need for waste disposal in areas of poor drainage or shallow soils), areas were moved into less restrictive categories, in effect "erasing" limitations from the capability map. Significant or large areas of managed forest were seen as valuable both economically and for their open space characteristics, and tended to be more restricted from development.

Because nearly 200,000 people lived within the park boundaries at the time the plan was designed, considerable conflict erupted over zoning near the 107 town and village centers. Early in the process, Adirondack Park Agency staff proposed differentiating between rural hamlets (which were relatively small and exhibited a traditional, rural character) and urban hamlets, such as Lake Placid, Lake George, and Saranac Lake (which had already undergone intensive development and commercialization). Rural hamlets were seen as an asset to the park's character and in need of protective zoning and design standards to preserve their small size and traditional style. Urban hamlets were seen as less in keeping with park character. Despite this negative perception of urban hamlets, some degree of continued intensive development without restrictions was considered necessary for community and economic stability. The distinction between rural and urban hamlets was dropped, however, by the Adirondack Park Agency commissioners due to the opposition voiced by politicians from the rural hamlets who were slated for more restrictive development.

The agency ultimately decided to treat all hamlets similarly, to impose no density restrictions within their boundaries, and to allow additional residential development in moderate- intensity and low-intensity use areas surrounding the hamlets. Low- and moderate-intensity use designation was assigned to areas with a high capacity for development if they fell within 3 to 5 miles of the hamlets. Any areas with high development capacity that were farther from existing centers of development were automatically zoned as rural use. Nearly all areas in the backcountry, with the exception of some lakeshores that were physically and biologically capable of withstanding intensive development, were automatically given a high level of protection for their declared open space values. In some cases, when hamlets had little or no adjacent area with high development capability, these criteria were applied with greater flexibility. Under these circumstances, sensitive resources were sometimes included within hamlet, moderate-intensity, or low-intensity use areas in order to provide all towns some room for expansion.

Even though the planning process back in the early 1970s was an attempt to integrate natural and social science information similar to what is needed as part of ecosystem management (see Chapter 1), much of the process was very subjective and was based largely on the Adirondack Park Agency staff's value judgments. Much of the scientific information used was fragmented and mainly focused on examining systems at large spatial scales, scales that are difficult to make relevant for answering questions at small site scales of a development project (see Section 3.2). For example, slope and elevation data were simply taken from topographic maps, while soils data were mapped from aerial photographs and not field checked, and scenic views (which were declared to be crucial to the open space character of the park) had no clearly defined standards for selection. Although natural features were sometimes mapped down to a resolution of 14 ha (35 acres), the land-use classification map clearly does not differentiate most small- and medium-sized waterways. It thus seems likely that many floodplains (which occur at roughly the same scale as streams and also tend to be linear features) were not incorporated into the map. Unfortunately the data that were used in these analyses and how these data were evaluated are poorly documented, so it is impossible to determine what type of information received the most attention in the development of the land-use maps for the Adirondack Park. Based on the type of information collected during the planning process (see Table 5-1), the approach used by the Adirondack Park Agency was probably successful in identifying those areas that were clearly going to be negatively affected by development (i.e., wetlands) and could be identified from examination of information obtained at coarse spatial scales. However, many of these areas were probably already obvious to the planners, and the mapping operation mainly verified and reinforced this grouping by making the information available in a visual form.

Spatial Decisionmaking in Mapping

One facet of planning that proved more difficult than anticipated earlier was the task of determining the boundaries among land-use categories. Mapping resolution was typically about 122 ha (300 acres) for features such as soil types. In some instances, more detailed inventories were available. For example, one of the original resource mappers recalled the minimum area mapped to have been approximately 18 ha (45 acres) or "the area on a minute quadrangle that would fit under a nickel" (G. Dupree, personal communication 1995). Based on the information in Table 5-1, inventory data covered a broad range of resolutions. The patterning that resulted on the land-use classification map clearly resulted from the mapping features at a scale much greater than 122 ha.

Boundaries eventually chosen among different zoning districts do not normally conform to the natural resource boundaries drawn on the original capability map because these natural boundaries were often considered to be too imprecise. Instead, identifiable boundaries (i.e., roads, rivers, and town and county lines) were chosen wherever possible. For example, where a zone included a roadside buffer or a floodplain, offset lines of 0.06 km (1/10th mile) or 0.16 km (1/4 mile) from roads or streams were used.

The relatively coarse resolution of mapping and the frequent use of arbitrary cultural boundaries instead of permanent natural boundaries (such as watershed divides or habitat boundaries) could at first glance be seen as grounds for criticism of the method, or at least the way in which it was implemented. Although the classification map showed little in the way of natural features, the predominance of straight lines and inorganic, geometrical shapes is clearly at odds with natural boundaries in most locations. Particularly noticeable on the map is the fact that zones show little conformity to watershed boundaries and drainage patterns. For example, many boundaries cut across streams at random angles.

Land ownership and development patterns appeared to have played a far more important role in determining allowable density of development than the resource capability analyses of the agency. As described earlier, the plan attempted to channel development toward existing town and village centers while restricting growth from outlying areas. Lands located adjacent to state forest preserve lands were restricted from intensive development. These state-owned lands gave the park an unmistakable visual perception of "wildness" and have been vigorously protected by New York conservationists for more than a century.

Those owning private lands adjacent to public lands reacted to highly restrictive zoning with outrage, claiming it was a state ploy to reduce land values and clearing the way for a less costly state land acquisition program. The jagged shape of many state land boundaries, along with the existence of many private land inholdings surrounded by state lands, imparts legitimacy to their claims. Still, park planners had little interest in fighting battles then faced by the National Park Service as it struggled to manage private growth adjacent to areas such as the Rocky Mountain, Yellowstone, and Grand Teton National Parks.

The planners still faced the difficult problem of deciding what density of development should be allowed, and how these density zones should be related to the diverse resource capabilities and cultural features of the 1.4 million ha of private lands existing in the park. Hamlets posed no such problem for the planners because it was decided that no density limits should be applied to community centers. In fact, a high density of development was to be encouraged to reduce pressures on more remote tracts of land. For other areas, the planners decided to assign rough allowable densities of development, which were to be described as the number of additional structures allowed per area of land. Density restrictions applied, with the following zones groupings being used: moderate intensity—requiring roughly 0.5 ha (1.3 acres) for a new structure [193 structures/km^2 (500/mi^2)]; low intensity—requiring 1.3 ha (3.2 acre) for a new structure [77 structures/km^2 (200/mi^2)]; rural use—3.4 ha (8.5 acres) [29 structures/km^2 (75/mi^2)] and resource management—the most stringent private land classification, requiring 17.3 ha (42.6 acres) [6 structures/km^2 (15/mi^2)] (APA Act 1972).

Using these broad zoning guidelines and the criteria described in Table 5-1, planners placed all private lands within one or another of the density categories. More detailed regulations and criteria for permit granting, such as the setback requirements from waterbodies and roads, were also associated with each of the zones. Planners calculated that the allowable development under the new density guidelines would permit the number of structures in the park to more than double.

In January 1973, after completion of the draft land-use classification map, meetings with local officials resulted in 800 proposed changes to boundaries and classifications. Nearly 500 of these resulted in boundary revisions, most involving expansion of hamlet boundaries (many hamlets were nearly doubled in size) and of low- and moderate-intensity use areas. In no cases were these areas reduced in size. Few if any changes were based on any understanding of ecological resilience or sensitivity.

Shoreline Zoning

The zoning of shorelines resulted in special problems for park planners. With hundreds of lakes, thousands of ponds, and several thousand kilometers of rivers, shorelines represented critical ecological, scenic, and recreational resources in the minds of those creating the zoning scheme.

The original plan for controlling development proposed by the Adirondack Park Agency would have protected shorelines through both land-use classifications and a set of shoreline restrictions that were to apply to all lakes, ponds, and navigable streams and rivers. Most shorelines would have received considerable protection (either rural use or resource management zoning) from the staff's recommendations because of the presence of several ecological and social features considered important to protect in the overlay mapping operation (i.e., floodplains, high water tables, designated wild and scenic rivers, marshes, and bogs). However, many shorelines (especially lakeshores accessible by road) had already become valuable for residential subdivisions and commercial uses. Because these shorelines had much greater value than upland or more remote areas, the planners were quite concerned about the possibility of challenges to the constitutionality of the entire zoning scheme based on a claim that regulation constituted an uncompensated taking of land by the State of New York.

The Adirondack Park Agency changed their originally strict protection of shorelines to allow more intensive development, especially around lakeshores. Protection of the shorelines was further reduced by the agency when maximum allowable densities for each classification zone were increased, first by the agency in the final version of the proposed plan and then by the legislature as it translated the plan into law. The legislature further increased shoreline development capacity by enacting a provision that allows shoreline clustering of as many as 33 buildings per linear kilometer (53/mile) in moderate-intensity use areas and 26 buildings per kilometer (42/mile) in low-intensity use areas. Clustering is permitted if sufficient contiguous land is owned behind the shore to meet the overall density guidelines for the land category in question. This series of compromises, driven by financial, legal, and political pressures, had an unusual effect on shorelines. The lower density requirements of other more remote areas served to channel growth pressure to the shorelines that were both undeveloped and zoned for more intensive development.

The original proposals for shoreline restrictions were among some 200 provisions that were weakened by Adirondack representatives in the state legislature the night before the final vote to adopt the plan in 1973. For example, the minimum lot widths and minimum setbacks established for each land-use classification are given in Table 5-3. Although these setback restrictions offered some protection of Adirondack shorelines, they must be considered in the context of density guidelines and specific site conditions before one could conclude that they offer effective protection to the biological, physical, aesthetic, and recreational values of the park.

TABLE 5-3   Minimum lot widths and minimum setbacks required for the five land-use classifications in the Adirondack Park

Land use classification	Minimum lot width	Minimum setback
Hamlet	15.2 m (50 ft)	15.2 m (50 ft)
Moderate intensity	30.5 m (100 ft)	15.2 m (50 ft)
Low intensity	38.1 m (125 ft)	22.9 m (75 ft)
Rural use	45.7 m (150 ft)	22.9 m (75 ft)
Resource management	61.0 m (200 ft)	30.5 m (100 ft)

5.5.3 Forms of Resistance

Passage of the Adirondack Park Land Use and Development Plan, containing the land-use restrictions just described, constituted a radical shift in legal authority to control land and resources. Rights formerly held by private landowners and only loosely regulated by local governments were now controlled by state regulators. The dominant strategy to conserve park resources was one of centralizing legal authority. The history of regulation may be told as one of resistance to this shift in authority, resistance that took many forms and is further discussed later.

Local Government Authority

It was originally expected that once the regional plan was adopted as a state zoning law, local governments would then begin the process of refining the boundaries through more detailed resource and development suitability analyses. The act passed in 1973 required that the state review and approve most development proposed on private lands within the park. However, the authority to review and approve projects would revert back to local governments if they completed their refinement of the resource capacity analyses, density boundaries, and zoning restrictions, and if they incorporated state standards. Class A Regional projects would always be reviewed by the state; however, smaller Class B projects would be reviewed by local planning and zoning boards.

Millions of dollars of state funds were allocated to local governments over the following two decades to support local planning efforts to refine the state plan and regulations. Among all 107 towns and villages, however, only 10 have sought approval from the state to secure the authority to review Class B development proposals. Many local governments remained hostile to state officials, resentful at their loss of control over private development. The poor Adirondack economy was perceived by them to be further threatened by the development restrictions, and their primary source of revenue was local property taxes. The highly restrictive density guidelines appeared to freeze development potential and local tax bases.

During the first decade following state adoption of the state plan, local refusal to adopt state standards was viewed as a form of resistance to the entire zoning scheme, and it became part of a larger strategy to repeal the law and abolish the Adirondack Park Agency. On more than one occasion, the agency's future was in question as legislative votes to repeal the statute were marshaled by a coalition of private development and local government interests. These challenges were successfully averted by changing representation among park agency commissioners. A shift in leadership from an environmentalist to a local official in 1977 deflated claims that the Adirondack Park Agency was insensitive to local economic conditions.

Judicial and Administrative Appeals

One of the most significant concerns of the original park planners was that courts might find the state density restrictions to be a violation of the federal Constitution's protection of private property rights. The Fifth Amendment requires that private property not be taken for public purposes without "just compensation." Because many parcels of land were severely restricted from development by the new regulations, the question of their constitutionality was prominent in the minds of developers, regulators, and environmentalists.

Throughout the 1970s several cases were brought before the state judiciary claiming that specific provisions of the agency's control amounted to a taking, and called either for compensation or repeal of the standards. In all of these instances, courts found that the underlying purposes of the regulations—protecting biological, physical, and even aesthetic resources—were legitimate uses of the state's police powers. By the early 1980s the weight of these decisions served to support public perception of the legitimacy of the state regulations, at least in the eyes of those with no intention of developing private lands in the park.

If a developer disagreed with the agency's density guidelines, they could appeal by requesting a map amendment. These requests were taken very seriously by the agency staff and commissioners, who recognized the amendments as the Achilles heel of zoning law. Proposals for map amendments were not permitted to be considered in the context of a specific site review. Instead they were encouraged to be part of local governments' refinements of the state plan. Still local delay in plan adoption discouraged this more comprehensive review, leaving the agency with the burden to judge many individual requests, the vast majority of which were denied.

Development Despite the Law

Most large developers confronted the state regulations directly with expertise from lawyers, planners, and natural scientists. Many Adirondack residents, however, simply developed or subdivided their lands as they always had planned to. Because these minor projects were difficult to recognize, especially single family home construction or two lot subdivisions, they often escaped state review. This problem was compounded by the complexity of regulations, which were difficult even for agency staff to interpret. Jurisdiction, for example, was often triggered by scientific interpretations of what soils or species defined a wetland, or the precise location of the development with respect to zoning boundaries. Because the need to obtain a development permit was often dependent upon site conditions or more careful analysis of the coincidence between property boundaries and zoning boundaries, this complexity quickly translated into substantial delays in state responses to seemingly simple requests for jurisdictional determinations.

The confusion of small landowners grew as local government officials often granted local building permits pursuant to local law, not knowing that state permits were also required. The result was mass misunderstanding of which projects required permits and what types of land uses were "grandfathered" by the state laws. These effects created an enormous monitoring and enforcement problem for the state, which did not allocate sufficient funds to staff the detective work required to ensure compliance. For the first 5 years of its existence, for example, the agency employed only one enforcement officer to cover an area roughly the size of Massachusetts. The resulting delays fueled anger among those who already perceived the newly centralized authority to be illegitimate.

Challenges to the Scientific Foundations of the Regulations

Perhaps the greatest challenge to the Adirondack Park plans was the gradual questioning of the scientific basis used in the development of the density regulations. Although the plan was originally characterized as one grounded in principles of sustainable resource management and sound ecological principles, developers and local officials slowly began to understood the uncertainties inherent in these concepts. There were many questions that arose that highlighted the inconsistencies and lack of clear ecological explanations for how decisions were made. For example, What logic would cause one tract of land to need 1.3 ha (3.2 acres) for a new structure, when seemingly identical ecological conditions across the road demanded 17.3 ha (42.6 acres)? What vegetative and soil characteristics define what can be classified as a wetland, a condition that normally initiates state jurisdiction over a new development? What is the adverse environmental effect of developing within a floodplain when the amount of fill required to raise a structure from the 100-year flood zone amounted to only one-millionth or less of the floodplain's volume?

Although these questions were common in the daily regulatory life of the young Adirondack Park Agency, they never became a basis for significant legislative reductions to the state's authority. Instead, a series of court interpretations support the precautionary policies and granted the agency broad discretion to interpret its legislated mandates, even if its choices were only loosely connected to rational interpretations of ecological evidence. The state was always able to play its trump card, the need to protect the open space character of the park declared to be of state interest in the Constitution and law for nearly a century.

Appeal to Technology and Mitigation Measures

The park agency has reviewed thousands of development proposals during the past 25 years. Developers were rarely denied permits. Instead, their proposals were reshaped by agency lawyers, planners, and ecologists to conform more closely to the broad objectives of the statute to protect the biological, physical, and open space resources of the park. Changes suggested by the agency might be as simple as reducing density, clustering development to avoid ecologically sensitive areas, realigning roads to diminish the disturbed area or natural drainage patterns, increasing setbacks from water bodies, and most commonly changes in sewage treatment technologies. These changes either took the form of alterations in the developers' proposals or the form of permit conditions.

In the case of relatively small subdivisions, the difference of being allowed to develop five to six lots determined the economic viability of a developer so that intense debates evolved over the capability of technology to control adverse environmental effects (i.e., such as sewage contamination of water bodies, pesticide runoff from golf courses, oil and gas contamination from marinas, or stream sedimentation) that would result from this development. In most early cases, the agency treated proposed mitigation measures with substantial pessimism, especially if they relied upon relatively new and untried technologies. Instead, they preferred a more cautious approach of simply reducing density or moving structures, roads, or other facilities back from waterfronts. The economic desirability of waterfront construction, along with the enormous length of Adirondack shorelines in private ownerships, resulted in the agency decisionmakers having to repeatedly deal with development conflicts along the shorelines.

In its infancy, the agency developed the policy of granting permits with an extensive set of conditions, carefully worded to prevent adverse environmental effects. Yet the absence of monitoring and enforcement staff have left the Adirondacks with thousands of approved projects, and little understanding of whether the permit conditions (often technologically dependent) have been implemented or have been effective in preventing damage. This failure has severely inhibited the agency's ability to learn from environmental responses to its past regulatory behavior. Worse, it leaves little ability to judge the overall effectiveness of an institutional structure that has cost well over $100 million dollars to implement. This conclusion seems especially important when considering what alternative level of protection might have been achieved had the state instead invested these funds to acquire additional forest preserve lands, which would offer more predictable insurance against environmental damage.

Civil and Uncivil Disobedience

Local resistance to the state regulations included both civil and uncivil disobedience. Some residents stated publicly that they refused to obtain development permits from the agency, justified by their perception that the law was illegitimate. Public rallies were held where young women—dressed in red, white, and blue costumes—danced in support of individual freedom, synonymous in their minds with private property rights. Horse manure was dumped on the entrance to the park agency's headquarters, along with a sign stating, "We've taken yours long enough, here's some of ours." Agency project review officers were evicted from private lands at gunpoint, and more than one agency vehicle returned to headquarters with bullet holes in fenders or doors. Late one night an arsonist was caught attempting to burn the agency building to the ground. Publicity given to those who defied the state's regulatory machinery created many local folk heroes.

These hostilities resulted from deeply felt frustration driven by economic insecurity. Because the agency was created in 1973, with the OPEC oil embargo, recession, and skyrocketing interest rates following quickly behind, it was understandable that Adirondack residents would blame local economic stagnation on the newly formed agency, rather than global and national economic trends.

Gradual political response to the hostility and outrage took the form of changed appointments to the commission. Permit review periods were shortened to a 21-day period for small subdivisions and single-family dwellings. Agency deliberations were finally opened to the public, and a new appeals process allowed project sponsors to contest decisions or permit conditions. Regional offices were established in parts of the park distant from headquarters. All of these changes were attempts to reduce the authoritarian image of the agency and to make it more responsive to local needs and conditions. The effect was gradual increases in acceptance of the rules and regulations, especially during the early 1980s.

Controlling Political Appointments

The Adirondack Park Agency is comprised of a group of commissioners and staff. The commissioners include three members of the governor's cabinet, five appointees from outside of the park, and three from within. This distribution was clearly intended to represent statewide interests more heavily than those of park residents. The staff of the agency has averaged roughly 50 employees over the past several decades. Both the director and the legal counsel are also appointed by the governor, while the remainder are appointed through normal competitive civil service procedures. This appointment process, both for commissioners and staff, provided substantial political control of agency decisions by the governor's office.

Although there has never been a clear correlation between any single political party and the preservationist interests of the governor and his appointees, there have been consistent swings of a pendulum between state and private interests. These shifts in direction were strongly resisted by a professional staff of ecologists, planners, and lawyers, who viewed the reforms as selling out to development interests. These conflicting viewpoints were managed by the chairman by removing staff authority for decision-making and vesting it among other commissioners who the chairman could more easily control.

The influence of preservationist or libertarian ideology among staff and commissioners may be exceptionally important in the interpretation of subtle and uncertain ecological information, which often provided the basis for permit or zoning decisions that had important economic implications. The balance of ideology between commissioners and staff was affected by the governor's appointment powers, but terms of appointment were rarely cut short, providing considerable stability.

5.5.4   Case Study of the *Oven Mountain Pond Estates* Development Project in the Adirondack Park

This section examines how science was used by the Adirondack Park Agency to determine whether to permit residential development at Oven Mountain Pond (Adirondack Park Agency. Permit: Project Summary, Project 91-110, May 18, 1995). This was a Class A Project, which required review and approval from the state. In this example, the Adirondack Park Agency used both quantitative and qualitative information to determine if the proposed development would detrimentally affect the proposed site or other park resources. The proposed developmental project was to occur in an undeveloped location, part of it classified for rural use, some as low-intensity use, some as wetlands, and some bordered lands classified as wild forest by the Adirondack Park Agency. The original proposal by the developer was to build a 66-lot residential area with 84% of the site to be retained as open space under the ownership of the homeowners' association.

The developers initially evaluated the development potential of this site using the method outlined by McHarg (1969) at the suggestion of the Adirondack Park Agency staff. Overlay composite maps were developed by the applicant that were used in the review process by the agency and several environmental groups. The applicant had to submit a significant volume of information to the agency related to the potential development site, revise the application based on interactions with the agency, and finally resubmit a complete application in December 1993. The developers were granted a permit in 1995, but the final approved project had changed considerably from the original version. For example, five lots had been eliminated (see later discussion of eutrophication), and nine lots had been reconfigured based on the assessment of ecological damage that would accrue to the sites from development.

As part of its review of this development project, the Adirondack Park Agency had to determine the impacts of development on water quality and eutrophication potentials, critical resource areas (i.e., wildlife habitat, wetlands), fish and wildlife populations directly, maintaining open space resources and vegetation cover, and aesthetics (i.e., scenic vistas and roads; Adirondack Park Agency Act 1991). The water quality issues were definitely of primary concern for the Adirondack Park Agency when assessing the impact of development in the proposed site. The water quality assessment was based on quantitative information (i.e., specific percolation rates of water through soil, specific quantities of phosphorous loading that could occur into a lake, etc.), which was combined with models to predict future effects in aquatic ecosystems. This contrasted with the wildlife information, which was relatively qualitative and was not based on any data that had been accumulated for the area. This was more based on having a coarse-scale understanding of habitat requirements for animals. The project area also is not inhabited by any known rare or endangered species so this was not as critical of an issue in the assessment of this land for development purposes. The data that had the least quantitative information and was almost entirely based on value judgments was the analysis of the open space and aesthetic issues associated with development. There was no science used in evaluating open space issues, but these issues did reinforce the stricter limitations that were made on the amount of development that was possible on this site.

Because the proposed project site contains several bodies of water (including an 8.5 ha pond) and streams, water quality issues had to be examined by the Adirondack Park Agency. The development site contains two watersheds. The site also exists over a portion of an unconfined surficial aquifer that does not have a protective layer of bedrock. Contaminants could migrate unimpeded into the aquifer through the soil. The impacts of development on water quality was therefore of considerable interest to the Adirondack Park Agency, and the developer had to prove that these bodies of water would not be adversely affected by the building activities and resident's future uses of the land (i.e., waste disposal systems, fertilization of lawns, etc.). The applicant had to collect data specific for the site to estimate water quality and average flow rates. The data collected had to be conducted by a qualified soil scientist.

The effects of sewage and fertilizers on water quality were also examined using a model to predict what levels of nitrate would be found in streams and the pond from knowing the number of dwellings and the number of people within each dwelling who would contribute N to the ecosystem. A U.S. Geological Survey ground water model was used to estimate nitrate loading into the unconfined aquifer. Based on these model runs and using the worse-case scenarios as requested by the Adirondack Park Agency staff, nitrate levels were predicted to be much lower than the 10 mg/l drinking water standard and were not considered to be a major problem in the Adirondack Park.

There was also concern for the potential future eutrophication of wetlands and the pond within the development site due to the additions of P from sewage disposal systems, and from the application of pesticides and fertilizers. Phosphorus is of concern because it is the limiting nutrient in the aquatic ecosystems in the Adirondacks. Two different models were used by both the agency and the applicant's consultants to assess the eutrophication potential of the water systems due to development. Some model runs predicted no effect of development on P levels in the pond, while other runs suggested the pond to have a high potential for eutrophication. The model results were used by the agency to recommend reducing the number of dwellings within the pond's watershed from the seven proposed by the builder to four due to the potential for eutrophication. The final revised proposal submitted by the developer had further reduced the number of dwellings to be built to three in this watershed. The adjustments made to how much development was going to be allowed was based on the utilization of quantitative data collected in the field and was analyzed for future potentials using models. In this case, where the outcome was not favorable to the builder, even though the applicant and the agency were generally satisfied with the model output, the applicant questioned and controversy occurred as to whether the models were accurate.

Eliminating development lots in the watershed of the pond was also driven by the need to maintain corridors for the wildlife using the wetlands around the pond. The applicant and the agency staff agreed that a corridor was needed between the wetlands and upland areas. This resulted in a recommendation by the agency to change the location of one of the lots in the watershed so that a corridor could be maintained. The applicant agreed to adjust the location of the lot. However, no case was made to show that this strip of land could function as an effective pathway for wildlife to continue to access the wetlands and pond. How this corridor would function in the landscape around the pond was not considered. Recently, the utility of corridors has been strongly questioned because it has become apparent that many animals do not use them (Simberloff 1993). There is a very strong possibility that this corridor by the pond will be insufficient to maintain wildlife not evolved to utilize human modified landscapes.

This raises an interesting point that development decisions are being made where the concerned parties are clearly aware of the environmental issues (i.e., fragmentation) associated with their activities and there is obvious interest in minimizing the impacts of development in the system (i.e., maintaining wildlife populations). However, the scientific understanding of how to effectively design the utilization of different landscape units to maximize the maintenance of wildlife populations or ecosystem functions are evolving so rapidly today that it is difficult to filter through all of the information and come to concrete solutions that can be used to develop policy. This is further compounded by the fact that many of the earlier ideas that were theory based and had not been tested in the field were adopted as tools to manage systems.

5.5.5 Reflections on Ecosystem Management and the Adirondack Experiment

The history of the Adirondack Park land-use and development plan has been an extraordinary political and ecological experiment. Although biological and physical resource information was surveyed during the creation of the plan, and this information base has been refined during the development of local land-use plans and during site development reviews, there has been no body of ecological theory or principles that has driven the spatial distribution of the different density zones nor the agency's thousands of permit decisions. Instead, a simpler vision drove the plan's design, one of attempting to protect certain values that are relatively easy to identify, such as water quality, wildlife habitats, shorelines, and the open space character of the park.

The legal framework for control of development involved three tiers: (1) the regional plan, which established the broad density limits for various zones; (2) local planning refinements to the regional plan; and (3) site plan review when new development was proposed. This legal structure gave regulators the latitude to protect critical ecological processes and functions, but these concepts remained undefined. Instead, the regulators concentrated on the management of ecosystem components, especially those that could be easily measured (i.e., such as species diversity or a few water quality parameters). Their behavior and attention was directed most by a precautionary philosophy, one that assumed somehow less human influence on the landscape was better than more. The private land zoning scheme described earlier created a gradient of acceptable development, ranging from urban cores to relatively undisturbed resource management areas requiring 18.6 ha (46 acres) per new structure. The management of the density of development was felt to relieve the agency from the need to conduct more technical analyses.

Ecosystem management is no different than any other form of management in having the broad purpose of human organization to produce desired values. Because of this, it is not immune to the contentious political conflicts that surround the allocation of other scarce resources, such as welfare, agricultural subsidies, and tax benefits and burdens. In the Adirondacks desired values include certain types of ecological outcomes, such as maximum biological diversity, least environmental contamination, sustained yield of certain resources such as timber, water, and wildlife, and perhaps of greatest importance to conservationists, wildness. The creation of the land use and development plan in 1973 might be interpreted as a process of imposing downstate conservation values on private landowners within the park. The mechanism used to apply these broader values was that of state-imposed zoning, which constrained the rights of private owners and reduced the authority of local governments to govern development.

The Adirondack Park is a great location in which to test ecosystem management as defined earlier in this book. The past approach used to define land-use boundaries in the park has created one of the problems that ecosystem management has to address today. It would be very informative to go the Adirondack Park today and to determine what the impacts of the imposition of these artificial boundaries have been on the resilience and resistance characteristics of these ecosystems to current human and anthropogenic stresses.

6 | Science and Management of Ecosystems
Synthesis

6.1　Ecosystem Management Framework

The ecosystem management paradigm has been chosen as the approach to be used by federal agencies to manage federal lands. Because of the difficulty that federal agencies and academicians have had in defining what these two words mean and because the perception is that this means more data collection than is currently occurring, ecosystem management has a strong potential to be rejected as the approach to use for managing ecosystems. This would be extremely unfortunate because this approach has elements that are crucial to retain for "good" management of ecosystems. The components of ecosystem management (articulated in earlier chapters) are fundamentally important if ecosystems are to be managed where they function within the range of states natural for that system and at the same time produce the values (i.e., species, timber, nontimber forest products, etc.) humans have identified as desirable. Ecosystem management should be used as a tool that identifies what the trade-offs are in accepting a particular management option. It is an approach that will move the management of our natural ecosystems from being solely based on human "values" to one balanced by scientific information helping to clarify the consequences of the chosen values.

The ecosystem management perspective should be used to aid setting priorities regarding the natural resources that need to be managed. As part of this perspective, it is understood that the management practices and decisions must be modified with the changing social, political, environmental, and scientific atmosphere. Ecosystem management is a conceptual framework that allows us to make choices and sacrifices with the best available information existing today.

Because the elements of ecosystem management are critical to maintain in any management activity, even if the name should change its principal elements should not be lost in a rush to adopt a new name. The use of the words *ecosystem management* to define a particular approach to management may become ineffective if individuals are unable to move beyond the point of trying to distinctly define the words. Currently, there exists a conflict of whether it is best to try to define these terms quite specifically or to keep them quite general so that they have broad-scale applications (see Chapter 2). Unfortunately, if ecosystem management is defined too strictly, it becomes a book of rules and totally negates what the ecosystem approach stands for. If ecosystem management is defined too generally to satisfy all interest groups, it has the problem of not being very useful; this approach is the one most frequently seen in the literature. One of the rationales for writing this book was to search behind the words and to identify the core attributes of ecosystem management, in this way influencing people to move beyond the stage of vague conjecturing about what they think it means. It may be that if people are unable to move out of the quagmire in which ecosystem management appears to be in, then ecosystem management becomes a placeholder until the next management approach is introduced. Whatever the ultimate fate of ecosystem management, the essential attributes that it encompasses must be retained. In this case, it means essentially stripping off the words and defining the underlying core principles of good natural resource management (defined here as the attributes of ecosystem management).

Ecosystem management also explicitly states that it is impossible to have all the values or products from a given unit of land. In the past it was felt that ecosystems could be managed so that all species and products could be simultaneously obtained from one ecosystem. Today it is apparent that multiple use is a dream that is highly desirable but unattainable because of the strong feedback loops that exist in ecosystems. The implications of changing one component or part of an ecosystem cannot be viewed in isolation because its effects can ramify throughout the system and modify how that system responds as a whole to a disturbance. This has implications for how ecosystems are managed for a specific value, such as species, timber, mushrooms, etc. When one value is emphasized in an ecosystem, other attributes may not be maintained at the same levels. For example, when a timber company manages their land to optimize tree productivity, they may selectively plant genetically superior tree species that are too effective at depleting site nutrient capitals. The initial response by management to decreased nutrient pools is to fertilize the site to maintain tree growth rates. However, fertilizing with nutrients feeds back to affect within-plant carbon allocation patterns so that the plant tissues become more palatable to herbivores. This may then require the use of pesticides to eradicate the increased herbivore populations. In this simple example, humans have had to replace two of the functional roles in this ecosystem. This particular scenario is actually not unusual when ecosystems are managed for one plant or animal species. Thus, it is difficult to optimize one variable without changing other attributes in ecosystems.

The definitions and applications of ecosystem management have been mainly driven by our values, and relatively little has occurred as a direct result of science itself. This fact is strongly highlighted by re-examining the historical timelines for ecology, management, and policy presented in Chapter 2. That timeline emphasized the paucity of scientific research resulting in the formulation of policy for forests; in most cases, science has been driven by public concern, which directed the allocation of funds to particular research areas relevant to the question. Interestingly, many of the definitions of ecosystem management, especially from the federal agencies (presented in Chapter 2), explicitly incorporated the role of the social and natural sciences as being integral to management. However, most of these definitions perceived that natural and social sciences only mesh at the point where two intercrossing circles overlap one another, an area defined as management (see Figure 1-1). The rest of each circle, one representing the social and the other the natural sciences, is depicted as separate units functioning in isolation from one another. The assumption is that only where the two fields intersect does a balance exist in how both fields contribute to management. In reality, there is no balance but a dominance by one field over the other, with natural sciences having a minor role in how decisions are made. The use of the two intersecting circles gives a false perception that most parts of the social and natural science circles can each function in isolation from one another. It is for this reason that the egg analogy is more appropriate (see Figure 1-1). It is important to make sure that, even if the natural sciences are entirely enmeshed by the social sciences, there is a core of natural sciences that contribute to value assessments.

The overlapping zone of the two circles where management occurs (see Figure 1-1) suggested that there is a balance in how the natural and social sciences contribute to management. In fact, ecosystem management at the implementation level does not incorporate the social and natural sciences in a balanced manner. The actual implementation approach is dictated by the disciplinary comfort zone of the individuals in charge. This management level is also being driven by policy being generated at higher levels, which is strongly controlled by individual or group values (i.e., general public, special interest groups, etc.). Because of the overriding influence of individual or group values in dictating policy, it appears that there is a perception that science is not really relevant to policy formulation. It has not helped that scientists have not been able produce the results or conclusions within the time frame preferred by policymakers because of the temporal displacement of the identification of a problem and the period needed to study the problem. A prime example of this was the NAPAP study (see Chapter 4), which was specifically mandated by the Clean Air Act to assess the scientific basis of acid deposition impacts on forests for policymakers. One of the main criticisms of this program was its inability to synthesize results from research conducted over a 10-year period in time for policymakers to utilize that information.

Ecosystem management seems distinguishable from some other forms of human organization and management in that its potential effectiveness is dependent on ecological knowledge. An ideal knowledge base would provide a clear vision of how biological, physical, and chemical processes would change in response to various management alternatives or human stresses. In this way, ecosystem management rests on faith in our rational capacity to understand the forces underlying any current ecological order, as a basis for creating policies that will guide the creation of a future order, one designed to produce some desired value set.

Limitations in our knowledge base obviously constrain our capacities to predict ecological response to management interventions. Normal scientific response to this problem has been to define problems narrowly in space, time, or variables considered. This form of reductionism, although necessary for implementing scientific methods, has led to surprises, such as the persistence and biological accumulation of DDT, extensive groundwater contamination from herbicides, and the ozone-depleting effect of chlorofluorocarbon aerosols. The probability of encountering these types of surprises is diminished by broadening the scale and variables considered in any management problem. But expanding the scope of inquiry increases uncertainty about causal relations and predicted outcomes. Under conditions of uncertainty, ecosystem management might best be thought of as experimentation, providing an opportunity to monitor the effects of resource management as a basis for adjusting later decisions.

At this point, public values are driving the direction that ecosystem management is moving. These social values are important and will probably always drive what and how natural resources are managed. However, ecosystem management by definition incorporates both social and natural sciences (see Chapter 2); the decision matrix specifically identifies the scientific implications of our decisions. In its basic form, ecosystem management really means that what is going to be managed is identified (the endpoints of risk assessment) using our social values and then science is used to outline the ecological ramifications of what humans want to manage. Ecosystem management gives a framework within which both sciences can be examined and where choices can be made as to what course of action should be taken. For example, if it is decided that management should focus on a particular species or a product from the ecosystem because a high value is placed on it, the ecosystem management framework should make us comprehend the variety of possible scenarios linked to the choice. Therefore, it is important to accept the implications and repercussions of our choices. At times this may require management to maintain the current equilibrium state of an ecosystem if there is a desire to maintain a particular species in the landscape that is dependent on a specific ecosystem state. However, at the same time this decision compels us to explicitly consider the fact that ecosystems are dynamic and maintaining a system in a particular equilibrium state may cause that system to become more susceptible to particular types of disturbances or potentially even degrade.

In earlier chapters the ecological roots of management were examined, which highlighted strongly how theory development in ecology has had really very little to do with how our natural resources are actually managed. What this means is that political factors are overriding and setting the guidelines of how managers actually implement management of forest ecosystems. This means that the ecological training silviculturalists have received is not considered when management plans are implemented. This is unfortunate because the scientific training of silviculturalists has been strongly grounded in ecology for a long time. This is highlighted by one of the earliest texts in silviculture being entitled, *Foundations of Silviculture upon an Ecological Basis* (Toumey 1928). Even back in 1921, Toumey published an ecological article discussing the damage occurring to forests due to smoke, ash, and fumes from manufacturing plants in Connecticut.

Much of our changing management paradigms have been driven by public perceptions of the decreased health of forest or even of rangeland ecosystems. This has been further complicated by science only recently developing tools to detect ecosystem resistance and resilience characteristics after a disturbance and to detect the effects of human activities separate from natural changes in ecosystems (see Chapter 4). These shifting values of how people perceived natural resources has also co-occurred with changing views of the use of public lands for harvesting timber or the use of public rangelands by ranchers.

6.2   What Should Be Monitored for Ecosystem Management

There exist many useful tools that can be adapted to measure the success of ecosystem management; many of these are discussed in Chapters 3 and 4. It is important to keep in mind that it is *not important to develop all new tools* to practice ecosystem management. Most of the approaches presented in Chapter 4 are adaptations of methods presented in Chapter 3.

For example, many of the spatial scales analyses (i.e., GIS) are very useful today but they need to be combined with other approaches to be effectively used in ecosystem management. It is very important that ecosystem management not be perceived as being a "cookbook" approach to management. It is crucial that managers and researchers *do not identify a set of specific tools that are universally transferred to all ecosystems.* It is also important that the perception that data have to be collected on all parts of an ecosystem be removed from our understanding of the approach. This perception has given rise to the conviction that ecosystem management is impossible to implement because it has such massive data needs—an incorrect assumption. It is important to use the appropriate ecological and silvicultural information to identify several attributes of ecosystems that are sensitive to monitor and allow us to detect when a system's state may be changing. Greater emphasis should also be placed on synthesizing the large amount of existing published literature to help identify how plant species or ecosystems can be indexed into functional groups based on their responses or sensitivities to various ecosystem variables (see Chapter 4). Detecting ecosystem level changes requires the selection of parameters that alert us to problems that would escape our attention if reliance was placed only on the use of models or to measuring individual components at lower levels of organization.

The type and quality of data collected as part of assessing ecosystem management practices will determine the kind of management that will occur; it is important to carefully consider what the data needs are for each system. Frequently the statement is made that monitoring is crucial for ecosystem management. Monitoring is important but *care should be given to what is monitored.* Monitoring programs should be evaluated so that a particular type of data collection does not occur just because that is what was collected in the past, especially if recent analyses have shown them to be of little utility. Long-term data monitoring is extremely important to maintain if we are going to be able to address many of our environmental problems that have longer temporal scales. Much of our longer term data has been collected by federal agencies, such as the Forest Service, because they have been able to maintain data collection over a long time scale. The first federal laboratory (i.e., Coweeta Hydrologic Laboratory) was established in 1934 (see Appendix A). Few organizations are as well structured to maintain baseline data collections as some of the United States Federal Laboratories. The U.S. National Science Foundation also established Long-term Ecological Research sites in the United States in 1980, with a significant portion of that program allocated to collecting baseline data for very different ecosystems, ranging from the arctic tundra to tropical forests. In fact, several of the LTERs are located at Forest Service field sites and are being jointly managed; these collaborations have been extremely successful.

There is a need to re-examine the necessity for transferring information collected at one scale to address questions occurring at another scale. It is important to keep in mind that *information collected at smaller scales should not always be scaled up.* For example, seedling studies conducted under controlled conditions are excellent for teasing apart the mechanisms underlying the processes of interest but do not readily transfer to the field situation (see Chapter 3). However, much research is still being conducted that assumes information collected on seedlings will automatically transfer to address questions on mature plants. Seedling studies should be used to initially identify the mechanisms; that information should then be used to design field experiments specifically focusing on the mechanisms identified. In these cases, the approaches or tools do not exist to allow us to sufficiently separate cause–effect relationships in the field experiments.

The Adirondack case study also highlighted the importance of the scale of analysis in determining how effectively information could be used to make management decisions (see Chapter 5). It also demonstrates the fact that there is a need to go to the relevant scale to collect information for management purposes, frequently using information from several different scales but not trying to transfer information directly between these scales. In the Adirondack case study, ecological information was collected at several scales—larger scale analyses more to index the landscape into sensitive and nonsensitive zones, followed by discrete analyses of one nutrient (i.e., P) at smaller scales. This was followed by site-specific analyses of the potential for P loading into a lake based on chemical analyses combined with the use of hydrologic models. The decision as to how much development to allow in this area was ultimately decided using the site-specific data.

Identification of ecosystem attributes specifically by scale will help explain how to best approach ecosystem measurement and how to obtain the minimum information needed to manage according to the ecosystem management ideal. It is necessary to reiterate that a *complete* understanding of ecosystem functions cannot be and should not be obtained using the methods outlined in Chapter 3 and 4. Each approach has a scale at which some functions can be more easily detected, with the sensitivity being lost when shifting to another scale of analysis. There must be a willingness to accept the inaccuracies caused by scaling and measurement errors and also by the scientific community's imperfect ecological knowledge base. This highlights the importance of adaptive management being integral to ecosystem management.

6.3   Impediments to Utilizing the Ecosystem Management Approach

Impediments to implementing ecosystem management exist not only at the level of developing new tools to assess management but also at the individual and group levels, where people have to feel comfortable with what this management approach entails. Ecosystem management requires people to be aware of information being generated in several different disciplines and to be able to integrate that information to develop management strategies. However, many people are not comfortable functioning within interdisciplinary frameworks because it forces them to move beyond the constraints of their disciplinary fields. This discomfort in functioning in interdisciplinary teams is further propagated by our graduate school training programs that confine students within their disciplinary boundaries. Furthermore, university level evaluations of faculty for permanent positions are based on their disciplinary strengths, which means that faculty do not encourage their students to cross disciplinary boundaries. In most educational institutions, students are actively discouraged from examining issues that may cross disciplinary boundaries. It is not uncommon to hear people say that students who cross disciplinary boundaries have a shallow knowledge of several fields and their knowledge is not founded in theory, a perception that has no validity. There is the perception that you cannot conduct *original* research for your doctoral degree program if you do not focus narrowly in a particular field. Graduate students are trained to become experts in some field, an impediment to implementing ecosystem management if the ecosystem approach is to be used due to its requirement for integrating many different fields.

What institutional structures are most appropriate for ecosystem management or how people's behavior controls their responses cannot be addressed in this book; that is another book to be written by someone else qualified to address these topics. However, these factors do have a significant impact on if and how ecosystem management will occur and therefore should be acknowledged because they can impede our ability to implement ecosystem management (see the National Forest Ecosystem Management case study). If flexibility is designed into management programs, it is easier to modify management activities as new information becomes available. Flexibility has to be seen as a distinct benefit and hence an objective of policy design, especially because it may be expensive to implement flexible management schemes. Trade-offs may have to be made between flexibility and efficiency. Woodhouse and Hamlett (1992) summarized four elements that they felt contributed to inflexibility in a program:

Lead time. A long period between initiating a program and getting convincing feedback about its inadequacies routinely interferes with error correction. It took about 25 years for information on DDT to be recognized.

Infrastructure . If a program depends on huge facilities, complicated networks, and other supporting facilities, change will be expensive.

Unit size. Large-scale endeavors, such as nuclear power plants, are generally harder to modify than smaller scale efforts.

Timing of Payment. If costs are primarily borne in advance, as in building a huge power plant, then learning from experience cannot occur in time to prevent high-cost mistakes.

The difficulty of managing ecosystems that are not delimited by their ecological boundaries is something that all the agencies and companies managing forests are currently struggling with (see case studies in Chapter 5). There is no easy solution to this problem. This problem has fostered the formation of new cooperatives between different federal agencies, private companies, and environmental groups to try to facilitate the management of a resource where the activities of one group may negatively impact some attribute of an ecosystem. It is not clear how successful these cooperative ventures will be for facilitating the management of a resource when people's values may limit their ability to accept the fact that not all outputs from an ecosystem can be maintained, especially when lands are being managed by companies whose business profits rely on a specific product. By definition, if it is accepted that there exists a desire for the maintenance of ecosystems in healthy states into the future, humans should not expect to have all outputs from an ecosystem. There will have to be flexibility in people's willingness to negotiate what is best for the ecosystem. It will also require different organizations to give up some of their control over a resource and may require the development of new infrastructures that are better designed to negotiate natural resource management under the umbrella of the ecosystem management paradigm.

6.4   Adaptive Management

Adaptive management is a term that occurs in most writings discussing ecosystem management (Everett et al. 1993). It is always stressed that adaptive management is important, but it is not really clear what this means. In most cases, adaptive management states that rules and management criteria must be flexible enough to changing biophysical events and changing human goals. Because a continual change in ecosystem processes is accepted, it is important to realize that no one single management practice is adequate. Adaptive management does explicitly accept the fact that variation is embedded within ecosystems. Adaptive management means establishing measurable objectives, both in ecosystem function and social desires, increasing current levels of data gathering by managing scientifically (with controls), monitoring, and adjusting management practices to meet changes in ecosystem capacity or social demands. The demands being placed on many systems may result in environmental changes occurring at a rate that exceeds the adaptive capacity of many species so that management has to be flexible to deal with new information as it becomes available and adjust accordingly.

It is extremely important that feedback loops are integrated into the ecosystem management structure and are kept flexible. It will help stabilize natural resource management by allowing incremental changes that are easier to work with, instead of major readjustments in practices every 30 years or so. This will require interdisciplinary integration to ensure many levels are monitored in an effort to measure the overall fit of practices (SAF Task Force Report 1992). Yet stating the facts is not enough and learning has to be a vital part of the process. To achieve our desired goals for ecosystem management requires the financial commitment to this type of program. Even though ecosystem management has financial costs associated with its implementation, the benefits from using such an approach will outweigh the costs if sustainable ecosystem management can be achieved in the future.

One way to achieve adaptive management is for management to use a trial-and-error approach because (1) even in the highly uncertain endeavors, it is possible to foresee and protect against some of the worst risks; and (2) it is impossible to know the likelihood or severity of hypothetical risks from new projects (Schmitt 1969); thus it makes sense to err on the side of caution. Well-designed trial-and-error processes are not purely reactive but are designed so that they prepare for error correction. This can be done by forming clear hypotheses about the probabilities associated with expected impacts and by establishing procedures for careful monitoring of initial trials to refine key uncertainties.

Decision analysis for ecosystem management might involve the production of a probability-based model indicating possible scenarios for management activity and associated outcomes. Probabilities can be thought of as expressions of the strength of our beliefs or knowledge (Solow 1994). If our knowledge is expressed in probabilities, then the basis exists for talking about what information does, gathering more information where needed, and measuring the change in our knowledge. Probabilities expressed as dependent on our current knowledge are sometimes called *conditional* probabilities. They depend integrally on the assumptions being make and the investigations conducted. Conditional probabilities are expressed as the "probability of an event A, given the conditions H." For example, when evaluating various management options, it may be necessary to express the probability of the spotted owl going extinct given a specific management option.

Though there have been many calls for adaptive management lately, few have noted specific methods for updating management decisions when more information becomes available. One way to formalize the updating of uncertain information is through Bayesian analysis. This type of analysis involves expressing uncertainty through probabilities of various outcomes, and re-evaluating the probabilities as more information becomes available. Sometimes managers are uncomfortable with the sound of probability estimation, but it is used every day when weather forecasts are given (i.e., the probability of rain tomorrow might be 50%). Similarly, the probability that the spotted owl will not go extinct, under various management scenarios, could be specified. Often there is a need to both express the uncertainty of outcomes as probabilities and to leave ourselves room to change them.

6.5   Principles of Ecosystem Management

Ecosystem management is a new and evolving *approach* to natural resource management. The following 10 principles or components have been identified that characterize the ecosystem approach. These principles include (they are not ranked in order of importance or chronology):

- Draft and implement their own formal working definition of ecosystem management that accounts for the specific characteristics of a given management unit and its philosophy. This definition should incorporate the concept that *humans are part of the ecosystem* and therefore should account for their values, impacts, and demands on the system.

- Identify management goals and objectives. Although this sounds obvious, case-study analysis showed that often management goals and objectives were not clearly identified (Schoeder and Keeler 1990). They should be identified and written into a formal management document, such as a management plan. In this fashion, goals and objectives are clearly highlighted and can be easily referred to.

- Define management unit and boundaries. In order to efficiently manage an area it is essential that boundaries are determined. This process of boundary designation forces the management entity to identify what it considers to be "the ecosystem." Both spatial and temporal scales should be examined when making this determination. Spatial scales include, for instance, legal property boundaries, range(s) of focus species, and hydrological regimes. Temporal scales include aspects such as disturbance patterns, population growth, and climatic change.

Again, depending upon the management entity, the criteria used to determine the management unit boundaries will vary. For example, in one instance large-scale disturbances such as fire may be driving the ecosystem. In this case this pattern of disturbance should be incorporated into the boundary delineation decisionmaking process. In another scenario, the range of a focus species, such as the grizzly bear, may influence management unit boundaries. Despite the difference in criteria that are used to make this designation, there is a commonality of guiding principles between areas that is recognized as the ecosystem approach.

- Develop and implement a management plan. Management plans have been mandated since 1976. They are famous for their often lengthy creation process and the tendency for completed plans to end up on a dusty shelf. Despite all of this, a clear and concise management plan is an effective tool in implementing the ecosystem approach. A management plan can serve several purposes: (1) to clarify goals and objectives, (2) to guide management actions, (3) to act as a checklist for implementation of action items, and (4) through revision, to promote adaptive management when situations and circumstances have changed.

- Identify policies, laws, and regulations that directly affect management activities. Often management entities are faced with a myriad of laws and regulations that guide their actions and powers. Oftentimes these laws and regulations can conflict with each other. It is necessary to identify these legal structures and to abide by their requirements.

- Carefully select and utilize ecosystem management tools and technologies. This area is rapidly changing and evolving. Ecosystem management tools and technologies can be broken down into scientific tools and consensus-building or communication tools. Scientific tools may include geographic information systems, monitoring techniques, social surveys, and economic forecasts. Consensus-building or communication tools are very important. Often one of the greatest obstacles to implementation of the ecosystem approach is the lack of support from an affected party. Newsletters, educational presentations, and meetings to generate information exchange between varied groups all promote understanding and communication. The importance of communication is an absolute necessity whether it is in regard to boundary determination, determining management goals and objectives, or coordinating land management activities with adjacent landowners.

- Collect, analyze, and integrate economic, social, and ecological information and make decisions using this science-based information. The ecosystem approach is different from past management schemes in that it promotes the integration of social, economic, and ecological data in order to make more holistic decisions. Integration of these three areas is evolving.

- Clearly identify ecological constraints or limits.

- Coordinate management activities with adjacent landowners, resource users and extractors, and other institutions or agencies that have an interest in or jurisdiction over the management unit. Coordination and communication is of paramount importance. Communication breakdowns consistently serve as obstacles to implementation of the ecosystem approach. Communication can lead to coordination of land-use activities, enforcement and support of management unit boundaries, and consensus on problem definition.

- Enable feedback mechanisms at all levels that promote adaptive management. Adaptive management is the name of the game. New technologies and information are constantly being generated, and circumstances (ecological, social, political, and economic) always change. Given this highly dynamic environment, management plans must be adjusted and monitoring schemes, among other things, must be updated and adjusted. In order for this to occur, feedback mechanisms must be present that promote this adjustment and change.

In contrast to the ecosystem approach, ecosystem management is management-unit specific. In other words, depending upon the goals, objectives, and philosophy of a given management entity, the actual definition of ecosystem management will change to reflect these values. For example, a timber company will frame their definition of ecosystem management in terms of fiber or timber production, a national park will describe ecosystem management in terms of recreational values, and the forest service in terms of biodiversity. The difference between these definitions should reflect the difference between the goals and objectives of these management entities. It is important to recognize that there is no one perfect way to implement ecosystem management.

The ecosystem approach is a means of taking the best information at hand and making the best possible decision in relation to management goals and objectives. The importance of a management-unit–specific definition of ecosystem management should not be underestimated. It became apparent when analyzing the range of case studies that often a major obstacle to implementation of ecosystem management was due to a lack of a clear operational definition.

6.6   Costs and Benefits of Instituting Ecosystem Management as a New Environmental Management Scheme

6.6.1   Potential Costs of Ecosystem Management

- *Potential for confusion.* The *ecosystem* concept and *ecosystem management* are not readily comprehensible terms to a wide audience. Indeed, these terms already mean different things to different groups and individuals. Ecosystem management, though a relatively new concept, already has been defined and practiced in a variety of ways (see Chapter 2). Both the lack of understanding of terms and the baggage carried by the variety of interpretations may cause confusion about what is being managed and how management should proceed. For example, the attention to ecosystem function is necessary, but it is at first glance less easy to define methods for management of ecosystem functions.

- *A new name doesn't necessarily mean better management.* Changing from multiple use to ecosystem management does not automatically solve management problems, nor does it provide a clear set of goals and associated framework for problem solving.

- *Borders issue.* There is a potential for simply changing management unit borders and assuming this will translate into comprehensive management. But the change from managing at the scale of an endangered species (using its range to delineate management borders) to looking

at a larger unit that is delineated as "the ecosystem" does not necessarily solve management problems. Rather, it is more important to
manage problems at the appropriate scale than to delineate borders,
and to declare that management is going to proceed at the "ecosystem level." Because of the confusion regarding what constitutes the
ecosystem as a physical entity, ecosystem management could occur at
just about any spatial scale.

- When the boundaries of ecological impacts do not coincide with *jurisdictional borders*, implementation of ecosystem management regimes
will be a challenge. It will suggest the development of consistent
policies covering land that is managed by a variety of users who
may have different ideas about how management should proceed.
The classical definition of ecosystem borders is based on characteristic energy-cycling patterns—temperate forests have different energy
cycling patterns than do wetlands or tropical forests. Because these
borders are both relatively intangible and because they do not cor-
respond to jurisdictional borders, there is some confusion as to how
best to delineate ecosystem management units. This may lead to disputes on management unit borders, which, when these arguments
are extreme, simply prevents addressing the underlying problems. Ultimately, management goals are (and should be) used to define system
borders, but ecological impacts are influenced by the biological, geological, and chemical processes. The mismatch between border definitions
can be a problem.

- Ecosystem management requires *integrated management*. The interactions between economic, social, and environmental aspects of
management issues must be examined. The complex interactions between human activities and ecological processes may be debated if
they are not obvious or well established. This can make management
difficult when there is no consensus as to what causes adverse impacts. For example, is overfishing the only reason for the decline in
commercial fish harvest, or are there important mediating factors that
have been overlooked? Fisherpeople would likely argue that there are
a number of factors involved—variability in population sizes could be
related to climate (El Niño) or simply natural fluctuations. Whatever
the reason, applying ecosystem management to fisheries would enable

the harvest to fluctuate with stocks and have everyone focusing on the sustainability of the fish population, not the income generated from fishing.

6.6.2 Benefits of Ecosystem Management

- Ecosystem management provides an opportunity to avoid crisis management. In fact, it may be the only way to get beyond the piecemeal approach taken in the past to environmental problems (such as treating the loss of species only when they are close to extinction).

- Ecosystem management should allow you to have a flexible management system that is able to incorporate the uncertainty associated with unpredictable disturbance cycles. It is important to build flexibility into ecosystem management plans.

- Ecosystem management is a placeholder that minimizes or reduces negative effects that can occur with single or multiple resource management approaches.

- Ecosystem management allows for a hierarchy approach to be used to examine process rates and ecosystem functions that may be connectable to management. By looking at the immediate problem as part of a nested hierarchy in the larger system, it is possible to get beyond the one symptom at a time approach.

- Ecosystem management allows for sustainable management of the ecosystem, and not just parts of the ecosystem, by focusing on both maintenance of system function as well as system outputs.

- Ecosystem management allows us to minimize future risk by addressing potential problems now, anticipating catastrophe, and maintaining functional ecosystems (biodiversity, gene pool preservation).

6.6.3 Summary

Again, it is important to reiterate that the term *ecosystem management* is less important than the practices that fall under it, but the name change from multiple use has signified the acknowledged need for better practices. Debate has been centering on what are the appropriate practices that should occur with ecosystem management. Because there is no one single definition of ecosystem management, appropriate management should be defined in the context of the particular ecosystem and the needs/wants of people.

Despite its limitations, ecosystem management is the only way to get past the piecemeal approach to environmental management. Traditional approaches tend to tackle the symptoms of one problem at a time, usually when the problems become crisis-like, when they adversely affect the economy, or when they attract widespread public attention. The piecemeal approach has led to different laws and sets of regulations aimed at specific symptoms of general problems. One set of regulations relates to building and development, one set for air pollution, one for water pollution, one for pesticides, still others for wildlife harvests and endangered species, etc. This has produced a fragmented structure of environmental protection agencies and has resulted in divisions within agencies working at cross-purposes.

Ecosystem management will get beyond the fragmented approach only if it entails

1. Assessment of environmental problems at their appropriate scales. Determining when issues are best treated on a small scale (local pollution problem) and when a regional approach should be taken.

2. Assessment of the state of an ecosystem. The need to ask, What is wanted by the management unit? What are the current conditions? What is the "normal" or expected range of conditions? How different are current conditions from the expected? Also, the need to try to establish the cause–effect relationships between human activity and resultant ecological impacts, and to evaluate the relative contribution of various activities. After an evaluation of which human activities contribute most to degradation and which can be changed to ameliorate impacts, management plans should reflect these options.

3. Clear presentation of trade-offs that are necessary for reducing impacts. This entails looking at the benefits and costs of human activities. For instance, the costs of agriculture include erosion of topsoil and reductions in local biodiversity. These have to be balanced with the benefits from growing the desired crops.

4. Acknowledgment of uncertainty and adaptive management. This involves accepting uncertainty and designing flexible management

systems that are able to incorporate the uncertainty associated with unpredictable events. Flexibility must be built in to management, and changes must be made when new information suggests such changes. This may cost more in the short term, as it will entail leaving room for unexpected change. However, the long-run costs will likely be lower than less flexible management because expensive, rigid programs will not be instituted, and ineffective management will not be kept around just because it has been instituted. In addition, long-term costs will be lower because the ecosystem will be less likely to be irreversibly degraded by management activities.

There should also be acknowledgment that current management has been more driven by social values than by our changing scientific knowledge base. That statement oddly contradicts the verbiage of the scientific community, which states that more ecology means better management. Many scientists assume that if new knowledge is produced, which would be a result of ecosystem management, then effective management systems can be implemented for our desired values. This does not address the fact that frequently what is managed for is contrary to the maintenance of the ecosystem itself. The public wants to protect their "values" and the legal system is responsive to the value changes. Frequently these values represent a coalition of different interest groups who effectively lobby to determine the type of policy that is formulated. Therefore institutional structures become responsive to the "values" instead of our knowledge base.

Ecosystem management may not survive as the management guideline for federal agencies. It conflicts with many groups who do not want to accept the fact that it is impossible to optimize all outputs from an ecosystem. If this philosophy is perpetuated, it will be impossible to do ecosystem management. Others perceive ecosystem management to be too complex and difficult to implement. People have to learn and begin to accept the complexity of our systems without assuming that no tools exist to analyze our systems. In this book, tools that do exist or can be adapted to assess ecosystem management practices are presented. Again, this is not a cookbook approach but necessitates that individual plans be developed for systems using ecological groups. Ecology has a fundamental role to play in developing the assessment criteria for managing ecosystems, but it should be remembered that silviculturalists have used the ecological approach for managing forests for a long time. It is important to recognize that fact and facilitate their ability to include ecology in management. It is also important that there is recognition of when the species approach is relevant to analyzing ecosystems and that the many cases for which this approach does not work are acknowledged.

References

Aber JD, Federer CA (1992) A generalized, lumped-parameter model of photosynthesis, evapotranspiration and net primary production in temperate and boreal forest ecosystems. Oecologia 92: 463–474.

Aber JD, Melillo JM (1982) Nitrogen immobilization in decaying hardwood leaf litter as a function of initial nitrogen and lignin content. Can. J. Bot. 60: 2263–2269.

Aber JD, Melillo JM, Nadelhoffer KJ, McClaugherty CA, Pastor J (1985) Fine root turnover in forest ecosystems in relation to quantity and form of nitrogen availability: a comparison of two methods. Oecologia (Berl.) 66: 317–321.

Abrahamsen G (1980) Effects of acid precipitation on soil and forest: methods of field experiments. pp. 190–191. In: (Drablos D, Tollan A, eds.) Ecological Impact of Acid Precipitation. SNSF Project. Oslo, Norway.

Abrams MD (1990) Adaptations and responses to drought in *Quercus* species of North America. Tree Physiol. 7: 227–238.

Adams JB, Kapos V, Smith MO, Filho A, Gillespie AR, Roberts DA (1990) A new Landsat view of land use in Amazonia. Int. Soc. Photogramm. Remote Sens. Manaus, Brazil 28: 177–185.

Aerts R, Caluwe HD (1994) Nitrogen use efficiency of CAREX species in relation to nitrogen supply. Ecology 75: 2362–2372.

Agee JK (1991) The historical role of fire in Pacific Northwest forests. pp. 25–38. In: Walstad JD, Radosevich SR, Sandberg DV, eds. Natural and Prescribed Fire in Pacific Northwest Forests. Oregon State University Press, Corvallis, Oregon.

Agee JK (1993) Fire and weather disturbances in terrestrial ecosystems of the Eastern Cascades. pp. 359–414. In: (P.F. Hessburg, compiler) Eastside Forest Ecosystem Health Assessment, Vol. III. USDA For. Serv. Pac. Northwest Res. Sta. PNW 93-0304.

Agee J, Johnson D (eds.) (1988) Ecosystem Management for Parks and Wilderness. University of Washington Press, Seattle, Washington. 237 pp.

Alfaro RI, MacDonald RN (1988) Effects of defoliation by the western false hemlock looper on Douglas-fir tree-ring chronologies. Tree-Ring Bull. 48: 3–11.

Allen MF (1991) The Ecology of Mycorrhizae. Cambridge Studies in Ecology. Cambridge University Press. Cambridge. 184 pp.

Allen TFH, Hoekstra TW (1990) The confusion between scale-defined levels and conventional levels of organization in ecology. J. Vegetat. Sci. 1: 5–12.

Allen TFH, Hoekstra TW (1992) Towards a Unified Ecology. Columbia University Press. New York. 384 pp.

Allen TFH, Hoekstra TW (1994) Toward a definition of sustainability. pp. 98–107. In: Covington WW, DeBano FL, technical coordinators. Sustainable Ecological Systems: Implementing an Ecological Approach to Land Management. U.S. Department of Agriculture, For. Serv. Rocky Mountain For. Range Exp. Sta. Gen. Techn. Rep. RM-247. Fort Collins, Colorado.

Allen TFH, Starr TB (1982) Hierarchy: Perspectives or Ecological Complexity. University of Chicao Press, Chicago. 310 pp.

Amaranthus MP, Perry DA (1987) Effect of soil transfer on ectomycorrhiza formation and the survival and growth of conifer seedlings on old, nonreforested clearcuts. Can. J. For. Res. 17: 944–950.

Amaranthus MP, Perry DA (1988) Mycorrhizal formation and growth of Douglas-fir seedlings in three southern Oregon vegetation types. In: Maintaining Long-Term Productivity of Pacific Northwest Forests. Proceedings of a Symposium, March 31–April 2, 1987, Corvalis, Oregon. Timber Press, Portland, Oregon.

American Forest and Paper Association (1994) Sustainable Forestry Principles and Implementation Guidelines. Am. For. Paper Assoc. Washington DC.

Anderson JE (1991) A conceptual framework for evaluating and quantifying naturalness. Conserv. Biol. 5: 347–352.

Andersson F (1970) Methods and preliminary results of estimation of biomass and primary production in a south Swedish mixed deciduous woodland. pp. 281–288. In: Proc. Brussels Symp. Productivity of Forest Ecosystems. UNESCO.

Andersson F (1989) Swedish forests in a changing environment—effects on soils, tree nutrition, and growth. In: Forest Health and Productivity. The Marcus Wallenberg Foundation Symposium Proceedings 5: 29–49. Falun.

Andersson F, Persson T (1988) Liming as a measure to improve soil and tree conditions in areas affected by air pollution. SNV Report 3518, Solna, 131 pp.

Antonovics J, Clay K, Schmitt J (1987) The measurement of small-scale environmental heterogeneity using clonal transplants of *Anthoxanthium odoratum* and *Danthonia spicata*. Oecologia 71: 601–607.

Aplet GH, Johnson N, Olson JT, Sample VA (eds.) (1993) Designing Sustainable Forestry. Island Press. Covelco, California.

Arno SF, Brown JK (1991) Overcoming the paradox in managing wildland fire. Western Wildlands 17: 40–46.

Arno SF, Ottmar RD (1993) Reducing hazard for catastrophic fire. pp. 17–18. In: Everett RL, compiler. Eastside Forest Ecosystem Health Assessment. Volume IV. Restoration of Stressed Sites, and Processes. United States Department of Agriculture. National Forest System. Forest Service Research April 1993.

Arnolds E (1991) Decline of ectomycorrhizal fungi in Europe. Agric Ecosyst Environ 35: 209–244.

Arnone JAI, Gordon JC (1990) Effect of nodulation, nitrogen fixation and CO_2 enrichment on the physiology, growth and dry mass allocation of seedlings of *Alnus rubra* Bong. New Phytol. 116: 55–66.

Ashton PMS, Berlyn GP (1992) Leaf adaptations of some *Shorea* species to sun and shade. New Phytol. 121: 587–596.

Aubertin GM, Patric JH(1974) Water quality after clear-cutting a small watershed in West Virginia. J. Environ. Qual. 3: 243–249.

Axelsson B (1985) Increasing forest productivity and value by manipulating nutrient availability. pp. 5–37. In: (Ballard R, Farnum P, Ritchie GA, Winjum JK, eds.) Weyerhaeuser Science Symposium. Forest Potentials. Productivity and Value. Weyerhaeuser. Tacoma, Washington.

Baker WL (1992) Effects of settlement and fire suppression on landscape structure. Ecology 73: 1879–1887.

Bailey R (1972) A Hierarchical Landscape Classification for Recreational Land Use Planning in the Finger Lakes Region of New York State. Unpublished Ph.D. Dissertation, Cornell University, New York.

Baldocchi DD (1993) Scaling water vapor and carbon dioxide exchange from leaves to a canopy: rules and tools. pp. 77–113. In: (Ehleringer JR, Field CB, eds.) Scaling Physiological Processes: Leaf to Globe. Academic Press. New York.

Balesdent J, Girardin C, Mariotti A (1993) Site-related 13C of tree leaves and soil organic matter in a temperate forest. Ecology 74: 1713–1721.

Ballard R (1979) Use of fertilizers to maintain productivity of intensively managed forest plantations. pp. 321–342. In: Proceedings Impact of Intensive Harvesting on Forest Nutrient Cycling. State University of New York, Syracuse, New York. Northeast For. Exp. Sta. USDA For. Serv. Broomall, Pennsylvania.

Barber K, Rodman S (1990) FORPLAN: a marvelous toy. J. For. 88: 26–30.

Barlow SW, Bridges JW, Calow DM, Conning RN, Curnow AD, Dayan IFH, Purchase CW, Suckling (1992) Toxicity, toxicology, and nutrition. Ch. 3. In: Risk: Analysis, Perception, Management. The Royal Society. London.

Barnard JE, Lucier AA, Johnson AH, Brooks RT, Karnosky DF, Richter DD. (1990) Changes in Forest Health and Productivity in the United States and Canada. NAPAP State of Science and Technology Report No. 16. Washington DC: National Acid Precipitation Assessment Program.

Barret JW (ed.) (1995) Regional Silviculture of the United States. 3rd Ed. John Wiley & Sons. New York.

Bartell SM, Gardner RH, O'Neill RV (1992) Ecological Risk Estimation. Lewis. Ann Arbor, Michigan. 252 pp.

Bartolome JW, Muick PC, McClaran MP (1987) Natural Regeneration of California Hardwoods. pp. 26–30. Gen. Tech. Rep. PSW-100, Berkeley, CA. Pacific Southwest For. Range Exp. Sta. Forest Service.

Baskin Y (1994) Ecologists dare to ask: How much does diversity matter? Science 264: 202–203.

Baskin Y (1995) Ecosystem function of biodiversity. BioScience 44: 657–660.

Basnet K, Scatena FN, Likens GE, Lugo AE (1993) Ecological consequences of root grafting in tabonuco (*Dacryodes excelsa*) trees in the Luquillo Experimental Forest, Puerto Rico. Biotropica 25: 28–35.

Bazzaz FA (1993) Scaling in biological systems: population and community perspectives. pp. 233–254. In: (Ehleringer JR, Field CB, eds.) Scaling Physiological Processes: Leaf to Globe. Academic Press. New York.

Bazzaz FA, Swipe TW (1987) Physiological ecology, disturbance, and ecosystem recovery. pp. 203–227. In: (Schulze ED, Zwolfer H, eds.) Potentials and Limitations of Ecosystem Analysis. Springer-Verlag. Berlin.

Beck MB, Jakeman AJ, McAleer MJ (1993) Construction and evaluation of models of environmental systems. Ch. 1. In: (Beck MB, Jakeman AJ, McAleer MJ, eds.) Modelling Change in Environmental Systems. John Wiley & Sons. New York.

Beissinger S (1994) Personal communication with Steve Beissinger, Marbled Murrelet Recovery Team.

Bell G, Lechowicz MJ (1994) Spatial heterogeneity at small scales and how plants respond to it. pp. 391–414. In: (Ehleringer JR, Field CB, eds.) Exploitation of Environmental Heterogeneity by Plants. Academic Press. New York.

Berendse F, Aerts R (1987) Nitrogen-use-efficiency: a biologically meaningful definition? Funct. Ecol. 1: 293–296.

Berendse F, Berg B, Bosatta E (1987) The effect of lignin and nitrogen on the decomposition of litter in nutrient-poor ecosystems: a theoretical approach. Can. J. Bot. 65: 1116–1120.

Berg B (1984) Decomposition of root litter and some factors regulating the process: long-term root litter decomposition in a Scots pine forest. Soil Biol. Biochem. 16: 609–617.

Berg B, Jansson P-E, Meentemeyer V (1984) Litter decomposition and climate—regional and local models. pp. 389–404. In: (Ågren GI, ed.) State and Change of Forest Ecosystems—Indicators in Current Research. Swed. Univ. Agric. Sci. Dept. Ecology & Environmental Research Rep. no. 13.

Billings WD (1985) The historical development of physiological plant ecology. pp. 1–15. In: (Chabot BF, Mooney HA, eds.) Physiological Ecology of North American Plant Communities. Chapman and Hall. New York.

Binkley C, Hagenstein P (1987) Economic analysis of the 1985 RPA program: a way to determine efficient allocations. J. For. 85: 1.

Binkley D (1992) Mixtures of nitrogen-fixing and non nitrogen fixing tree species. pp. 99–123. In: (Cannell MGR, Malcolm DC, Robertson PA, eds.) The Ecology of Mixed Species Stands of Trees. Blackwell Scientific. London.

Binkley D, Cromack K Jr, Fredriksen RL (1982) Nitrogen accretion and availability in some snowbrush ecosystems. For. Sci. 28: 720–724.

Bloom AJ, Chapin FSI, Mooney HA (1985) Resource limitation in plants—an economic analogy. Ann. Rev. Ecol. Syst. 16: 363–392.

Bloomfield J, Vogt KA, Vogt DJ (1993) Decay rate and substrate quality of fine roots and foliage of two tropical tree species in the Luquillo Experimental Forest, Puerto Rico. Plant Soil 150: 233–245.

Bloomfield J, Vogt KA, Wargo P (1996) Tree root turnover and Senescence. pp. 363–381. In: (Waisel Y, Eshel A, Kafkafi U, eds.) Plant Roots: the Hidden Half. 2nd Ed. Marcel Dekker. New York.

Boerner REJ (1982) Fire and nutrient cycling in temperate ecosystems. BioScience 32: 187–192.

Boerner REJ (1984) Foliar nutrient dynamics and nutrient use efficiency of four deciduous tree species in relation to site fertility. J. Appl. Ecol. 21: 1029–1040.

Bolstad PV, Gower ST (1990) Estimation of leaf area index in fourteen southern Wisconsin forest stands using a portable radiometer. Tree Physiol. 7: 115–124.

Bond WJ (1993) Keystone species. pp. 237–253. In: (Schulze ED, Mooney HA, eds.) Biodiversity and Ecosystem Function. Ecological Studies 99. Springer-Verlag. Berlin.

Bond WJ, Midgley J, Vlok J (1988) When is an island not an island? Insular effects and their causes in fynbos shrublands. Oecologia 77: 515–521.

Bonnett M, Zimmerman K (1991) Politics and preservation: the Endangered Species Act and the northern spotted owl. Ecol. Law Qu. 18: 105–171.

Bormann BT, Brookes MH, Ford ED, Kiester AR, Oliver CD, Weigand JF (1993) A broad, strategic framework for sustainable-ecosystem management. Eastside Forest Ecosystem Health Assessment. Volume V. U.S. Department of Agric. For. Serv. 62 pp.

Bormann FH (1985) Air pollution and forests: an ecosystem perspective. BioScience 35: 434–441.

Bormann FH (1990) Air pollution and temperate forests: creeping degradation. pp. 25–44. In: (Woodwell GM, ed.) The Earth in Transition. Patterns and Processes of Biotic Impoverishment. Cambridge University Press. Cambridge.

Bormann FH, Likens GE (1979) Pattern and Process in a Forested Ecosystem. Disturbance, Development and the Steady State Based on the Hubbard Brook Ecosystem Study. Springer-Verlag. New York. 253 pp.

Bortone SA, Davis WP (1994) Fish intersexuality as an indicator of environmental stress. BioScience 44: 165–172.

Bourgeron PS, Jensen ME (1993) An overview of ecological principles for ecosystem management. In: (Jensen ME, Bourgeron PS, eds.) Ecosystem Management: Principles and Applications. Vol. II. Eastside Forest Health Assessment. U.S. Department of Agric. For. Serv. Washington DC.

Bowen GD (1984) Tree roots and the use of soil nutrients. pp. 147–180. In: (Bowen GD, Nambiar EKS, eds.) Nutrition of Plantation Forests. Academic Press. Orlando, Florida.

Boyce RL, Friedland AJ, Chamberlain CP (1994) Changes in canopy uptake and allocation of 15-NH^{4+} and 15-NO^{3-} during the growing season in red spruce. 74th Annual Meeting of the Ecol. Soc. Am. Knoxville, Tennessee. August 7–11, 1994. Bull. Ecol. Soc. Am. Vol 70: 21.

Brady NC (1990) The Nature and Properties of Soils. 10th Ed. Macmillan Publishing. New York. 621 pp.

Bray JR (1963) Root production and the estimation of net productivity. Can. J. Bot. 41: 65–72.

Bray JR, Gorham E (1964) Litter production in forests of the world. Adv. Ecol. Res. II: 101–157.

Brewer G (1986) Methods for Synthesis: Policy Exercises. In: (Clark WC, Munn RE, eds.) Sustainable Development of the Biosphere. Cambridge. University Press. 491 pp.

Brewer G, DeLeon P (1983) Foundations of Policy Analysis. Dorsey Press. New York.

Brix H (1971) Effects of nitrogen fertilization on photosynthesis and respiration in Douglas-fir. For. Sci. 17: 407–414.

Brokaw NVL (1985) Gap-phase regeneration in a tropical forest. Ecology 66: 682–687.

Brown GW, Gahler AR, Marston RB (1973) Nutrient losses after clear-cut logging and slash burning in the Oregon Coast Range. Water Resource Res. 9: 1450–1453.

Brown JH, Heske EJ (1990) Control of a desert grassland transition by a keystone rodent guild. Science 250: 1705–1707.

Brown S, Iverson LR (1992) Biomass estimates for tropical forests. World Res. Rev. 4: 366–384.

Brown S, Lugo AE (1982) The storage and production of organic matter in tropical forests and their role in the global carbon cycle. Biotropica 14: 161–187.

Brunson M (1993) "Socially acceptable" forestry: what does it imply for ecosystem management? Western J. Appl. For. 8: 116–119.

Bryant JP, Chapin FS II, Klein DR (1983) Carbon-nutrient balance of boreal plants in relation to vertebrate herbivory. Oikos 40: 357–368.

Bryant JP, Chapin FS III, Reichardt PB, Clausen T (1985) Adaptation to resource availability as a determinant of chemical defense strategies in woody plants. Recent Adv. Phytochem. 19: 219–237.

Bunnel FL, MacLean SF, Brown J (1975) Barrow, Alaska, USA. pp. 73–124. In: (Rosswall T, Heal OW, eds.) Structure and Function of Tundra Ecosystems. Ecological Bull. 20. Swedish Natural Science Research Council. Stockholm, Sweden.

Burch W Jr (1976) Who participates? A sociological interpretation of natural resource decisions. Natural Resources Journal. Symposium issue on NEPA.

Burch W (1984) Much ado about nothing—some reflections on the wider and wilder implications of social carrying capacity. Leisure Sci. 6: 487.

Burch W Jr (1987) Ties that bind: the social benefits of recreation provision. President's Commission on American's Outdoors: Literature Review.

Burns RC, Hardy RWF (1975) Nitrogen fixation in bacteria and higher plants. Springer-Verlag. Berlin.

Cairns J Jr (1979) Biological monitoring—concept and scope. pp. 3–20. In: (Cairns J Jr, Patil GP, Waters WE, eds.) Environmental Biomonitoring, Assessment,

Prediction, and Management. International Co-operative Publishing House. Fairland, Maryland.

Cairns J Jr (1986) The myth of the most sensitive species. BioScience 36: 670–672.

Caldwell MM, Richards JH (1989) Hydraulic lift: water efflux from upper roots improves effectiveness of water uptake by deep roots. Oecologia 19: 1–5.

Campbell R (1989) Pearl of the low country. SC Wildl. Mag. 36: 36–43.

Cannell MGR (1982) World Forest Biomass and Primary Production Data. Academic Press. London. 391 pp.

Carey C (1993) Hypothesis concerning the causes of the disappearance of boreal toads from the mountains of Colorado. Conserv. Biol. 6: 355–362.

Carpenter SR (1982) Comparison of equations for decay of leaf litter in tree-hole ecosystems. Oikos 39: 17–22.

Carter HR, Eriksen RA, Sander TG (1988) Status of the marbled murrelet in California. Pacific Seabird Group Bull. 15:25–26.

Chabot BF, Mooney HA (eds.) (1985) Physiological Ecology of North American Plant Communities. Chapman and Hall. New York. 351 pp.

Champion International Corporation (1995) Setting the Standard. Caring for our Forests in the Northeast. Champion International. Bucksport, Maine. 9 pp.

Chapin FS I (1983) Patterns of nutrient absorption and use by plants from natural and man-modified environments. pp. 175–187. In: (Mooney HA, Godron M, eds.) Disturbance and Ecosystems: components of Response. Springer-Verlag. Berlin.

Chapin FS III (1980) The mineral nutrition of wild plants. Ann. Rev. Ecol. Syst. 11: 233–260.

Chapin FS III (1990) The ecology and economics of storage in plants. Ann. Rev. Ecol. Syst. 21: 423–447.

Chapin FS III (1991a) Integrated responses of plants to stress. BioScience 41: 29–36.

Chapin FS III (1991b) Effects of multiple environmental stresses on nutrient availability and use. pp. 67–88. In: (Mooney HA, Winner WE, Pell EJ, eds.) Response of Plants to Multiple Stresses. Academic Press. London.

Chapin FS III (1993) Functional role of growth forms in ecosystem and global processes. pp. 287–312. In: (Ehleringer JR, Field CB, eds.) Scaling Physiological Processes: Leaf to Globe. Academic Press. New York.

Chapin FS III, Kedrowski RA (1983) Seasonal changes in nitrogen and phosphorous fractions in evergreen and deciduous taiga trees. Ecology 64: 376–391.

Chapin FS III, Bloom AJ, Field CB, Waring RH (1987) Plant responses to multiple environmental factors. BioScience 37: 49–57.

Chapin FS III, Jefferies RL, Reynolds JF, Shaver GR, Svoboda J (1992) Arctic Ecosystems in a Changing Climate. Academic Press. San Diego, California.

Chapin FS, Giblin AE, Nadelhoffer KJ, Laundre JA (1995) Responses of arctic tundra to experimental and observed changes in climate. Ecology 76: 694–711.

Chen J, Franklin JF, Spies TA (1992) Vegetation responses to edge environments in old-growth Douglas-fir forests. Ecol. Appl. 2: 387–396.

Chen J, Franklin JF, Spies TA (1995) Growing-season microclimatic gradients from clearcut edges into old-growth Douglas-fir forests. Ecol. Appl. 5: 74–86.

Chiariello NR, Gulmon SL (1991) Stress effects on plant reproduction. pp. 161–188. In: (Mooney HA, Winner WE, Pell EJ, eds.) Responses of Plants to Multiple Stresses. Academic Press. New York.

Christensen NL, Agee JK, Brussard PF, Hughes J, Knight DH, Minshall GW, Peek JM, Pyne SJ, Swanson FJ, Thomas JW, Wells S, Williams SE, Wright HA (1989) Interpreting the Yellowstone fires of 1988. BioScience 39: 678–685.

Clark TW, Minta SC (1994) Greater Yellowstone's Future. Homestead Publishing. Moose, Wyoming.

Clark TW, Amato ED, Whittemore DG, Harvey AH (1991) Policy and programs for ecosystem management in the greater Yellowstone ecosystem: an analysis. Conserv. Biol. 5: 412–422.

Clements FE (1916) Plant Succession: An Analysis of the Development of Vegetation. Carnegie Institute Washington Publ. # 242.

Cole DW, Rapp M (1981) Elemental cycling in forest ecosystems. pp. 341–409. In: (Reichle DE, ed.) Dynamic Properties of Forest Ecosystems. Cambridge University Press. Cambridge.

Comstock JP, Ehleringer JR. (1992) Correlating genetic variation in carbon isotopic composition with complex climatic gradients. Proc. Natl. Acad. Sci. USA 89: 7747–7751.

Connell JH, Sousa WP (1983) On the evidence needed to judge ecological stablity or persistence. Am. Natural. 121: 789–824.

Conservation International, Ecotrust, Pacific GIS (1995) The rain forests of home: an atlas of people and place. Part I: natural forests and native languages of the coastal temperate rain forest.

Cornell University Department of Agronomy/US Soil Conservation Service (1972) Soil Survey Interpretation of Soils in New York State. Cornell University. Ithaca, New York.

Costanza R (1992) Toward an operational definition of ecosystem. pp. 239–256. In: (Costanza R, Norton BG, Haskell BD, eds.) Ecosystem Health: New Goals for Environmental Management. Island Press. Washington DC.

Covington WW, Everett RL, Steele R, Irwin LL, Daer TA, Auclair AND (1994) Historical and anticipated changes in forest ecosystems of the inland west of the

United States. pp. 13–64. In: (Sampson RN, Adams DL, eds.) Assessing Forest Ecosystem Health in the Inland West. Haworth Press. New York.

Cowling EB (1992) The performance and legacy of NAPAP. Ecol. Appl. 2: 111–116.

Crunkilton DD, Pallardy SG, Garrett HE (1992) Water relations and gas exchange of northern red oak seedlings planted in central Missouri clearcut and shelterwood. For. Ecol. Manag. 53: 117–129.

Cuevas E, Medina E (1986) Nutrient dynamics within Amazonian forest ecosystems. I. Nutrient flux in fine litter fall and efficiency of nutrient utilization. Oecologia 68: 466–472.

Dahl R (1982) Dilemmas of Pluralist Democracy: Autonomy vs. Control. Yale University Press. New Haven, Connecticut.

Dahlgren RA, Vogt KA, Ugolini FC (1991) The influence of soil chemistry on fine root aluminum concentrations and root dynamics in a subalpine Spodosol, Washington State, USA. Plant Soil 133: 117–129.

Dawson J (1983) Dinitrogen fixation in forest ecosystems. Can. J. Microbiol. 29: 979–992.

Dawson TE, Chapin SF III (1993) Grouping plants by their form-function characteristics as an avenue for simplification in scaling between leaves and landscapes. pp. 313–319. In: (Ehleringer JR, Field CB, eds.) Scaling Physiological Processes. Leaf to Globe. Academic Press. San Diego, California.

Dawson TE, Ehleringer JR (1991) Streamside trees that do not use stream water. Nature 350: 335–337.

Dawson TE, Ehleringer JR (1993) Isotopic enrichment of water in the "woody" tissues of plants: implications for plant water source, water uptake, and other studies which use the stable isotopic composition of cellulose. Geochimi. Cosmochimi. Acta 57: 3487–3492.

DeBonis J (1989) The Inner Voice. 1: 4–5. (Occasional newsletter).

De Pietri DE (1992) The search for ecological indicators: is it possible to biomonitor forest system degradation caused by cattle ranching activities in Argentina? Vegetatio 101: 109–121.

DeAngelis DL, Gardner RH, Shugart HH (1981) Productivity of forest ecosystems studies during the IBP: the woodlands data set. pp. 567–672. In: (Reichle DE, ed.) Dynamic Properties of Forest Ecosystems. Cambridge University Press. Cambridge.

DeCola (1989) Fractal analysis of a classified Landsat scene. Photogr. Eng. Remote Sensing 35: 601–610.

Delwiche CC, Steyn PL (1970) Nitrogen isotope fractionation in soils and microbial reactions. Environ. Sci. Tech. 4: 929–935.

Denslow JS (1980) Gap partitioning among tropical rainforest trees. Trop. Succession 12 supplement 47–55.

Dietrich W (1992) The Final Forest: The Battle for the Last Great Trees of the Pacific Northwest. Simon and Schuster. New York. 303 pp.

Dighton J, Mason PA (1985) Mycorrhizal dynamics during forest tree development. pp. 117–139. In: Developmental Biology of Higher Fungi (Moore D, Casselton LA, Wood DA, Frankland JC, eds.) Cambridge University Press. New York.

Dixon RK, Meldahl RS, Ruark GA, Warren WG (eds.) (1990) Process Modeling of Forest Growth Responses to Environmental Stress. Timber Press. Portland, Oregon. 441 pp.

Dobereiner J, Campello AB (1977) Importance of legumes and their contribution to tropical agriculture. pp. 191–220. In: (Hardy RWF, Gibson AH, eds.) A Treatise on Dinitrogen Fixation. IV: Agronomy and Ecology. John Wiley & Sons. New York.

Eagar C, Adams MB (eds.) (1992) Ecology and Decline of Red Spruce in the Eastern United States. Springer-Verlag. New York.

Edmonds RL (1982) Analysis of Coniferous Forest Ecosystems in the Western United States. US/IBP Synthesis Series 14. Hutchinson Ross Publ. Co. Stroudsburg, Pennsylvania. 19 pp.

Egler FE (1954) Vegetation science concepts. I. Initial floristic composition, a factor in old-field vegetation development. Vegetatio 4: 412–417.

Ehleringer JR (1990) Correlations between carbon isotope discrimination and leaf conductance to water vapor in common beans. Plant Physiol. 93: 1422–1425.

Ehleringer JR, Cooper TA (1988) Correlations between carbon isotope ratio and microhabitat in desert plants. Oecologia 76: 562–566.

Ehleringer JR, Dawson TE (1992) Water uptake by plants: perspectives from stable isotope composition. Plant Cell Environ. 15: 1073–1082.

Ehleringer JR, Field CB (1993) Scaling Physiological Processes. Leaf to Globe. Academic Press. San Diego, California. 388 pp.

Ehleringer JR, Osmond CB (1991) Stable isotopes. pp. 281–300. In: (Pearcy RW, Ehleringer JR, Mooney HA, Rundel PW, eds.) Plant Physiologica Ecology Field Methods and Instrumentation. Chapman and Hall. London.

Ehleringer JR, Hall AE, Farquhar GD (1993) Stable Isotopes and Plant Carbon-Water Relations. Academic Press. New York.

Elsik CH, Flagler RB, Boutton TW (1993) Carbon isotope composition and gas exchange of Loblolly and Shortleaf pine as affected by ozone and water stress. pp. 227–242. In: (Ehleringer JR, Hall AE, Farquhar GD, eds.) Stable Isotopes and Plant Carbon-Water Relations. Academic Press. New York.

Emmingham WH, Waring RH (1977) An index of photosynthesis for comparing forest sites in western Oregon. Can. J. For. Res. 7: 165–174.

EPA (1987) Unfinished business: A Comparative Assessment of Environmental Problems. Office of Policy Analysis, US EPA. Washington DC.

EPA (1990) Ecological Indicator Report for the Environmental Monitoring and Assessment Program. February 1990. US EPA. Office of Research and Development. Research Triangle Park, North Carolina.

EPA (1992a) Framework for ecological risk assessment. Office of Health and Environmental Assessment, US EPA. Washington DC.

EPA (1992b) Report on the Ecological Risk Assessment Guidelines Strategic Planning Workshop. Risk Assessment Forum. US EPA, Washington DC.

Everett R, Oliver C, Saveland J, Hessburg P, Diaz N, Irwin L (1993) Adaptive ecosystem management. In: (Jensen ME, Bourgeron PS, eds.) Ecosystem Management: Principles and Applications, Vol. II. Eastside Forest Health Assessment. U.S. Dept. Agric., For. Serv. Washington DC.

Ewel JJ (1986) Invasibility: lessons from south Florida. pp. 214–230. In: (Mooney HA, Drake JA, eds.) Ecology of Biological Invasions of North American and Hawaii. Springer-Verlag. New York.

Fahey TJ (1982) Nutrient dynamics of aboveground detritus in lodgepole pine ecosystems, Wyoming. Ecol. Monogr. 53: 51–72.

Fahey TJ, Birk EM (1991) Measuring internal distribution and resorption. pp. 226–241. In: (Lassoie JP, Hinckley TM, eds.) Techniques and Approaches in Forest Tree Ecophysiology. CRC Press. Boca Raton, Florida.

Fahey TJ, Knight DH (1986) The lodgepole pine forest ecosystem. BioScience 36: 610- -617.

Falk DA (1990) The theory of integrated conservation strategies for biological diversity. pp. 5–10. In: (Mitchell RS, Cheviak CJ, Leopold DJ, eds.) Ecosystem Management: Rare Species and Significant Habitats. New York State Museum Bull. 471.

Farquhar GD, O'Leary MH, Berry JA (1982) On the relationship between carbon isotope discrimination and the intercellular carbon dioxide concentration in leaves. Austral. J. Plant Physiol. 9: 121–137.

Feeney (1989) The Pacific Northwest's ancient forests: ecosystems under siege. In: (Chandler WJ, ed.) Audubon Wildlife Report 1989/1990. Academic Press. San Diego, California.

FEMAT (Forest Ecosystem Management Assessment Team) (1993) Forest Ecosystem Management: An Ecological, Economic and Social Assessment. Washington DC. U.S. Government Printing Office, no. 1993-793-071.

Fernow BE (1907) A Brief History of Forestry, in Europe and the United States and Other Countries. University Press. Toronto, Ontario. 438 pp.

Field CB (1991) Ecological scaling of carbon gain to stress and resource availability. pp. 35–65. In: (Mooney HA, Winner WE, Pell EJ, eds.) Response of Plants to Multiple Stresses. Academic Press. London.

Field C, Mooney HA (1986) The photosynthesis-nitrogen relationship in wild plants. pp. 25–55. In: (Givnish TJ, ed.) On the Economy of Plant Form and Function. Cambridge University Press. Cambridge.

Field C, Merino J, Mooney HA (1983) Compromises between water-use efficiency and nitrogen-use efficiency in five species of California evergreens. Oecologia 60: 384–389.

Findlay S, Jones CG (1990) Exposure of cottonwood plants to ozone alters subsequent leaf decomposition. Oecologia 82: 248–250.

Flanagan LB, Johnson KH (1995) Genetic variation in carbon isotope discrimination and its relationship to growth under field conditions in full-sib families of *Picea mariana*. Can. J. For. Res. 25: 39–47.

Fogel R, Hunt G (1979) Fungal and arboreal biomass in a western Oregon Douglas-fir ecosystem: distribution patterns and turnover. Can. J. For. Res. 9: 245–256.

Fogel R, Hunt G (1983) Contribution of mycorrhizae and soil fungi to nutrient cycling in a Douglas-fir ecosystem. Can. J. For. Res. 13: 219–232.

Forman RTT, Godron M (1986) Landscape Ecology. John Wiley & Sons. New York. 619 pp.

Forsman ED, Meslow EC, Wight HR (1980) Distribution and biology of the spotted owl in Oregon. Wildlife Monogra. 87: 1–64.

Fortin M-J, Drapeau P (1995) Delineation of ecological boundaries: comparison of approaches and significance tests. Oikos 72: 323–332.

Franco AA (1978) Contribution of the legume-Rhizobium symbiosis to the ecosystem and food production. pp. 65–74. In: (Dobereiner J, Burris RH, Hollaender A, eds.) Limitations and Potentials for Biological Nitrogen Fixation in the Tropics. Plenum Press. New York.

Frank AB (1885) Ueber die auf Wurzelsymbiose beruhende Ernährung gewisser Baume durch unterirdische Pilze. Berichte der Deutsche Bot. Gesellschaft 3: 128–145.

Frankel OH, Soulé ME (1984) Conservation and Evolution. Cambridge University Press, Cambridge.

Frankel OH, Brown AHD, Burdon JJ (1995) The Conservation of Plant Biodiversity. Cambridge University Press. Cambridge. 299 pp.

Frankland JC (1992) Mechanisms in fungal succession. pp. 383–401. In: (Carroll GC, Wicklow DT, eds.) The Fungal Community. Its Organization and Role in the Ecosystem. 2nd Ed. Marcel Dekker. New York.

Franklin JF (1989) Toward a new forestry. Am. Forests 95: 37–44.

Franklin JF (1993) Preserving biodiversity: species, ecosystems, or landscapes. Ecol. Appl. 3: 202–205.

Franklin JF, Forman RTT (1987) Creating landscape patterns by forest cutting: Ecological consequences and principles. Landscape Ecol. 1: 5–18.

Franklin JF, Spies TA (1991) Composition, function, and structure of old-growth Douglas-fir forests. pp. 71–79. In: (Ruggiero LF, Aubry KB, Carey AB, Huff MH, eds.) Wildlife and Vegetation of Unmanaged Douglas-Fir Forests. USDA For. Serv. Pac. Northwest Res. Sta. Gen. Tech. Rep. PNW-GTR-285.

Franklin JF, Bledsoe CS, Callahan JT (1990) Contributions of the Long-Term Ecological Research program. BioScience 40: 509–523.

Franklin JF, Cromack K Jr, Denison W, McKee A, Maser C, Sedell J, Swanson F, Juday G (1981) Ecological characteristics of old-growth Douglas-fir forests. USDA For. Serv. Pac. Northwest For. Range Exp. Sta. Gen. Techn. Rep. PNW-118. 48 pp.

Freedman B (1981) Intensive forest harvest: a review of nutrient budget considerations. Information Report M-X-121. Maritime Forest Research Centre, Canadian Forestry Service, Department of the Environment. Fredericton, New Brunswick, Canada.

Frissell CA (1992) Cumulative effects of land use on salmon habitat in southwest Oregon coastal streams. Ph.D. Dissertation, Oregon State University, Corvallis, Oregon.

Frissell CA (1993) The shrinking range of the Pacific Salmon: status and distribution of anadromous salmonids in the Pacific Northwest and California. Report for the Wilderness Society. Seattle, Washington.

GAO (1994) Ecosystem Management. Additional Actions Needed to Adequately Test a Promising Approach. GAO/RCED-94-111. United States General Accounting Office. Report to Congressional Requesters. 86 pp.

Gao B-C, Goetz AFH (1990) Column atmospheric water vapor and vegetation liquid water retrievals from airborne imaging spectrometer data. J. Geophys. Res. 95: 3549–3564.

García-Montiel DC, Scatena FN (1994) The effect of human activity on the structure and composition of a tropical forest in Puerto Rico. For. Ecol. Manag. 63: 57–78.

Garden CT Jr., Van Miegroet HV (1994) Relationships between soil nitrogen dynamics and natural 15-N abundance in plant foliage from Great Smoky Mountains National Park. Can. J. For. Res. 24: 1636–1645.

Gardes M, White TJ, Fortin JA, Bruns TD, Taylor JW (1991) Identification of indigenous and introduced symbiotic fungi in ectomycorrhizae by amplification of nuclear and mitochondrial ribosomal DNA. Can. J. Bot. 69: 180–190.

Georgia-Pacific (1995) Forests for the Future. Georgia-Pacific's Commitment to Sustainable Forestry. Georgia-Pacific Corporation. Corporate Communications, Forest Resources. Atlanta, Georgia. 24 pp.

Geber MA, Dawson TE (1990) Genetic variation in and covariation between leaf gas exchange, morphology, and development in *Polygonum arenastrum*, an annual plant. Oecologia 85: 53–58.

Gerlach LP, Bengston DN (1994) If ecosystem management is the solution, what's the problem? Eleven challenges for ecosystem management. J. For. 92: 18–21.

Gholz HL (1982). Environmental limits on aboveground net primary production, leaf area, and biomass in vegetation zones of the Pacific Northwest. Ecology 63: 469–481.

Gholz HL, Hendry LC, Cropper WP Jr. (1986) Organic matter dynamics of fine roots in plantations of slash pine (*Pinus elliotti*) in north Florida. Can. J. For. Res. 16: 529–538.

Gholz HL, Ewel KC, Teskey RO (1990) Water and forest productivity. For. Ecol. Manage. 30: 1–18.

Gleason HA (1939) The individualistic concept of the plant association. Am. Midl. Nat. 21: 92–110.

Glinka KD (1927) The great soil groups of the world and their development. (Transl. by C.F. Marbut.) Edwards. Ann Arbor, Michigan.

Goldstein B (1992) The struggle over ecosystem management at Yellowstone. BioScience 42: 183–187.

Golley F (1993) A History of the Ecosystem Concept in Ecology: More Than the Sum of the Parts. Yale University Press. New Haven, Connecticut.

Gordon DR, Welker JM, Menke JW, Rice KJ. (1989) Competition for soil water between annual plants and blue oak (*Quercus douglassi*) seedlings. Oecologica. 79: 533–541.

Gordon JC (1992) Ecosystem Management: An Idiosyncratic Overview. Prepared for "Defining Sustainable Forestry" (January 13–15, 1992) Reston, Virginia, USA. 4 pp.

Gordon J (1993) Ecosystem Management: An Idiosyncratic Overview. pp. 240–244. In: (Aplet GH, Johnson N, Olson JT, Sample VA, eds.) Defining Sustainable Forestry. Island Press. Covelco, California.

Gordon JC (1994) From vision to a policy vision: a role for foresters. J. For. 92: 17–19.

Gordon JC, Dawson JO (1979) Potential uses of nitrogen fixing trees and shrubs in commercial forestry. Bot. Gaz. 140 (Suppl.): 588–590.

Gordon JC, Bormann BT, Kiester AR (1992) The physiology and genetics of ecosystems: a new target or "forestry contemplates an entangled bank." pp. 1–14.

In: (Colombo SJ, Hogan G, Wearn V, eds.) 12th North American Forest Biology Workshop. Sault Ste. Marie, Ontario, Canada. Ontario Ministry of Natural Resources.

Gorham E, Vitousek PM, Reiners WA (1979) The regulation of chemical budgets over the course of terrestrial ecosystem succession. Ann. Rev. Ecol. Syst. 10: 53–84.

Gorte R (1986) The Forest Service's 1980 RPA Program: Comprison with Accomplishments. Congressional Research Service, Library of Congress. Report 86-902 ENR.

Gosz JR (1992) Gradient analysis of ecological change in time and space: implications for forest management. Ecol. Appl. 2: 248–261.

Gosz JR (1993) Ecotone hierarchies. Ecol. Appl. 3: 369–376.

Gosz JR, Sharpe PJH (1989) Broad-scale concepts for interactions of climate, topography and biota at biome transitions. Landscape Ecol. 3: 229–243.

Gosz JR, Holmes RT, Likens GE, Bormann FH (1981) The flow of energy in a forest ecosystem. Sci. Am. 238: 92–102.

Gower ST, Vogt KA, Grier CC (1992) Carbon dynamics of Rocky Mountain Douglas-fir: influence of water and nutrient availability. Ecol. Monogr. 62: 43–65.

Gower ST, Isebrands JG, Sheriff DW (1995) Carbon allocation and accumulation in conifers. pp. 217–254. In: (Smith WK, Hinckley TM, eds.) Resource Physiology of Conifers: Acquisition, Allocation, and Utilization. Academic Press. New York.

Gower ST, Running SW, Gholz HL, Haynes BE, Hunt JER, Ryan MG, Waring RH, Cropper JWP (1996) Influence of climate and nutrition on carbon allocation and net primary production of four conifer forests. Tree Physiology (in press).

Graham BF Jr, Bormann FH (1966) Natural root grafts. Bot. Rev. 32: 255–292.

Graham RT (1994) Silviculture, fire and ecosystem management. pp. 339–351. In: (Sampson RN, Adams DL, eds.) Assessing Forest Ecosystem Health in the Inland West.

Graveland J, van der Wal R, van Balen JH, van Noordwijk AJ (1994) Poor reproduction in forest passerines from decline of snail abundance on acidified soils. Nature 368: 446–448.

Grier CC (1975) Wildfire effects on nutrient distribution and leaching in a coniferous ecosystem. Can. J. For. Res. 5: 599–607.

Grier CC, Waring RH (1974) Conifer foliage mass related to sapwood area. For. Sci. 20: 205–206.

Grier CC, Logan RS (1977) Old-growth *Pseudotsuga menziesii* communities of a western Oregon watershed: biomass distribution and production budgets. Ecol. Monogr. 47: 373–400.

Grier CC, Running SW (1976) Leaf area of mature north-western coniferous forests; relation to site water balance. Ecology 58: 893–899.

Grier CC, Gower ST, Vogt KA (1990) Douglas-fir productivity: a conceptual model of its regulation and nutrient availability. pp. 257–273. In: (Gessel, SP, Lacate DS, Weetman GF, Powers RF, eds.) Sustained Productivity of Forest Soils. Proceedings of the 7th North American Forest Soils Conference. University of British Columbia. Vancouver, British Columbia, Canada.

Grier CC, Vogt KA, Keyes MR, Edmonds RL (1981) Biomass distribution and above- and below-ground production in young and mature *Abies amabilis* zone ecosystems of the Washington Cascades. Can. J. For. Res. 11: 155–167.

Griffin JR (1977) Oak woodland. pp. 383–415. In: (Barbur MG, Majors J, eds.) Terrestrial Vegetation of California. John Wiley & Sons. New York.

Grime JP (1977) Evidence for the existence of three primary strategies in plants and its relevance to ecological and evolutionary theory. Am. Nat. 111: 1169–1194.

Grumbine RE (1990) Protecting biological diversity through the greater ecosystem concept. Natural Areas 10: 114–120.

Grumbine RE (1992) Ghost Bears: Exploring the Biodiverstiy Crisis. Island Press. Washington DC. 294 pp.

Grumbine RE (1994) What is ecosystem management? Conserv. Biol. 8: 27–38.

Halpern CB (1988) Early successional pathways and the resistance and resilience of forest communities. Ecology 69: 1703–1715.

Hambidge G (1938) Soils and men—a summary. pp. 1–44. In: Soils and Men—Yearbook of Agriculture 1938. USDA US Govt. Printing Office. Washington DC. 1232 pp.

Hamburg SP, Sanford RL Jr (1986) Disturbance, *Homo sapiens*, and ecology. Bull. Ecol. Soc. Am. 67: 169–171.

Hansen AJ, Spies TA, Swanson FJ, Ohmann JL (1991) Conserving biodiversity in managed forests. BioScience 41: 382–392.

Hansen AJ, Garman SL, Marks B, Urban DL (1993) An approach for managing vertebrate diversity across multiple-use landscapes. Ecol. Appl. 3: 481–496.

Harley JL (1969) The Biology of Mycorrhizae. 2nd Ed. Leonard Hill. London.

Harley JL, Smith SE (1983) Mycorrhizal Symbiosis. Academic Press. London. 483 pp.

Harmon ME, Franklin JF, Swanson FJ, Sollins P, Gregory SV, Lattin JD, Anderson NG, Cline SP, Aumen NG, Sedell JR, Lienkaemper GW, Cromack K, Cummins KW (1986) Ecology of coarse woody debris in temperate ecosystems. Adv. Ecol. Res. 15: 133–302.

Harris WF, Kinerson RS, Edwards NT (1978) Comparisons of belowground biomass of natural deciduous forest and loblolly pine plantations. Pedobiology 17: 369–381.

Harvey AE (1994) Integrated roles for insects, diseases and decomposers in fire dominated forests of the inland western United States: past, present and future forest health. pp. 211–220. In: (Sampson RN, Adams DL, eds.) Assessing Forest Ecosystem Health in the Inland West. Food Products Press. Haworth Press. New York.

Harvey AE, Larsen MJ, Jurgensen MF (1976) Distribution of ectomycorrhizae in a mature Douglas-fir/larch forest soil in western Montana. For. Sci. 22: 393–398.

Haskell BD, Norton BG, Costanza R (1992) What is ecosystem health and why should we worry about it? pp. 3–20. In: (Costanza R, Norton BG, Haskell BD, eds.) Ecosystem Health: New Goals for Environmental Management. Island Press. Washington DC.

Haynes RW (1990) Analysis of the timber situation in the United States: 1989–2040. Gen. Tech. Rep. RM-199. Fort Collins, CO. USDA Forest Service. Rocky Mountain Forest and Range Exp. Sta. Ft. Collins, Colordo.

Haynes RW, Brooks DJ (1991) Wood and timber availability from a Pacific rim perspective. In: Proceedings of the Annual Convention of the Society of American Foresters, San Francisco.

Hedin LO, Granat L, Likens GE, Buishand TA, Galloway JN, Butler TJ, Rodhe H (1994) Steep declines in atmospheric base cations in regions of Europe and North America. Nature 367: 351–354.

Hendrey GR (ed) (1992) FACE: Free-air CO^2 enrichment for plant research in the field. Criti. Rev. Plant Sci. 11(2/3): 75–83.

Hessburg PF, Mitchell RG, Filip GM (1993) Historical and current roles of insects and pathogens in eastern Oregon and Washington forested landscapes. pp. 489–535. In: (Hessburg PF, compiler) Eastside Forest Ecosystem Health Assessment. Vol. III. Assessment. USDA For. Serv. Pac. Northwest Res. Sta. PNW 93-0304. Portland, Oregon.

Heywood VH, Baste I, Gardner KA (1995) Introduction. pp. 1–19. Global Biodiversity Assessment. UNEP. Cambridge University Press. Cambridge.

Hillel DJ (1991) Out of the earth—Civilization and the life of the soil. The Free Press Macmillan. New York. 321 pp.

Hiltner L (1896) On the importance of root nodules of *Alnus glutinosa* for nitrogen fixation by that plant. Landwirtsch. Vers.-Stn. 46: 153–161.

Hogberg P, Johannisson P, Hogbomg M, Nasholm T, Hallgren J-E (1995) Measurements of abundances of 15-N and 13-C as tools in retrospective studies of N balances and water stress in forests: a discussion of preliminary results. Plant Soil 168/169: 125–133.

Holling CS (1973) Resilience and stability of ecological systems. Ann. Rev. Ecol. Syst. 4: 1–24.

Holling CS (1986) The resilience of terrestrial ecosystems: local surprise and global change. pp. 289–317. In: (William CC, Munn RE, eds.) Sustainable Development of the Biosphere. IISASA. Cambridge University Press. Cambridge.

Hollins J (1993) The second law: a basis for ecological indicators. Unpublished manuscript.

Hornbeck JW, Swank WT (1992) Watershed ecosystem analysis as a basis for multiple-use management of eastern forests. Ecol. Appl. 2: 238–247.

Houck OA (1995) Reflections on the endangered species act. Environ. Law 25: 689–702.

Howarth FG, Sohmer SH, Duckworth WD (1988) Hawaiian natural history and conservation efforts. BioScience 38: 232–237.

Huettle RF (1989) "New types" of forest damages in central Europe. pp. 22–74. In: (MacKenzie JJ, Ellll-Ashry MT, eds.) Air Pollution's Toll on Forests & Crops. Yale University Press. New Haven, Connecticut.

Hunt ERJ, Martin FC, Running SW (1991) Simulating the effects of climatic variation on stem carbon accumulation of a Ponderosa pine stand: comparison to annual growth increment data. Tree Physiol. 9: 161–171.

Hunter MC (1990) Wildlife, Forests, and Forestry. Principles of Managing Forests for Biological Diversity. Prentice Hall. New Jersey. 370 pp.

Ingestad T, Ågren GI (1992) Theories and methods on plant nutrition and growth. Physiol. Plant 84: 177–184.

Ingham RE, Trofymow JA, Ingham ER, Coleman DC (1985) Interactions of bacteria, fungi, and their nematode grazers: effects on nutrient cycling and plant growth. Ecol. Monogr. 55: 119–140.

International Paper (1994) Forest Stewardship. Issue Briefs 6.94. June 1994. 22 pp.

István S, Riedi RH (1994) Application of multifractals to the analysis of vegetation pattern. J. Veg. Sci. 5: 489–496.

Iverson D, Alston R (1986) The genesis of FORPLAN: a historical and analytical review of Forest Service Planning models. Intermountain Research Station. Gen. Tech. Rep. INT-214. Boise, Idaho.

Janos DP (1980) Mycorrhizae influence tropical succession. Biotropica 12: 56–64.

Jarvis PG, Leverenz JW (1983) Productivity of temperate, deciduous and evergreen forests. pp. 233–280. In: (Lange OL, Nobel PS, Osmond CB, Ziegler H, eds.) Encyclopedia of Plant Physiology. Vol. 12D. Springer. New York.

Jenny H (1941) Factors of soil formation. McGraw-Hill. New York.

Jenny H (1980) The Soil Resource. Springer-Verlag. New York. 377 pp.

Joffe JS (1936) Pedology. Rutgers University Press. New Brunswick, New Jersey.

Johnsen KH, Major JE (1995) Gas exchange of 20-year-old black spruce families displaying a genetic × environmental interaction in growth rate. Can. J. For. Res. (in press).

Johnson AH, Siccama TG (1989) Decline of red spruce in the high-elevation forests of the northeastern United States. pp. 191–243. In: (MacKenzie JJ, El-Ashry MT, eds.) Air Pollution's Toll on Forests and Crops. Yale University Press. New Haven, Connecticut.

Johnson DR, Agee JK (1988) Introduction to ecosystem management. pp. 3–14. In: (Agee JK, Johnson DR, eds.) Ecosystem Management for Parks and Wilderness. University of Washington Press. Seattle, Washington.

Johnson DW (1994) Effects of forest management on soil carbon storage. Water Air Soil Poll. 64: 83–120.

Johnson DW, Lindberg SE (eds.) (1992) Atmospheric Deposition and Forest Nutrient Cycling: A Synthesis of the Integrated Forest Study. Springer-Verlag. New York. 707 pp.

Johnson DW, Todd DE (1983) Relationships among iron, aluminum, carbon, and sulfate in a variety of forest soils. Soil Sci. Soc. Am. J. 47: 792–800.

Johnson DW, Todd DE (1987) Nutrient export by leaching and whole-tree harvesting in a loblolly pine and mixed oak forest. Plant Soil 102: 99–109.

Johnson J (1994) Conversation, February 21, 1994.

Johnson KH, Vogt KA, Clark HJ, Schmitz OJ, Vogt DJ (1996) Biodiversity and the productivity and stability of ecosystems. TREE 11: 372–377.

Johnson KN, Franklin JF, Thomas JW, Gordon J (1991) Alternatives for management of late successional forests of the Pacific Northwest: a report to the Agriculture Committee and the Merchant Marine and Fisheries Committee of the U.S. House of Representatives. Washington D.C.

Jones CG, Coleman JS (1991) Plant stress and insect herbivory: toward an integrated perspective. pp. 249–274. In: (Mooney HA, Winner WE, Pell EJ, eds.) Response of Plants to Multiple Stresses. Academic Press. New York.

Jørgensen SE, Halling-Sørensen B, Nielsen SN (1996) Handbook of Environmental and Ecological Modeling. CRC Press. Lewis Publishers. Boca Raton, Florida. 672 pp.

Karieva P (1994) Diversity begets productivity. Nature 368: 686–687.

Karr JR, Dudley DR (1981) Ecological perspective on water quality goals. Environ. Manag. 5: 55–68.

Kauffmann JB, Martin RE (1987) Effects of fire and fire suppression on mortality and mode of reproduction of California Black Oak (*Quercus kelloggii*). Gen. Tech. Rep. PSW-100: 122–126. San Francisco, CA.

Kay JJ, Schneider E (1994) Embracing complexity: the challenge of the ecosystem approach. Alternatives 20: 33–39.

Keddy PA, Lee HT, Wisheu IC (1993) Choosing indicators of ecosystem integrity: wetlands as a model system. pp. 61–79. In: (Woodley S, Kay J, Francis G, eds.) Ecological Integrity and the Management of Ecosystems. St. Lucie Press. Ottawa, Canada.

Kiester AR, Ford ED, Weinstein DA, Chen C, Dougherty P, Gherini S, Webb C, Droessler TD, Ladd L, Cothern S, Avery A, Beloin R, Gomez L, Teskey R, Barton C, Burkhart H, Amateis R, Jarvis P, Bassow S, Ford R (1990) Development and Use of Tree and Forest Response Models. National Acid Precipitation Assessment Program (NAPAP) Report 17. Government Printing Office. Washington DC. 250 pp.

Kimmins JP (1977) Evaluation of the consequences for future tree productivity of the loss of nutrients in whole tree harvesting. For. Ecol. Manag. 1: 169–183.

Kimmins JP (1987) Forest Ecology. MacMillan. New York. 531 pp.

Kimmins JP (1988) Community organization: Methods of study and prediction of the productivity and yield of forest ecosystems. Can. J. Bot. 66: 2654–2672.

Kimmins JP, Comeau PG, Kurz W (1990) Modelling the interactions between moisture and nutrients in the control of forest growth. For. Ecol. Manag. 30: 361–380.

King AW (1993) Considerations of scale and hierarchy. In: (Woodley S, Kay J, Francis G, eds.) Ecological Integrity and the Management of Ecosystems. St. Lucie Press. Delray Beach, FL.

Klute A (ed.) (1986) Methods of Soil Analysis: Physical and Mineralogical Methods—Part 1, 2nd Ed. Agron. Monograph #9. ASA and SSSA, Madison, Wisconsin. 1216 pp.

Knight DH (1991) The Yellowstone fire controversy. pp. 87–103. In: (Keiter RB, Boyce MS, eds.) The Greater Yellowstone Ecosystem. Redefining America's Wilderness Heritage. Yale University Press. New Haven, Connecticut.

Knight DH, Fahey TJ, Running SW (1985) Water and nutrient outflow from lodgepole pine forests in Wyoming. Ecol. Monogr. 55: 29–48.

Knowles P, Grant MC (1981) Genetic patterns associated with growth variability in Ponderosa pine. Am. J. Bot 68: 942–946.

Körner C (1993) Scaling from species to vegetation: the usefulness of functional groups. pp. 117–140. In: (Schulze E, Mooney HA, eds.) Biodiversity and Ecosystem Function. Springer-Verlag. Berlin.

Körner C, Farquhar GD, Wong SC (1991) Carbon isotope discrimination by plants follows latitudinal and altitudinal trends. Oecologia 88: 30–40.

Kotliar NB, Wiens JA (1990) Multiple scales of patchiness and patch structure: a hierarchical framework for the study of heterogeneity. Oikos 59: 253–260.

Kowalski S (1987) Mycotrophy of trees in converted stands remaining under strong pressure of industrial pollution. Angew. Botanik 61: 65–83.

Kramer PJ, Kozlowski TT (1960) Physiology of Trees. McGraw-Hill. New York. 642 pp.

Kramer PJ, Kozlowski TT (1979) Physiology of Woody Plants. Academic Press. New York. 811 pp.

Kremen C (1992) Assessing the indicator properties of species assemblages for natural areas monitoring. Ecol. Appl. 2: 203–217.

Krummel JR, Gardner RH, Sugihara G, O'Neill RV, Coleman PR (1987) Landscape pattern in a disturbed environment. Oikos 48: 321–324.

Krupa SV, Kickert RN (1987) The acid deposition program: an analysis of numerical models of air pollutant exposure and vegetation response. Acid Deposition Program. Calgary, Canada.

Lackey RT (1995) Seven pillars of ecosystem management. Draft. March 13, 1995.

Lajtha K, Whitford WG (1989) The effect of water and nitrogen amendments on photosynthesis, leaf demography, and resource-use efficiency in *Larrea tridentata*, a desert evergreen shrub. Oecologia 80: 341–348.

Lam NS-N (1990) Description and measurement of Landsat TM images using fractals. Photogramm. Engin. Remote Sensing 56: 187–195.

Lambers H, Poorter H (1992). Inherent variation in growth rate between higher plants: a search for physiological causes and ecological consequences. Adv. Ecol. Res. 23: 187–261.

Landres PB, Verner J, Thomas JW (1988) Ecological uses of vertebrate indicator species: a critique. Conserv. Biol. 2: 316–328.

Landsberg JJ, Prince SD, Jarvis PG, McMurtrie RE, Luxmoore R, Medlyn BE (1996) Energy conversion and use in forests: the analysis of forest production in terms of radiation utilisation efficiency. In: (Gholz HL, Nakare K, Shimoda H, eds.) The Use of Remote Sensing in the Modelling of Forest Productivity at Scales From the Stand to the Globe. Kluwer Academic. (in press).

Landsberg JJ, Linder S, McMurtrie RE (1995) A strategic plan for research on managed forest ecosystems in a globally changing environment. pp. 1–17. In: Global Change and Terrestrial Ecosystems GCTE Report No. 4. GCTE Activity 3.5: Effects of Global Change on Managed Forests. Implementation Plan.

Lapeyrie FF, Chilvers GA (1985) An endomycorrhiza-ectomycorrhiza succession associated with enhanced growth of *Eucalyptus dumosa* seedlings planted in a calcareous soil. New Phytol. 100: 93–104.

Larcher W (1975) Physiological Plant Ecology. Springer-Verlag. Berlin. 252 pp.

Lassoie JP, Hinckley TM (1991) Techniques and Approaches in Forest Tree Ecophysiology. CRC Press. Florida. 599 pp.

Last FT, Mason PA, Ingleby K, Fleming LV (1984) Succession of fruitbodies of sheathing mycorrhizal fungi associated with *Betula pendula*. For. Ecol. Manage. 9: 229–234.

Leffer JW (1978) Ecosystem responses to stress in aquatic microcosms. pp. 14–29. In: (Thorp JH, Gibbons JW, eds.) Energy and Environmental Stress in Aquatic Ecosystems. U.S. Dept. of Energy. National Tech. Inform. Center. Springfield, Virginia.

Leopold A (1949) A Sand County Almanac. Oxford University Press. New York. 295 pp.

Leverenz JW, Hallgren J-E (1991) Measuring photosynthesis and respiration of foliage. pp. 303-328. In: (Smith WK, Hinckley TM, eds.) Resource Physiology of Conifers: Acquisition, Allocation, and Utilization. Academic Press. New York.

Levin S (1989) Challenges in the development of a theory of ecosystem structure and function. pp. 242–255. In: (Roughgarden J, May RM, Levin SA, eds) Perspectives in Ecological Theory. Princeton University Press, Princeton, New Jersey.

Levin SA (1992) Orchestrating environmental research and assessment. Ecol. Appl. 2: 103–106.

Levin SA (1993) Concepts of scale at the local level. pp. 7–19. In: (Ehleringer JR, Field CB, eds.) Scaling Physiological Processes: Leaf to Globe. Academic Press. San Diego.

Levin SA, Harwell MA, Kelly Jr, Kimball KD (1986) Ecotoxicology: Problems and Approaches. Springer-Verlag. New York.

Lichtman P, Clark TW. (1990) Rethinking the "Vision" exercise in the Greater Yellowstone Ecosystem. 29 pp. Unpublished manuscript.

Lindblom C. (1977) Politics and markets. Basic Books pp. 313–329. Harper Collins. New York. 403 pp.

Lindeman RL (1942) The trophic dynamic aspect of ecology. Ecology 23: 399–418.

Liroff RA, Davis GG (1982) Protecting green space: Land use control in the Adirondack Park. Cambridge, Massachusetts. Ballinger.

Loope LL, Hamann O, Stone CP (1988) Comparative conservation biology of oceanic archipelagoes. BioScience 38: 272–282.

Lotka AJ (1925) The Elements of Physical Biology. Williams & Wilkins, Baltimore, Maryland.

Lubchenco J, Olson AM, Brubaker LB, Carpenter SR, Holland MM, Hubbell SP, Levin SA, MacMahon JA, Matson PA, Melillo JM, Mooney HA, Peterson CH, Pulliam HR, Real LA, Regal PJ, Risser PG (1991) The sustainable biosphere initiative: an ecological research agenda. Ecology 72: 371–412.

Lucier AA (1994) Criteria for success in managing forested landscapes. J. For. 92: 20–24.

Ludwig D, Hilborn R, Walters C (1993) Uncertainty, resource exploitation, and conservation: lessons from history. Ecol. Appl. 3: 547–549.

Lugo AE (1987) Are island ecosystems different from continental ecosystems? Acta Cienti. 1: 48–54.

Lugo AE (1988) Ecological aspects of catastrophes in Caribbean islands. Acta Cienti. 2: 24–31.

Lugo AE, Scatena FN (1995) Ecosystem-level properties of the Luquillo Experimental Forest with emphasis on the Tabonuco Forest. pp. 59–108. In: (Lugo AE, Lowe C, eds.) Tropical Forests: Management and Ecology. Ecological Studies 112. Springer-Verlag. New York.

Lugo AE, Sanchez MJ, Brown S (1986) Land use and organic content of some subtropical soils. Plant Soil 96: 185–196.

Lyons J, Knowles D (1988) The Forest and Rangeland Renewable Resource Planning Act: Congressional staff perspective. In: (Binkley C, Brewer G, Sample VA, eds.) Redirecting the RPA. Yale School of Forestry and Environ. Studies Bull. 95. 229 pp.

MacArthur RH (1955) Fluctuations of animal populations and a measure of community stability. Ecology 36: 533–536.

MacArthur RH, Wilson EO (1963) An equilibrium theory of insular zoogeography. Evolution 17: 373–387.

MacCleery DW (1993) American Forests, A History of Resiliency and Recovery. Forest History Society. Durham, North Carolina. 58 pp.

MacMahon JA, Phillips DL, Robinson JV, Schimpf DJ (1978) Levels of biological organization: an organism-centered approach. BioScience 28: 700–704.

Mandelbrot BB (1983) The Fractal Geometry of Nature. WH Freeman. New York.

Marks GC, Kozlowski TT (1973) Ectomycorrhizae. Their Ecology and Physiology. Academic Press. New York. 444 pp.

Marsh AS, Siccama TG (1996) Use of formerly plowed land in New England to monitor the vertical distribution of lead, zinc and copper in mineral soil. Water Air Soil Poll. (in press).

Marsh GP (1864) Man and Nature: Physical Geography as Modified by Human Action. Scribner. New York.

Marshall DB (1987) Status of the marbled murrelet in North America: with special emphasis on the population in California, Oregon, and Washington. U.S. Fish and Wildlife Service Biological Report 88. 19 pp. Washington D.C.

Marshall DB (1989) The marbled murrelet. Audubon Wildlife Report 1989: 435–455.

Maser C, Trappe JM, Ure DC (1978) Fungal-small mammal interrelationships with emphasis on Oregon forests. Ecology 59: 799–809.

Matson PA, Vitousek PM (1990) Remote sensing and trace gas fluxes. pp. 97–108. In: (Hobbs RJ, Mooney HA, eds.) Remote Sensing of Biosphere Functioning. Wiley. New York.

Mattson (1994) Grizzly Bear-Human Interactions in the Greater Yellowstone Ecosystem. (unpublished manuscript and figures).

May RM (1973) Stability and Complexity in Model Ecosystems. Princeton University Press. Princeton, New Jersey.

McClaugherty CA, Pastor J, Aber JD, Melillo JM (1985) Forest litter decomposition in relation to soil nitrogen dynamics and litter quality. Ecology 66: 266–275.

McHarg I (1969) Design with Nature. Natural History Press. Garden City, New Jersey.

McLaughlin SB, Downing (1995) Interactive effects of ambient ozone and climate measured on growth of mature forest trees. Nature 374. 252–254.

McLaughlin SB, Anderson CP, Edwards MT, Roy WK, Layton PA (1990) Seasonal patterns of photosynthesis and respiration of red spruce saplings from two elevations in declining southern Appalachian stand. Can. J. For. Res. 20: 485–595.

McNaughton SJ (1993) 17 Biodiversity and function of grazing ecosystems. pp. 361–383. In: (Schulze E, Mooney HA, eds.) Biodiversity and Ecosystem Function. Springer-Verlag. Berlin.

McNaughton SJ, Ruess RW, Seagle SW (1988) Large mammals and process dynamics in African ecosystems. BioScience 38: 794–800.

Meadows D, Robinson JM (1985) The Electronic Oracle: Computer Models and Social Decisions. John Wiley. New York.

Medina E, Montes G, Cuevas E, Roksandic Z (1986) Profiles of CO_2 concentration and (^{13}C values in tropical rainforests of the Upper Rio Negro Basin, Venezuela. J. Trop. Ecol. 2: 207–217.

Meentemeyer V (1978) Macroclimate and lignin control of litter decomposition rates. Ecology 59: 465–472.

Meentemeyer V, Box EO (1987) Scale effects in landscape studies. pp. 15–34. In: (Turner MG, ed.) Landscape Heterogeneity and Disturbance. Springer-Verlag. New York.

Meiners TM, Smith DW (1984) Soil and plant water stress in an Appalachian oak forest in relation to topography and stand age. Plant Soil 80:171–179.

Mikola P (1973) Mycorrhizae and feeder root diseases. pp. 351–382. In: (Marks GC, Kozlowski TT, eds.) Ectomycorrhizae. Their Ecology and Physiology. Academic Press. New York.

Miller DL, Leonard PM, Hughes RM, Karr JR, Moyle PB, Schrader LH, Thompson BA, Daniels RA, Fausch KD, Fitzhugh GA, Gammon JR, Haliwell DB, Angermeier PA, Orth DJ (1988) Regional applications of an index of biotic integrity for use in water resource management. Fisheries (Bethesda) 13: 12–20.

Mills SL, Soulé ME, Doak DF (1993) The keystone-species concept in ecology and conservation. BioScience 42: 219–224.

Milne BI (1992) Spatial aggregation and neutral models in fractal landscapes. Am. Natural. 139: 32–57.

Mladenoff DJ, Pastor J (1993) Sustainable forest ecosystems in the northern hardwood and conifer forest region: concepts and management. pp. 145–180. In: (Aplet GH, Johnson N, Olson JT, Sample VA, eds.) Defining Sustainable Forestry. Island Press. Covelco, California.

Mladenoff DJ, White MA, Pastor J, Crow TR (1993) Comparing spatial pattern in unaltered old-growth and disturbed forest landscapes. Ecol. Appl. 3: 294–306.

Molina R, O'Dell T, Luoma D, Amaranthus M, Castellano M, Russell K (1993) Biology, Ecology, and Social Aspects of Wild Edible Mushrooms in the Forests of the Pacific Northwest: A Preface to Managing Commercial Harvest. USDA For. Serv. Gen. Tech. Rep. PNW-GTR-309. Portland, Oregon.

Mooney HA (1972) The carbon balance of plants. Ann. Rev. Ecol. Syst. 3: 315–346.

Mooney HA, Drake JA (eds.) (1986) Ecology of Biological Invasions of North American and Hawaii. Springer-Verlag. New York.

Mooney HA, Kummerow J (1971) The comparative water economy of representative evergreen sclerophyll and drought deciduous shrubs of Chile. Bot. Gaz. 132: 245–252.

Mooney HA, Winner WE (1991) Partitioning response of plants to stress. pp. 129–141. In: (Mooney HA, Winner WE, Pell EJ, eds.) Response of Plants to Multiple Stresses. Academic Press. London.

Mooney HA, Drake BG, Luxmoore RJ, Oechel WC, Pitelka LF (1991) Predicting ecosystem responses to elevated CO_2 concentrations. BioScience 42: 96–104.

Moran EF (1987) Monitoring fertility degradation of agricultural lands in the lowland tropics. pp. 69–91. In: (Little PD, Horowitz MM, Nyerges AE, eds.) Lands at Risk in the Third World: Local-Level Perspectives. Westview Press. Boulder, Colorado.

Morrison ML (1986) Bird populations as indicators of environmental change. pp. 429-451. In: (Johnston RF, ed.) Current Ornithology. Vol. 3. Plenum Press. New York.

Morrison PH (1988) Old Growth in the Pacific Northwest: A Status Report. The Wilderness Society. Washington DC. 58 pp.

Morrison PH (1990) Ancient Forest on the Olympic National Forest: Analysis from a Historical and Landscape Perspective. The Wilderness Society. Seattle, Washington.

Muckenfuss E (1994) Cooperative Ecosystem Management in the ACE Basin. J. For. 92: 35–36.

Mueller-Dombois D, Bridges KW, Carson HL (eds.) (1981) Island Ecosystems. Biological Organization in Selected Hawaiian Communities. Hutchinson Ross. Stroudsburg, Pennsylvania.

Myers N (1988) Tropical forests and their species: going, going...? In: (Wilson EO, Peter FM, eds.) Biodiversity. National Academy Press. Washington DC.

Naeem S, Thompson LJ, Lawler SP, Lawton JH, Woodfin RM (1994) Declining biodiversity can alter the performance of ecosystems. Nature 368: 734–737.

Naiman RJ, Beechie TJ, Benda LE, Bisson PA, McDonald LH, Conner MD, Olson PL, Steel EA (1992) Fundamental elements of ecologically healthy watersheds in the Pacific Northwest coastal ecoregions. In: (Naiman RJ, ed.) Watershed Management: Balancing Sustainability and Environmental Change. Springer-Verlag. New York.

NAPAP (National Atmospheric Precipitation Assessment Program) (1993) pp. 37–49. In: 1992 Report to Congress. Ch. 4. NAPAP. Washington DC.

NCASI (1992) Impacts on private forestry from conservation strategies for threatened and endangered species. New York. National Council Paper Ind. on Air Stream Imp. Special Report. 92-07. 10 pp.

Nehlsen W, Williams JW, Lichatowich JA (1991) Pacific salmon at the crossroads: stocks at risk from California, Oregon, Idaho and Washington. Fisheries 16: 4–21.

Nobbe F von, Hiltner L (1904) On the nitrogen fixing abilities of Alnus and Eleagnus. Naturwiss. Z. Forst. Landwirtsch. 2: 366–369.

Norby RJ, O'Neill EG, Luxmoore RJ (1986) Effects of atmospheric CO_2 enrichment on the growth and mineral nutrition of *Quercus alba* seedlings in nutrient-poor soils. Plant Physiol. 82: 83–89.

Norman JM (1993) Scaling processes between leaf and canopy levels. pp. 41–74. In: (Ehleringer JR, Field CB, eds.) Scaling Physiological Processes: Leaf to Globe. Academic Press. New York.

Norse EA (1990) Ancient Forests of the Pacific Northwest. Island Press. Washington DC.

Norse EA (1991) Global Marine Biodiversity: A Strategy for Building Conservation into Decision-Making. Island Press, Washington DC.

Norse EA (1993) Global marine biodiversity. pp. 41–43. In: (Sandlund OT, Schei PJ, eds.) Proceedings of the Norway/UNEP (United Nations Environment Programme) Expert Conference on Biodiversity. Norwegian Institute of Nature Research. Trondheim, Norway.

Norstedt HO (1982) Nitrogen fixation by free-living microorganisms in the soils of a mature oak stand in Upland, Sweden. Holarctic Ecol. 5: 20–26.

Northrup RR, Yu Z, Dahlgren RA, Vogt KA (1995) Polyphenol control of nitrogen release from pine litter. Nature 377: 227–229.

Noss R (1990) Indicators for monitoring biodiversity: a hierarchical approach. Conserv. Biol. 4: 355–364.

NRC (National Research Council) (1986) Ecological Knowledge and Environmental Problem-Solving: Concepts and Case Studies. National Academy Press. Washington DC.

NRC (National Research Council) (1990) Forestry Research: A Mandate for Change. National Academy Press. Washington DC. 84 pp.

NRC (National Research Council) (1992) Review of EPA's Environmental Monitoring and Assessment Program (EMAP). Interim Report. National Research Council. Washington DC.

NRC (National Research Council) (1994) Rangeland Health: New Methods to Classify, Inventory, and Monitor Rangelands. National Academy Press. Washington DC. 180 pp.

O'Dell TE, Luoma DL, Molina RJ (1992) Ectomycorrhizal fungal communities in young, managed, and old-growth Douglas-fir stands. Northwest Environ. J. 8: 166–167.

O'Laughlin J (1993) Exploring the definition of forest health. pp. 9–14. In: Proceedings. Forest Health in the Inland West. USDA For. Serv. Boise National For. Dept. For. Resources. University of Idaho. Moscow, Idaho.

O'Laughlin J, Livingston RL, Thier R, Thornton J, Toweill DE, Morelan L (1994) Defining and measuring forest health. pp. 65–86. In: (Sampson RN, Adams DL, eds.) Assessing Forest Ecosystem Health in the Inland West. Haworth Press. New York.

O'Neill RV (1976) Ecosystem persistence and heterotrophic regulation. Ecology 57: 1244–1253.

O'Neill RV, DeAngelis DL, Waide JB, Allen TFH (1986) A Hierarchical Concept of Ecosystems. Princeton University Press. Princeton, New Jersey. 253 pp.

O'Neill RV, Johnson AR, King AW (1989) A hierarchical framework for the analysis of scale. Landsc. Ecol. 3: 193–205.

Odum EP (1953) Fundamentals of Ecology. WB Saunders. Philadelphia, Pennsylvania.

Odum EP (1959) Fundamentals of Ecology. 2nd Ed. WB Saunders. Philadelphia, Pennsylvania.

Odum EP (1969) The strategy of ecosystem development. Science 164: 262–270.

Odum EP (1971) Fundamentals of Ecology. 3rd. Ed. WB Saunders. Philadephia, Pennsylvania.

Odum EP (1985) Basic Ecology. WB Saunders. Philadephia, Pennsylvania. 613 pp.

Odum HT (ed.) (1970) A Tropical Rain Forest. A Study of Irradiation and Ecology at El Verde, Puerto Rico. US Atomic Energy Commission. Division of Technical Information Extension. Oak Ridge, Tennessee.

Oliver CD (1981) Forest development in North America following major disturbances. For. Ecol. Manag. 3: 153–168.

Oliver CD, Larson BC (1990) Forest Stand Dynamics. McGraw-Hill. New York. 467 pp.

Olsen ER, Ransey RD, Winn DS (1993) A modified fractal dimension as a measure of landscape diversity. Photogr. Eng. Remote Sensing 59: 1517–1520.

Omerik JM, Gallant AL (1986) Ecoregions of the Pacific Northwest. U.S. Environmental Protection Agency. Washington DC. 39 pp.

Osenberg CW, Schmitt RJ, Holbrook SJ, Abu-Saba KE, Flegal AR (1994) Detection of environmental impacts: natural variability, effect size, and power analysis. Ecol. Appl. 4: 16–30.

Overbay JC (1992) Ecosystem management. pp. 3–15. In: Proceedings of the National Workshop: Taking an Ecological Approach to Management. April 27–30, 1992, Salt Lake City, Utah. US Dept. Agric. For. Serv.

Owen CN, Sweeney JM (1995) Managing for Diversity: Champion International's Approach. J. For. 93: 12–14.

Page AL, Miller RH, Keeney DR (eds.) (1982) Methods of Soil Analysis: Chemical and Microbiological Properties—Part 2. 2nd Ed. Agron. Monogr. #9. ASA. SSSA, Madison, Wisconsin. 1184 pp.

Paine RT (1966) Food web complexity and species diversity. Am. Natural. 100: 65–75.

Pallardy SG, Pereira JS, Parker WC (1991) Measuring the state of water in tree systems. pp. 27–76. In: (Lassoie JP, Hinckley TM, eds.) Techniques and Approaches in Forest Tree Ecophysiology. CRC Press. Boca Raton, Florida.

Pallardy SG, Cermák J, Ewers FW, Kaufmann MR, Parker WC, Sperry JS (1995) Water transport dynamics in trees and stands. pp. 301–373. In: (Smith WK, Hinckley TM, eds.) Resource Physiology of Conifers: Acquisition, Allocation, and Utilization. Academic Press. New York.

Palmiotto PA (1993) Initial response of *Shorea* wildlings transplanted in gap and understory microsites in a lowland rain forest. J. Trop. For. Sci. 5: 403–415.

Parsons DJ, DeBenedetti SH (1979) Impact of fire suppression on a mixed-conifer forest. For. Ecol. Manag. 2: 21–33.

Parton WJ, Schimel DS, Cole CV, Ojima DS (1987) Analysis of factors controlling soil organic matter levels in great plains grasslands. Soil Sci. Soc. Am. J. 51: 1173–1179.

Parton WJ, Stewart JWB, Cole CV (1988) Dynamics of C, N, P and S in grassland soils: a model. Biogeochemistry 5: 109–131.

Pastor JJ, Aber JD, McClaugherty CA, Melillo JM (1984) Aboveground production and N and P cycling along a nitrogen mineralization gradient on Blackhawk Island, Wisconsin. Ecology 65: 245–268.

Pastor J, Naiman RJ, Dewey B, McInnes P (1988) Moose, microbes, and the boreal forest. BioScience 38: 770–777.

Patterson MT, Rundel PW (1993) Carbon isotope discrimination and gas exchange in ozone-sensitive and -resistant populations of Jeffrey Pine. pp. 213–225. In: (Ehleringer JR, Hall AE, Farquhar GD, eds.) Stable Isotopes and Plant Carbon-Water Relations. Academic Press. New York.

Pearson DL, Cassola F (1992) World-wide species richness patterns of Tiger Beetles (*Coleopers cicindelidae*): indicator taxon for biodiversity and conservation studies. Conserv. Biol. 6: 376–391.

Perlin J (1989) A Forest Journey, the Role of Wood in the Development of Civilization. Harvard University Press. Cambridge, Massachusetts. 445 pp.

Perry DA, Amaranthus MP, Borchers JG, Borchers SL, Brainerd RE (1989) Bootstrapping in ecosystems. BioScience 39: 230–237.

Perry DA, Borchers JG, Borchers SL (1990) Species migrations and ecosystem stability during climate change: the belowground connection. Conserv. Biol. 4: 266–274.

Persson H (ed.) (1990) Above and Below-ground Interactions in Forest Trees in Acidified Soils. Air Pollution Research Report 32. Commission of the European Communities. Directorate-General for Sciences, Research and Development. 258 pp.

Pfeffer J, Salancik G (1978) The External Control of Organizations: A Resource Dependence Perspective. Harper and Row. New York. 300 pp.

Pickett STA, Bazzaz FA (1978) Organization of an assemblage of early successional species on a soil moisture gradient. Ecology 59: 1248–1255.

Pickett STA, White PS (1985) The Ecology of Natural Disturbance and Patch Dynamics. Academic Press. Orlando, Florida. 472 pp.

Pierce LL, Running SW, Riggs GA (1990) Remote detection of canopy water stress in coniferous forests using the S001 Tematic Mapper Simulator and the Thermal Infrared Multispectral Scanner. Photogr. Eng. Rem. Sensing 56: 579–586.

Pimm SL (1991) The Balance of Nature? Ecological Issues in the Conservation of Species and Communities. The University of Chicago Press. Chicago, Illinois.

Pinedo-Vasquez M (1995) Human Impact on Varzea Ecosystems in the Napoo-Amazon, Peru. PhD Dissertation. Yale University. New Haven, Connecticut.

Pironzynski KA, Malloch DW (1975) The origins of land plants: a matter of mycotrophism. Biosystems 6: 153–164.

Pomeroy W (1961) Pomeroy Commission on the Adirondack Park. State of New York. Albany, N.Y.

Powell DS, Faulkner JL, Darr DR, Zhu Z, MacCleery DW (1992) Forest resources of the United States. USDA For. Serv. Washington, DC.

Power TM (1991) Ecosystem preservation and the economy in the Greater Yellowstone Area. Conserv. Biol. 5: 412–422.

Premo D (1993) Sustainability-based forest management in Michigan's Upper Peninsula. Unpublished paper. White Water Associates. Amasa, Michigan.

Premo D, Schwandt D (1993) A landscape managment coalition: The Mullikin Creek riparian management area. In: (Premo D, Premo B, Premo K, Rogers E, Tiller D, eds.) Total Ecosystem Management Strategies (TEMS). 1993 Annual Report: A Year of Progress, Results, and Recognition. White Water Associates. Amasa, Michigan.

Prendergast JR, Quinn RM, Lawton JH, Eversham BC, Gibbons DW (1993) Rate species, the coincidence of diversity hotspots and conservation strategies. Nature 365: 335–337.

Preston FW (1962) The canonical distribution of commonness and rarity: Parts 1 and 2. Ecology 43: 185–215, 410–432.

Publicover DA, Vogt KA (1991) A comment on "Forests are not just swiss cheese: canopy stereogeometry of non-gaps in tropical forests." by Lieberman et al. Ecology 72: 1507–1510.

Rapport DJ, Regier HA, Hutchinson TC (1985) Ecosystem behavior under stress. Am. Natural. 125: 617–640.

Rastetter EB, King AW, Cosby BJ, Hornberger GM, O'Neill RV, Hobbie JH (1992) Aggregating fine-scale ecological knowledge to model coarser-scale attributes of ecosystems. Ecol. Appl. 2: 55–70.

Read DJ (1993) Keystone species. pp. 181–209. In: (Schulze ED, Mooney HA, eds.) Biodiversity and Ecosystem Function. Ecological Studies 99. Springer-Verlag. Berlin.

Reich PB, Schoettle AW (1988) Role of phosphorus and nitrogen in photosynthetic and whole plant carbon gain and nutrient use efficiency in eastern white pine. Oecologia (Heidelberg) 77: 25–33.

Reich PB, Walters MB, Ellsworth DS (1992) Leaf life-span in relation to leaf, plant, and stand characteristics among diverse ecosystems. Ecol. Monogr. 62: 365–392.

Remmert H (1980) Arctic Animal Ecology. Springer-Verlag. Berlin.

Rey JR (1981) Ecological biogeography of arthropods on Spartina islands in northwest Florida. Ecol. Monogr. 51: 237–265.

Rey-Benayas JM, Pope KO (1995) Landscape ecology and diversity patterns in the seasonal tropics from Landsat TM imagery. Ecol. Appl. 5: 386–394.

Reynolds JF, Hilbert DW, Kemp PR (1993) Scaling ecophysiology from the plant to the ecosystem: a conceptual framework. pp. 127–140. In: (Ehleringer JR, Field CB, eds.) Scaling Physiological Processes: Leaf to Globe. Academic Press. San Diego, California.

Rice R (1989) Wilderness Society testimony before Subcommittee on Environment, Energy and Natural Resources. Committee on Government Operations. US House of Representatives. November 16, 1989.

Richards JH, Caldwell MM (1987) Hydraulic lift: substantial nocturnal water transport between soil layers by *Artemisia tridentata* roots. Oecologia 73: 486–489.

Richardson CJ, Lund JA (1975) Effects of clearcutting on nutrient losses in aspen forests on three soil types in Michigan. pp. 673–686. In: (Howell FG, Gentry JB, Smith MH, eds.) Mineral Cycling in Southeastern Ecosystems. ERDA Sympos. Ser. CONF-740513. Washington DC.

Richter DL, Bruhn JN (1993) Mycorrhizal fungus colonization of *Pinus resinosa* Ait transplanted on northern hardwood clearcuts. Soil Biol. Biochem. 25: 355–369.

Ricklefs RE (1990) Ecology. 3rd Ed. WH Freeman. New York. 896 pp.

Ricklefs RE, Schluter D (eds.) (1993) Species Diversity in Ecological Communities. Historical and Geographical Perspectives. University of Chicago Press. Chicago, Illinois. 414 pp.

Roberts L (1983) Is acid deposition killing west german forests? BioScience 33: 302–305.

Robertson D (1987) Hearing testimony. Subcommittee on Forests, Family Farms, and Energy of the Committee on Agriculture, House of Representatives. April 8, 1987. Serial no. 100-38. Washington, DC.

Robertson D (1992) Memorandum issued to regional foresters and research-station directors. June 4, 1992. Washington, DC.

Rokeach M (1979) Understanding Human Values. Free Press. New York.

Romme WH, Knight DH (1981) Fire frequency and subalpine forest succession along a topographic gradient in Wyoming. Ecology 62: 319–326.

Romme WH, Knight DH, Yavitt JB (1986) Mountain pine beetle outbreaks in the Central Rocky Mountains: effects on primary production. Am. Natural. 127: 484–494.

Roper Organization, Inc. (1992) Natural resource conservation: Where environmentalism is headed in the 1990s. Times Mirror Magazines Conservation Council. Washington, DC.

Roughgarden J (1995) Vertebrate patterns on islands. pp. 51–56. In: (Vitousek PM, Adsersen H, Loope LL, eds.) Islands. Biological Diversity and Ecosystem Function. Springer. Berlin.

Roughgarden J, Running SW, Matson PA (1991) What does remote sensing do for ecology? Ecology 72: 1918–1922.

Rounick JS, Winterbourn MJ (1986) Stable carbon isotopes and carbon flow in ecosystems. BioScience 36: 171–177.

Roush (1995) When rigor meets reality. Science 269: 313–315.

Rundel PW, Sharifi MR (1993) Carbon isotope discrimination and resource availability in the desert shrub *Larrea tridentata*. pp. 173–184. In: (Ehleringer JR, Hall AE, Farguhar GD, eds.) Stable Isotopes and Plant Carbon-Water Relations. Academic Press. New York.

Running SW (1980) Environmental and physiological control of water flux through *Pinus contorta*. Can. J. For. Res. 10: 82–91.

Running SW, Coughlan JC (1988) A general model for forest ecosystem processes for regional applications. I. Hydrologic balance, canopy gas exchange, and primary production processes. Ecol. Modelling 42: 125–154.

Running SW, Hunt ER Jr (1993) Generalization of a forest ecosystem process model for other biomes, BIOME-BGC, and an application for global-scale models. In: (Ehleringer JR, Field CB, eds.) Scaling Physiological Processes. Leaf to Globe. Academic Press. San Diego, California.

Running SW, Nemani RR (1988) Relating seasonal patterns of the AVHRR vegetation index to simulated photosynthesis and transpiration of forests in different climates. Remote Sens. Environ. 24: 347–367.

Russell M (1992) Lessons from NAPAP. Ecol. Appl. 2: 107–110.

Ryan MR, Linder S, Vose JM, Hubbard RM (1994) Respiration of pine forests. Ecol. Bull. 43: 50–63.

Rygiewicz PT, Andersen CP (1994) Mycorrhizae alter quality and quantity of carbon allocated below ground. Nature 369: 58–60.

Sackett S, Haase S, Harrington MG (1993) Restoration of southwestern Ponderosa pine ecosystems with fire. pp. 115–121. In: (Covington WW, DeBano LF, tech. coord.) Sustainable Ecological Systems: Implementing an Ecological Approach to Land Management. Gen. Tech. Rep. RM-247. Fort Collins, Colorado. USDA For. Serv. Rocky Mtn. For. Exp. Sta.

SAF Task Force (1992) Sustaining Long-Term Forest Health and Productivity. Society of American Foresters. Bethesda, Maryland.

Sage RF, Pearcy RW (1987) The nitrogen use efficiency of C3 and C4 plant. II. Leaf nitrogen effects on the gas exchange characteristics of *Chenopodium album* (L.) and *Amaranthus retroflexus* (L.) Plant Physiol. 84: 959–963.

Sagoff M (1992) Has nature a good of its own? Ch. 3. In: (Costanza R., Norton BG, Haskell BD, eds) Ecosystem Health: New Goals for Environmental Management. Island Press. Covina, California.

Salwasser H, Schonewald-Cox C (1987) The role of interagency cooperation in managing for viable populations. pp. 159–174. In: (Soulé ME, ed.) Viable Populations for Conservation. Cambridge University Press. New York.

Sample VA (1989) National Forest policy-making and program planning: the role of the President's Office of Management and Budget. J. For. 87: 17–25.

Sample VA (1990) Conservation Foundation Testimony: Joint Hearing Before the Committee on Agriculture, Nutrition, and Forestry and the Committee on Energy and Natural Resources. US Senate. October 25, 1989. S. Hrg. 101–553: 131–14.

Sample VA, Johnson N, Aplet GH, Olson JT (1993) Introduction. pp 3–8. In: (Aplet GH, Johnson N, Olson JT, Sample VA, eds.) Defining Sustainable Forestry. Island Press. Covina, California.

SAT (Scientific Analysis Team) USDA (1993) Viability Assessments and Management Considerations for Species Associated with Late Successional and Old-Growth Forests of the Pacific Northwest. USDA National Forest System. For. Serv. 530 pp.

Saunders DA, Hobbs RJ, Margules CR (1991) Biological consequences of ecosystem fragmentation: a review. Conserv. Biol. 5: 18–32.

Savage JM (1995) Systematics and the biodiversity crisis. BioScience 45: 673–679.

Saxena VK, Stogner RE, Hendler AH, Defelice TP, Yeh RJ-Y, Lin N-H (1989) Monitoring the chemical climate of the Mt. Mitchell State Park for evaluation of its impact on forest decline. Tellus 41B: 92–109.

Schimel DS (1993) New technologies for physiological ecology. pp. 359–365. In: (Ehleringer JR, Field CB, eds.) Scaling Physiological Processes: Leaf to Globe. Academic Press. New York.

Schindler DW (1992) A view of NAPAP from north of the border. Ecol. Appl. 2: 124–130.

Schlesinger WH, DeLucia EH, Billings WD (1989) Nutrient-use efficiency of woody plants on contrasting soils in the western Great Basin, Nevada. Ecology 70: 105–113.

Schluter D, Ricklefs RE (1993) Species diversity: An introduction to the problem. pp. 1–10. In: (Ricklefs RE, Schluter D, eds.) Species Diversity in Ecological Communities. University of Chicago Press. Chicago, Illinois.

Schmitt SA (1969) Measuring Uncertainty: An Elementary Introduction to Bayesian Statistics. Addison-Wesley. Reading, Massachusetts.

Schneider K (1993) Loggers listen to what Michigan forests say. New York Times, National Section, July 25, 1993. p. 20.

Schoeder JM, Keeler ME (1990) Setting objectives, a prerequisite of ecosystem management. In: (Mitchell RS, Cheviak CJ, Leopold DJ, eds.) Ecosystem Management: Rare Species and Significant Habitats. New York State Museum Bull. 471: 1–4.

Schoonmaker P, McKee A (1988) Species composition and diversity during secondary succession of coniferous forests in the Western Cascade Mountains of Oregon. For. Sci. 34: 960–979.

Schowalter TD (1989) Canopy arthropod community structure and herbivory in old-growth and regenerating forests in western Oregon. Can. J. For. Res. 19: 318–322.

Schowalter TD (1993) An ecosystem-centered view of insect and disease effects on forest health. pp. 189–195. In: (Covington WW, DeBano LF, technical coordinators) Sustainable Ecological Systems: Implementing and Ecological Approach to Land Management. Rocky Mountain For. & Range Exp. Sta. Gen. Tech. Rep. RM-247. Fort Collins, Colorado.

Schowalter TD, Hargrove WW, Crossley DA Jr (1986) Herbivory in forested ecosystems. Ann. Rev. Entomol. 31: 177–196.

Schullery P (1989) The fires and fire policy. BioSciene 39: 686–694.

Schulze ED, Chapin FSC III (1987) Plant specialization for environments of different resource availabilitiy. pp. 120–148. In: (Schulze ED, Zwolfer H, eds.) Potentials and Limitations of Ecosystem Analysis. Springer-Verlag. Berlin.

Schulze ED, Mooney HA (1993) Ecosystem function of biodiversity: a summary. pp. In: (Schulze ED, Mooney HA, eds.) Biodiversity and Ecosystem Function. Springer-Verlag. New York.

Schuster WSF, Phillips SL, Sandquist DR, Ehleringer JR (1992) Heritability of carbon isotope discrimination in *Gutierrezia microcephala*. Am. J. Bot. 79: 216–221.

Schutt P, Cowling EB (1985) Waldsterben, a general decline of forests in central Europe: symptoms, development, and possible causes. Plant Dise. 69: 548–558.

Seemann JR, Sharkey TD, Wang JL, Osmond CB (1987) Environmental effects of photosynthesis, nitrogen-use efficiency, and metabolite pools in leaves of sun and shade plants. Plant Physiol. 84: 96–802.

Sellers PJ, Mintz Y, Sud YC, Dalcher A (1986) A simple biosphere model (S.B) for use wiht general circulation models. J. Atmos. Sci. 43: 505–531.

Senge PM (1990) The Fifth Discipline: The Art and Practice of the Learning Organization. Doubleday Currency. New York. 424 pp.

Servheen C (1986) The threatened grizzly bear (*Ursus arctos horriblis*) in the conterminous United States. In: (Wilcox BA, Brussard PF, Marcot BG, eds.) The Management of Viable Populations: Theory, Applications and Case Studies. Stanford University. Stanford, California.

Shaver GR, Melillo JM (1984) Nutrient budgets of marsh plants: efficiency concepts and relation to availability. Ecology 65: 1491–1510.

Sheriff DW, Nambiar EKS, Fife DN (1986) Relationship between nutrient status, carbon assimilation, and water use efficiency in *Pinus radiata* (D. Don) needles. Tree Physiol. 2: 73–88.

Sheriff DW (1995) Gas exchange of field-grown *Pinus radiata*: Relationships with foliar nutrition and water potential, and with climatic variables. Australian J. Plant Physicology 22: 1015–1026.

Shortle WC, Smith KT, Minocha R, Alexeyev VA (1995) Similar patterns of change in stemwood calcium concentration in red spruce and siberian fir. J. Biogeogr. 22: 467–473.

Shugart HH (1984) A Theory of Forest Dynamics, the Ecological Implications of Forest Succession Models. Springer. New York. 304 pp.

Shugart HH (1987) Dynamic ecosystem consequences of tree birth and death patterns. BioScience 37: 596–602.

Siccama TG, Bliss M, Vogelmann HW (1982) Decline of red spruce in the Green Mountains of Vermont. Bull. Torrey Bot. Club 109: 162–168.

Silbergeld EK (1993) Revising the risk assessment paradigm: limits on the quantitative ranking of environmental problems. Ch. 5. In: (Richard Cothern, ed.) Comparative Environmental Risk Assessment. Lewis. Ann Arbor, Michigan.

Silen RR (1982) Nitrogen, corn, and forest genetics. The agricultural yield strategy— implications for Douglas-fir management. USDA For. Serv. Gen. Tech. Rep. PNW-137. 20 pp. Portland, Oregon.

Silvester WB (1977) Dinitrogen fixation by plant associations excluding legumes. pp. 141–190. In: (Hardy RWF, Gibson AH, eds.) A Treatise on Dinitrogen Fixation. IV. Agronomy and Ecology. John Wiley & Sons. New York.

Simberloff D (1987) The spotted owl fracas: mixing academics, applied and political ecology. Ecology 68: 766–771.

Simberloff D (1993) How forest fragmentation hurts species and what to do about it. pp. 85–90. In: (Covington WW, DeBano LF, Tech. Coord.) Sustainable Ecological Systems: Implementing an Ecological Approach to Land Management. USDA FS Rocky Mtn. For. Range Exp. Sta. Gen. Tech. Rep. RM-247. Fort Collins, Colorado.

Simberloff DS, Wilson EO (1969) Experimental zoogeography of islands. The colonization of empty islands. Ecol. 50: 278–296.

Slocombe DS (1993) Implementing ecosystem-based management. BioScience 43: 612–622.

Smith DM (1986) The Practice of Silviculture. 8th Ed. John Wiley & Sons. New York. 527 pp.

Smith JL, Halvorson JJ, Papendick RI (1993) Using multiple-variable indicator kriging for evaluating soil quality. Soil Sci. Soc. Am. J. 57: 743–749.

Smith RL (1990a) Ecology and Field Biology. 4th Ed. Harper and Row. New York. 922 pp.

Smith LZ (1990b) The spotted owl: a case study of conservation in the nineties. Hawk Migration Association 16: 15–16.

Smith WH (1991) Air pollution and forest damage. Chemical and Engineering News. November 11, 1991: 30–43.

Smith WK, Hinckley TM (1995) Resource Physiology of Conifers. Acquisition, Allocation and Utilization. Academic Press. San Diego, California. 396 pp.

Soil Survey Division (1995) Soil quality. National Committee Report. National Cooperative Soil Survey Conference on Soil Quality/Soil Health in San Diego, California. July 10–14, 1995.

Solbrig OT (1993) Plant traits and adaptive strategies: their role in ecosystem function. pp. 97–116. In: (Schulze E, Mooney HA, eds.) Biodiversity and Ecosystem Function. Springer-Verlag. Berlin.

Sollins P, Grier CC, McCorison FM, Cromack K, Fogel R, Fredriksen RL (1980) The internal element cycles of an old-growth Douglas-fir ecosystem in western Oregon. Ecol. Monogr. 50: 261–285.

Solow AR (1994) On the Bayesian estimation of the number of species in a community. Ecology 75: 2139–2142.

Son Y, Gower ST (1991) Aboveground nitrogen and phosphorus use by five plantation-grown trees with different leaf longevities. Biogeochemistry 14: 167–191.

Soulé ME (1980) Thresholds for survival: maintaining fitness and evolutionary potential. pp. 151–170. In: (Wilcox BA, Soulé ME, eds.) Conservation Biology: An Evolutionary-Ecological Perspective. Sinauer Associates. Sunderland, Massachusetts.

Soulé ME (1991) Conservation: tactics for a constant crisis. Science 253: 744–749.

Soulé M (1993) Normative conflicts and obscurantism in the definition of ecosystem management. pp. 20. In: (Covington WW, DeBano LF, tech. coord.) Sustainable Ecological Systems: Implementing an Ecological Approach to Land Management. USDA FS Gen. Tech. Rep. RM-247. Fort Collins, Colorado.

Sparks JP, Ehleringer JR (1995) Trends in leaf carbon isotope discrimination and nitrogen content for three riparian tree species along elevational transects in Utah. Annual Meeting of the Ecological Society of American on the Transdiciplinary Nature of Ecology. Snowbird, Utah, USA. Bull. Ecol. Soc. Am 76: 250.

Specht RL (1991) Changes in the Eucalypt forests of Australia as a result of human disturbance. pp. 177–197. In: (Woodwell GM, ed.) The Earth in Transition. Patterns and Processes of Biotic Impoverishment. Cambridge University Press. Cambridge.

Specht RL, Rayson P, Jackman ME (1958) Dark Island Health (Ninety-Mile Plain, south Australia). VI. Pyric succession; changes in composition, coverage, dry weight and mineral status. Austral. J. Bot. 6: 59–88.

Spies TA (1991) Plant species diversity and occurrence in young, mature, and old-growth Douglas-fir stands in western Oregon and Washington. pp. 111–121.

In: (Ruggiero LF, Aubry KB, Carey AB, Huff MH, eds.) Wildlife and Vegetation of Unmanaged Douglas-fir Forests. USDA F.S. Pacific Northwest Res. Sta. Gen. Tech. Rep. PNW-GTR-285. Portland, Oregon.

Spies TA, Franklin JF (1991) The structure of natural young, mature, and old-growth forests in Washington and Oregon. pp. 90–109. In: (Ruggiero LF, Aubry KB, Carey AB, Huff, MH, eds.) Wildlife and Vegetation of Unmanaged Douglas-fir Forests. USDA For. Serv. Gen. Tech. Rep. PNW-GTR-285. Portland, Oregon.

Spies TA, Franklin JF, Thomas TB (1988) Coarse woody debris in Douglas-fir forests of western Oregon and Washington. Ecology 69: 1689–1702.

Sprugel DG (1989) The relationship of evergreenness, crown architecture, and leaf size. Am. Natural. 133: 465–479.

Sprugel DG, Ryan MK, Brooks JR, Vogt KA, Martin TA (1994) Respiration from the organ-level to the stand—a model system for the application of scaling techniques. pp. 255–299. In: (Smith WK, Hinckley TM, eds.) Resource Physiology of Conifers. Acquisition, Allocation, and Utilization. Academic Press. San Diego, California.

Stahl E (1900) Der Sinn der mycorrhizenbildung. Jahrbucher für wissenscahftliche Botanik 34: 539–668.

Stahle DW, Cleavelenad MK, Hehr JG (1988) North Carolina climate change reconstructed from tree rings: A.D. 372 to 1985. Science 240: 1517–1519.

Stanley TR Jr. (1995) Ecosystem management and the arrogance of humanism. Conserv. Biol. 9: 255–262.

Steele JH (1991) Marine functional diversity. BioScience 41: 470–482.

Stone EL, Kalisz PJ (1991) On the maximum extent of tree roots. For. Ecol. Manag. 46: 59–102.

Sultan SE (1987) Evolutionary implications of phenotypic plasticity in plants. pp. 127–178. In: (Hecht MK, Wallace B, Prance GT, eds.) Evolutionary Biology. Vol. 21. Plenum Press. New York.

Suter GW (1988) Endpoints for regional ecological risk assessments. Environmental Management 14: 9–23.

Suter GW (1994) Ecological risk assessment prospects for the end of the 1990s. Canadian Technical Report of Fisheries and Aquatic Sciences 0 (1989): 288. Twentieth Annual Aquatic Toxicity Workshop. Quebec City, Quebec, Canada.

Swank WT (1988) 25. Stream chemistry responses to disturbances. pp. 339–357. In: (Swank WT, Crossley DA Jr, eds.) Forest Hydrology and Ecology at Coweeta. Ecological Studies. Vol. 66. Springer-Verlag. New York.

Swank WT, Crossley DA Jr (1988) Forest Hydrology and Ecology at Coweeta. Ecological Studies. Vol. 66. Springer-Verlag. New York.

Swanson FJ, Franklin JF (1992) New forestry principles from ecosystem analysis of Pacific Northwest Forests. Ecol. Appl. 2: 262–274.

Swift MJ, Heal OW, Anderson JM (1979) Decomposition in Terrestrial Ecosystems. Blackwell Scientific. Oxford.

Swift LW Jr, Cunningham GB, Douglass JE (1988) 3. Climatology and hydrology. pp. 35–55. In: (Swank WT, Crossley DA Jr, eds.) Forest Hydrology and Ecology at Coweeta. Ecological Studies. Vol. 66. Springer-Verlag. New York.

Szabolcs I (1989) Salt-Affected Soils. CRC Press. Boca Raton, Florida.

Tamm CO (1976) Acid precipitation: biological effects in soil and on forest vegetation. AMBIO 5: 235–238.

Tamm CO (1991) Nitrogen in plants and soils: Physiological and microbiological background for biological nitrogen turnover. In: Nitrogen in Terrestrial Ecosystems. Springer-Verlag. Berlin.

Tansley AG (1935) The use and abuse of vegetational concepts and terms. Ecology 42: 237–245.

Tarrant RF, Trappe JM (1971) The role of Alnus in improving the forest environment. Plant Soil (Special Vol.) 35–348.

Temple SA (1977) Plant-animal mutualism: coevolution with Dodo leads to near extinction of plant. Science 197: 885–886.

Temporary Study Commission of the Future of the Adirondacks. 1970. The Future of the Adirondack Park Technical Reports. Albany, New York.

Thomas J, Forsman E, Lint J, Meslow E, Noon B, Verner J (1990) A Conservation Strategy for the Northern Spotted Owl. Report of the Interagency Scientific Committee to Address the Conservation of the Northern Spotted Owl. USDA Forest Service, USDI Bureau of Land Management, USDI Fish and Wildlife Service, USDI Park Service. Portland, Oregon.

Thompson DB, Brown JH, Spencer WD (1991) Indirect facilitation of granivouous birds by desert rodents: experimental evidence from foraging patterns. Ecology 72: 852–863.

Thorburn P, Walker G, Hatton T (1992) Are river red gums taking water from soil, groundwater or streams? Catchments of green. National Conference on Vegetation and Water Management, Adelaide. pp. 37–42.

Tilman D (1988) Plant Strategies and the Dynamics and Structure of Plant Communities. Princeton University Press. Princeton, New Jersey.

Tilman D, Downing JA (1994) Biodiversity and stability in grasslands. Nature 367: 363–365.

Tingey DT (1994a) Project: Effects of CO_2 and Climate Change on Forest Trees. Experimental Tasks and Facilities. US Environmental Protection Agency. Environ. Res. Lab. Corvallis, Oregon. May 1994.

Tingey DT (1994b) Project: Effects of CO_2 and Climate Change on Forest Trees. Field Studies. US Environmental Protection Agency. Environ. Res. Lab. Corvallis, Oregon. May 1994.

Torgersen TR (1993) Maintenance and restoration of ecological processes regulating forest-defoliating insects. pp. 33–36. In: (Everett RL, compiler) Eastside Forest Ecosystem Health Assessment. Volume IV. Restoration of Stressed Sites, and Processes. United States Department of Agriculture. National Forest System. Forest Service Research. April 1993. Washington, D.C.

Toumey JW (1921) Damage to forests and other vegetation by smoke, ash, and fumes from manufacturing plants in Naugatuck Valley. Conn. J. Forestry 19: 267–373.

Toumey JW (1928) Foundations of Silviculture upon an Ecological Basis. John Wiley & Sons. New York. 438 pp.

Trappe JM (1962) Fungus associates of ectotrophic mycorrhizae. Bot. Rev. 28: 538–606.

Turner MG, Gardner RH, O'Neill RV (1995) Ecological dynamics at broad scales: ecosystems and landscapes. BioScience (Suppl.) s29–s35.

Twight BW, Lyden FJ (1989) Measuring forest service bias. J. For. 87: 35–41.

Ulrich B (1987) Stability, elasticity, and resilience of terrestrial ecosystems with respect to matter balance. pp. 8–47. In: (Schulze E, Zwolfer H, eds.) Potentials and Limitations of Ecosystem Analysis. Springer-Verlag. Berlin.

Ulrich B (1989) Effects of acid precipitation on forest ecosystems in Europe. pp. 181–212. In: Acid Precipitation. Vol. 2. Biological Ecological Effects. Springer-Verlag. Berlin.

Unger DG (1994) The USDA Forest Service perspective on ecosystem management. Symposium on Ecosystem Management and Northeastern Area Association of State Foresters Meeting. Burlington, Virginia. United States Government Printing Office. Washington DC.

USDA (United States Department of Agriculture) Forest Service (1992) Final Environmental Impact Statement on Management for the Northern Spotted Owls in the National Forests. USDA Forest Service. Washington, DC.

USDA USDODI (United States Department of Agriculture and United States Department of the Interior) (1994a) Record of Decision and Standards and Guidelines for Management of Habitat for Late-Successional and Old-Growth Related Species Within the Range of the Northern Spotted Owl. United States Department of Agriculture, U.S. Forest Service and United States Department of the Interior, Bureau of Land Management. Washington, DC.

USDA USDODI (United States Department of Agriculture and United States Department of the Interior) (1994b) Final Supplemental Environmental Impact Statement for Management of Habitat for Late-Successional and Old-Growth Related Species Within the Range of the Northern Spotted Owl. USDA For. Serv. and United States Department of the Interior, Bureau of Land Management. Washington, DC.

USDI (United States Department of the Interior) Fish and Wildlife Service (1992) Draft Recovery Plan for the Northern Spotted Owl. USDA Fish and Wildlife Service. Washington, DC.

USDOI BLM (1993) Final Supplemental Environmental Impact Statement for Management of Habitat for Late-Successional and Old-Growth Related Species Within Range of the Northern Spotted Owl. U.S. Forest Service, and Bureau of Land Management.

USFS (United States Forest Service) (1974) The Rocky Mountain Timber Situation 1970. Research Bull. INT-10: 25–26. Cited in Wilkinson and Anderson (1987) op. cit. at 154.

USFS (United States Forest Service) (1986a) A Recommended Renewable Resources Program. 1985–2030. USDA/FS-400.

USFS (United States Forest Service) (Old-Growth Definition Task Force) (1986b) Interim definitions for old-growth Douglas-fir mixed conifer forests in the Pacific Northwest and California. USDA For. Serv. Res. Note PNW-447. Pac. Northwest For. Range Exp. Sta. Portland, Oregon.

USFS (United States Forest Service) (1990) Forest Service Response to Questions for Senator Hatfield. Joint Hearing Before the Committee on Agriculture, Nutrition, and Forestry and the Committee on Energy and Natural Resources, US Senate. October 25, 1989. S. Hrg. 101-553: 125–126. Washington, DC.

Ustin SL, Wessman CA, Curtiss B, Kasischke E, Way J, Vanderbilt VC (1991) Opportunities for using the EOS imaging spectrometers and synthetic aperture radar in ecological models. Ecology 72: 1934–1945.

Ustin S, Smith MO, Adams JB (1993) Remote sensing of ecological processes: a strategy for developing and testing ecological models using spectral mixture analysis. pp. 339–355. In: (Ehleringer JR, Field CB, eds.) Scaling Physiological Processes: Leaf to Globe. Academic Press. New York.

Valentini R, Anfodillo T, Ehleringer JR (1994) Water sources and carbon isotope composition (213 C) of selected tree species of the Italian Alps. Can. J. For. Res. 24: 1575–1578.

Valentini R, Mugnozza GES, Ehleringer JR (1992) Hydrogen and carbon isotope ratios of selected species of a mediterranean macchia ecosystem. Funct. Ecol. 6: 627–631.

Van Breemen N, van Dijk HFG (1988) Ecosystem effects of atmospheric deposition of nitrogen in the Netherlands. Environm. Polluti. 54: 249–274.

Van Cleve K, Chapin FS III, Flanagan PW, Viereck LA, Dyrness CT (eds.) (1986) Forest Ecosystems in the Alaskan Taiga: A Synthesis of Structure and Function. Springer-Verlag. New York.

Van Cleve K, Chapin FS III, Dyrness CT, Viereck LA (1991) Element cycling in taiga forests: state-factor control. BioScience 41: 78–88.

van der Maarel E (1990) Ecotones and ecoclines are different. J. Veget. Sci. 1: 135–138.

Van Remortel RD (1994) 10. Soil classification and physiochemistry. pp. 10.1–10.28. In: Forest Health Monitoring. Southeast Loblolly/Shortleaf Pine Demonstration Interim Report. Environmental Monitoring and Assessment Program. EPA/620/R-94/006. April 1994. Washington, DC.

Van Miegroet H, Cole DW (1984) The impact of nitrification on soil acidification and cation leaching in a red alder ecosystem. J. Environ. Quality 13: 586–590.

Verner J (1984) The guild concept applied to management of bird populations. Environ. Manag. 8: 1–14.

Vig NJ, Kraft ME (1994) Environmental Policy in the 1990's. Congressional Quarterly Press, Washington, DC, 422 pp.

Virginia RA, Delwiche CC (1982) Natural 15-N abundance of presumed N^2-fixing and non-N^2-fixing plants from selected ecosystems. Oecologia 54: 317–325.

Virginia RA, Jarrell WM, Rundel PW, Shearer G, Kohl DH (1988) The use of variation in the natural abundance of ^{15}N to assess symbiotic nitrogen fixation by woody plants. pp. 375–394. In: (Rundel PW, Ehleringer JR, Nagy KA, eds.) Stable Isotopes in Ecological Research. Ecol. Studies 68. Springer-Verlag. New York.

Virkkala R (1993) Ranges of northern forest passerines: a fractal analysis. Oikos 67: 218–226.

Vitousek P (1982) Nutrient cycling and nutrient use efficiency. Am. Natural. 119: 553–572.

Vitousek PM (1990) Biological invasions and ecosystem processes: towards an integration of population biology and ecosystem studies. Oikos 57: 7–13.

Vitousek PM, Howarth RW (1991) Nitrogen limitation on land and in the sea: how can it occur? Biogeochemistry 13: 87–115.

Vitousek PM, Reiners WA (1975) Ecosystem succession and nutrient retention: a hypothesis. BioScience 25: 376–381.

Vitousek PM, Walker LR (1989) Biological invasion by Myrica faya in Hawaii: plant demography, nitrogen fixation, and ecosystem effects. Ecol. Monogr. 59: 247–265.

Vitousek PM, Gosz JR, Grier CC, Melillo JM, Reiners WA, Todd RL (1979) Nitrate losses from disturbed ecosystems. Science 204: 469–474.

Vitousek PM, Field CB, Matson PA (1990) Variation in foliar delta-carbon-13 in Hawaiian Metrosideros—Polymorpha, a case of internal resistance. Oecologia 84: 362–370.

Vitousek PM, Adsersen H, Loope LL (eds.) (1995) Islands. Biological Diversity and Ecosystem Function. Springer. Berlin. 238 pp.

Vogelman JE (1995) Assessment of forest fragmentation in southern New England using remote sensing and geographic information systems technology. Conserv. Biol. 9: 439–449.

Vogelmann H, Marvin J, McCormack M (1969) Ecology of the Higher Elevations in the Green Mountains of Vermont. Mimeographed report to the Governor's Commission on Environmental Control. State of Vermont, Montpelier, VT.

Vogt DJ (1987) Douglas-Fir Ecosystems in Western Washington: Biomass and Production as Related to Site Quality and Stand Age. PhD Dissertation, University of Washington, Seattle, Washington.

Vogt KA (1991) Carbon cycling in forest ecosystems. Tree Physiol. 9: 69–86.

Vogt KA, Persson H (1992) Measuring growth and development of roots. pp. 477–501. In: (Hinckley T, Lassoie J, eds.) Techniques and Approaches in Forest Tree Ecophysiology. CRC Press. Boca Raton, Florida.

Vogt KA, Edmonds RL, Grier CC (1981) Biomass and nutrient concentrations of sporocarps produced by myocorrhizal and decomposer fungi in *Abies amabilis* stands. Oecologia 56: 170–175.

Vogt KA, Grier CC, Meier CE, Edmonds RL (1982) Mycorrhizal role in net primary production and nutrient cycling in *Abies amabilis* ecosystems in western Washington. Ecology 63: 370–380.

Vogt KA, Grier CC, Vogt DJ (1986) Production, turnover, and nutrient dynamics of above- and belowground detritus of world forests. Adv. Ecol. Res. 15: 303–377.

Vogt KA, Dahlgren R, Ugolini F, Zabowski D, Moore EE, Zasoski R (1987) Aluminum, Fe, Ca, Mg, K, Mn, Cu, Zn and P in above- and belowground biomass. II. Pools and circulation in a subalpine *Abies amabilis* stand. Biogeochemistry 4: 295–311.

Vogt KA, Vogt DJ, Gower ST, Grier CC (1990) Carbon and nitrogen interactions for forest ecosystems. pp. 203–235. In: (Persson H, ed.) Above- and Below-Ground Interactions in Forest Trees in Acidified Soils. Air Pollution Report 32. Commission of the European Communities. Directorate-General for Sciences, Research and Development. Environment Research Programme. Brussels, Belgium.

Vogt KA, Publicover DA, Vogt DJ (1991) A critique of the role of ectomycorrhizas in forest ecology. Agric. Ecosyst. Environ. 35: 81–87.

Vogt KA, Bloomfield J, Ammirati JF, Ammirati SR (1992) Sporocarp production by basidiomycetes, with emphasis on forest ecosystems. pp. 563–581. In: (Carroll GC, Wicklow DT, eds.) The Fungal Community. Its Organization and Role in the Ecosystem. 2nd Ed. Marcel Dekker. New York.

Vogt KA, Publicover DA, Bloomfield J, Perez JM, Vogt DJ, Silver WL (1993) Belowground responses as indicators of environmental change. Environ. Exp. Bot. 33: 189–205.

Vogt KA, Vogt DJ, Asbjornsen H, Dahlgren RA (1995a) Roots, nutrients and their relationship to spatial patterns. Plant Soil 168–169: 113–123.

Vogt K, Vogt D, Brown S, Tilley J, Edmonds R, Silver W, Siccama T (1995b) Dynamics of forest floor and soil organic matter accumulation in boreal, temperate, and tropical forests. pp. 159–178. Ch. 14. In: (Lal R, Kimble J, Levine E, Stewart B, eds.) Soil Management and Greenhouse Effect. CRC Press. Lewis. Boca Raton, Florida.

Vogt K, Asbjornsen H, Ercelawn A, Montagnini F, Valdes M (1996a) Ecosystem integration of roots and mycorrhizas in plantations. In: (Nambiar S, Brown A, Cossalter C, eds.) Management of Soil, Water and Nutrients in Tropical Plantation Forests. Australian Centre for International Agricultural Research (ACIAR) (in press).

Vogt KA, Vogt DJ, Boon P, Covich A, Scatena FN, Asjbornsen H, O'Hara JL, Perez J, Siccama TG, Bloomfield J, Ranciato JF (1996b) Litter dynamics along stream, riparian and upslope areas 1, 2, and 5 years following Hurricane Hugo, Luquillo Experimental Forest, Puerto Rico. Biotropica (in press).

Vogt KA, Vogt DJ, Palmiotto PA, Boon P, O'Hara J, Asbjornsen H (1996c) Factors controlling the contribution of roots to ecosystem carbon cycles in boreal, temperate and tropical forests. Plant Soil (in press).

Waide RB, Lugo AE (1992) A research perspective on disturbance and recovery of a tropical montane forest. In: (Goldammer JG, ed.) Tropical Forests in Transition: Ecology of Natural and Anthropogenic Disturbance Processes. Birkhauser Verlag. Switzerland.

Walker BH (1992) Biodiversity and ecological redundancy. Conserv. Biol. 6: 18–23.

Wargo J (1996) Our Children's Toxic Legacy. Yale University Press. New Haven, Connecticut. 380 pp.

Wargo PM, Bergdahl DR, Tobi DR, Olson CW (1993) Root vitality and decline of red spruce. In: (Fuhrer E, Schutt P, eds.) Contributiones Biologiae Arborum Vol. 4. Ecomed, Landsberg/Lech. Germany. 134 pp.

Waring RH (1983) Estimating forest growth and efficiency in relation to canopy leaf area. Adv. Ecol. Res. 13: 327–354.

Waring RH (1987) Nitrate pollution: a particular danger to boreal and subalpine coniferous forests. pp. 93–105. In: (Fujimore T, Kimura M, eds.) Human Impacts and Management of Mountain Forests. Forestry and Forest Products Research Institute. Ibaraki. Japan.

Waring RH, Franklin JF (1979) Evergreen coniferous forests of the Pacific Northwest. Science 204: 1380–1386.

Waring RH, Pitman GB (1985) Modifying lodgepole pine stands to change susceptibility to mountain pine beetle attack. Ecology 66: 889–897.

Waring RH, Schlesinger WH (1985) Forest Ecosystems: Concepts and Management. Academic Press. Orlando, Florida. 340 pp.

Waring RH, McDonald AJS, Larsson S, Ericsson T, Wiren A, Arwidsson E, Ericsson A, Lohammar T (1985) Differences in chemical composition of plants grown at constant growth rates with stable mineral nutrition. Oecologia 66: 157–160.

Warkentin BP (1995) The changing concept of soil quality. J. Soil Water Conserv. 50: 226–228.

Watson RB, Muraoka DD (1992) The northern spotted owl controversy. Society of Natural Resources 5: 85–90.

Webb WL, Lauenroth WK, Szarek Sr, Kinerson RS (1983) Primary production and abiotic controls in forests, grasslands, and desert ecosystems in the United States. Ecology 64: 134–151.

Welker JM, Menke JW (1990) The influence of simulated browsing on tissue water relations, growth, and survival of *Quercus douglasii* (Hook and Arn.) seedlings under slow and rapid rates of soil drought. Functional Ecol. 4: 807–817.

Wessman CA, Aber JD, Peterson DL, Melillo JM (1988) Foliar analysis using near IR reflectance spectroscopy. Can. J. For. Res. 18: 6–11.

West P (1987) Social benefits of outdoor recreation: sociological perspectives and implications for planning and policy. President's Commission on American Outdoors. USDA For. Serv. A Recommended Renewable Resources Assessment and Programming: Executive Summary. USDA For. Serv. Gen. Tech. Rep. PSW-2. San Francisco, California.

Westman WE (1985) Ecology, Impact Assessment, and Environmental Planning. John Wiley & Sons. New York. 532 pp.

Weyerhaeuser (1995) Handbook: Best Management Practices for Weyerhaeuser Forestry in Arkansas and Oklahoma. Arkansas/Oklahoma Forest Council. Weyerhaeuser. Federal Way, Washington.

Whittaker RH (1953) A consideration of climax theory: the climax as a population pattern. Ecol. Monogr. 23: 41–78.

Whittaker RH (1975) Communities and Ecosystems. Macmillan. New York.

Wiens JA (1976) Population responses to patchy environments. Annu. Rev. Ecol. Syst. 7: 81–120.

Wiens JA (1989) Spatial scaling in ecology. Funct. Ecol. 3: 385–397.

Wiens JA, Stenseth NC, Horne BV, Ims RA (1993) Ecological mechanisms and landscape ecology. Oikos 66: 369–380.

Wilde SA (1941) Forest Soils: Origin, Properties, Relation to Vegetation, and Silvicultural Management. University of Wisconsin, Madison, Wisconsin. 384 pp.

Wilderness Society (Natural Resources Defense Council and National Wildlife Federation) (1993) A Critique of the Clinton Forest Plan. The Wilderness Society. Washington DC.

Wilkins DA (1991) The influence of sheathing (ecto-) mycorrhizas of trees on the uptake and toxicity of metals. Agric. Ecosyst. Environ. 35: 209–244.

Wilkinson C, Anderson M (1987) Land and Resource Planning in the National Forests. Island Press. Washington, DC.

Williams M (1989) Americans and their Forests. A Historical Geography. Cambridge University Press. New York.

Williamson M (1981) Island Populations. Oxford University Press. Oxford.

Wilson EO (1988) The current state of biological diversity. pp. 3–18. In: (Wilson EO, ed.) Biodiversity. National Academy Press. Washington DC.

Wilson JL, Tkacz BM (1993) Status of insects and diseases in the southwest: implications for forest health. pp. 196–203. In: (Covington WW, DeBano LF, technical coordinators) Sustainable Ecological Systems: Implementing an Ecological Approach to Land Management. USDA Forest Service General Technical Report RM-247. Ft. Collins, Colorado.

Wittig R (1993) General aspects of biomonitoring heavy metals by plants. pp. 3–27. In: (Markert B, ed.) Plants as Biomonitors: Indicators for Heavy Metals in the Terrestrial Environment. Weinheim, New York.

Woodhouse EJ, Hamlett PW (1992) Decision making about biotechnology: the costs of learning from error. In: The Social Response to Environmental Risk: Policy Formulation in an Age of Uncertainty. Kluwer Academic Norwell, Massachusetts.

Woodley S (1993) Monitoring and measuring ecosystem integrity in Canadian national parks. pp. 155–176. In: (Woodley S, Kay J, Francis G, eds.) Ecological Integrity and the Management of Ecosystems. St. Lucie Press. Ottawa, Canada.

Woodward FI (1987) Climate and Plant Distribution. Cambridge University Press. Cambridge.

Wren CD (1986) Mammals as biological monitors of environmental metal levels. Environm. Monitor. Assess. 6: 127–144.

Yakir D, Berry JA, Giles LJ, Osmond CB, Thomas RB (1993) Applications of stable isotopes to scaling biospheric photosynthetic activities. pp. 323–335. In: (Ehleringer JR, Field CB, eds.) Scaling Physiological Processes: Leaf to Globe. Academic Press. New York.

Appendix:
Ecological, Management and Legal Development of
Concepts Relevant to Ecosystems Through Time

TIME	ECOLOGICAL TIMELINE	MANAGEMENT TIMELINE	LEGAL TIMELINE
55,000 BP–1400	• Indians present in the Americas; little known about their contribution to ecosystems; use of fire to create "prairies."		
1400s		• Wood scarcity in Europe	
1543			• Parliment restricts cutting of English timber.
1584		• *Discourse Concerning Western Planting* by Richard Hakluyt. Talks about list of commodities from New England. The search for commodities reflected in explorers attitudes towards nature (i.e., mercantile possibilities).	
1789			• U.S. Constitution established authority over land policy and interstate commerce. These provisions have been used in matters of jurisdiction over wildlife issues.
late 1700s		• Firewood shortages in Philadelphia, Pennsylvania.	
1803			• Louisiana Purchase: U.S. doubles in size.
mid to late 1800s		• Economic importance of forests grows in U.S. for the construction of new cities, furniture, railroad ties. • Clear that wildlife resources were not infinitely abundant and human activity causing declining populations (ie., passenger pigeon becoming extinct; buffalo populations declining [between 1872-1874 more than 3 million killed annually with populations decreasing from 13 million in 1867 to only 200 in 1883])	

TIME	ECOLOGICAL TIMELINE	MANAGEMENT TIMELINE	LEGAL TIMELINE
1832			• First U.S. National Park established in Hot Spring, Arkansas.
1842(+−)		• Arrival of immigrants to "untamed" wilderness. Conversion from forest to farm and grazing.	
1840–70s		• Taming of the western wilderness by missionaries, surveyors, military men, railroad builders.	• **Preemption Act**—1841: heads of family could purchase 65 hectares cheaply. • The New York Association for the Protection of Game created in 1844. Earliest known "modern" sportsmen's group. • U.S. Department of the Interior established. Managed the General Land Office and the Bureau of Indian Affairs.
1862			• **Homestead Act:** broader than 1841 Act. Increased settlement in the west after the end of the Civil War.
1864	• George Perkins Marsh publishes *Man and Nature*, suggests wise management could mitigate soil erosion problem in New England.		
1866	• Haeckel defines œcology.		
1871			• U.S. Fish Commission established. First attempt by federal government to manage and protect fish and wildlife.
1872		• Franklin Hough speech to the American Association for the Advancement of Science states duty of the government to protect forest resources.	• Yellowstone National Park established by Congress.

TIME	ECOLOGICAL TIMELINE	MANAGEMENT TIMELINE	LEGAL TIMELINE
1873	• Great Chicago Fire, Peshtigo Fire.	• Timber supplies obtained from Wisconsin, Minnesota, & Michigan. Millions of acres were cut and transported to Mills along the Mississippi River and the edges of the Great Lakes.	• **Timber Culture Act**: donated 65 hectares of land to anyone who would plant 16 hectares to trees and maintain then for 10 years
1880			• Congress passed a bill to make forest investigations a regular part of the agriculture department.
1882			• 1st American Forest Congress: fears of timber famine, civic leaders looked toward the future.
1885	• Frank published article that coined the word mycorrhizas and identified this to be a symbiotic association between a fungus and a plant.		• Adirondack Preserve created in New York.
1887			• **Hatch Act**: Provided for financial assistance to states for agricultural experiment stations including forestry research activities.
1886–1918		• Poaching of large animals and administrative needs led to Yellowstone National Park management by U.S. Calvary. • Fear of "timber famine" fueled new ideas regarding use of resources	
1890		• Frontier declared "closed."	
1891			• Over-exploitation of public forests and wildlife resulted in establishment of first forest reserve near Yellowstone National Park. • **Forest Reserve Act** (formed basis of national forest system). Established in response to removal

TIME	ECOLOGICAL TIMELINE	MANAGEMENT TIMELINE	LEGAL TIMELINE
			of millions of acres of timber land in the East and Great Lakes, U.S.
1894		• Vigorous local opposition arose to creation of forest reserves: sheep/cattle grazing occurring in the Cascade Mountains, Rocky Mountains, Sierra Nevada Mountains in the U.S.	• Passage of the 1894 Trespass Act against sheep.
1896	• Hiltner concluded that elemental atmospheric N_2 was being fixed by nodules on the roots of the plant *Alnus glutinosa*.		• **Cullom-White Act**. Regulated the interstate shipment of wild game.
1897			• Some relaxation of 1894 act. Some grazing permitted. • Passage of **Organic Administration Act**: allowed for creation of administrative forest ranger force within Dept. of Interior. Set primary goals for new forest reserves: "No national forest shall be established except to improve and protect the forest within the boundaries, or for the purpose of securing favorable conditions of water flows, and to furnish a continuous supply of timber for the use and necessities of citizens of the United States." Drives Forest Service management still.
1900		• Timber in Great Lake region is depleted, started looking towards South and West for new supplies. • Scientific Forestry beginning to take hold with Bernard Fernow, Gifford Pinchot, President Roosevelt. Forestry and conservation became linked. Forestry = controlling access	• **Lacey Act** (first federal wildlife management law). Prohibited interstate transportation of game killed in violation of local laws to stop large scale commerce in wild game. Assisted states in protecting depleted wildlife resources at the federal level.

TIME	ECOLOGICAL TIMELINE	MANAGEMENT TIMELINE	LEGAL TIMELINE
		to national forests, regulating certain practices, protecting from fire, studying and applying best scientific information.	
1905		• Management of Forest Service transferred from Dept. of Interior to Dept. of Agriculture. • Recreational use of forests became important management objective. Long-term leases to private citizens encouraged for summer homes.	• U.S. Forest Service created. • 2nd American Forest Congress, call for fire protection.
1907		• Term "Conservation" coined by WJ McGee. • Gifford Pinchot developed concepts of interrelatedness of natural resources (soil, water, wildlife, minerals etc.)	
1908		• 1st Forest Experiment Station, Flagstaff Arizona, U.S.	
1910		• Forest Products Laboratory established in Madison, Wisconsin, U.S.	
1911			• **Weeks Act of 1911**—allowed cooperation in fire fighting between U.S. federal and state forest fire protection agencies, created National Forest Reservation Commission to create national forests in the eastern part of the U.S.
1913			• **Weeks-McLean Migratory Bird Act.** Protected all migratory game and insectivorous birds through the Department of Agriculture.
1914	• Extinction of the passenger pigeon. Last bird died in captivity.		

TIME	ECOLOGICAL TIMELINE	MANAGEMENT TIMELINE	LEGAL TIMELINE
1916	• Clements published *Plant Succession: an analysis of the the development of vegetation*. He was the first to develop the concept of plant communities being dynamic as a formal ecological theory.		
1918			• Park Service established.
1917–1978		• In Yellowstone National Park region land use management for fiber, livestock, wildlife, water, minerals. Forests over cut, livestock overgrazed, wildlife overexploited, watersheds damaged, minerals mined without environmental concerns. All of this spurred the creation of environmental legislation.	
1920		• Private forester David T. Mason advocates long-term sustained yield operations. • Aldo Leopold worked out fundamentals of game management: 1) recognized prohibiting hunting on all federal lands aggravated species survival problems; 2) advocated establishment of numerous, small, temporary refuges in national forests.	• **Federal Water Power Act.** In response to danger of dams on fish populations recommended that federally sponsored dams include devices to allow fish to swim around dams (fish ladders).
1925	• Lotka published *The Elements of Physical Biology*. He was the first to treat populations and communities as thermodynamic systems.		
1924			• **Clarke-McNary Act:** augmented cooperative federal and state fire suppression.

TIME	ECOLOGICAL TIMELINE	MANAGEMENT TIMELINE	LEGAL TIMELINE
1928	• Toumey publishes *Foundation of Silviculture Upon An Ecological Basis* stresses that ecology was born out of siviculture.		• **McSweeny-McNary Act**: expanded forestry research.
1930s		• Depression Era, planting of millions of trees to protect private farm land from loss of topsoil from devastating droughts of late 1920s-early 1930s. Done by Workers Progress Administration. Later in 1942 Soil Conservation Service continued this work. Civilian Conservation Corps created.	
1932		• Ecological Society of America's Committee for the Study of Plant and Animal Communities said U.S. needs a nature sanctuary system, to protect species of concern, represent wide range of ecosystem types, manage for ecological fluctuations (disturbances), use a core reserve/buffer zone approach (Shelford 1933). Mentioned interagency cooperation, and environmental education.	
1934	• USFS Coweeta Experiment Forest hydrology lab opened, North Carolina, U.S. First watershed level studies established.		• **Article X of the Lumber Code** enacted, to control logging on private lands. Was advocated by Mason, Pinchot, and Bob Marshall. Designed to have strict regulatory controls over logging plans in private sector. Law struck down in Supreme Court after less than one year. Regardless, it resulted in closer cooperation between the timber industry and the U.S. Forest Service.

TIME	ECOLOGICAL TIMELINE	MANAGEMENT TIMELINE	LEGAL TIMELINE
1935	• The word Ecosystem first coined by Tansley. First time presented concept that animals and plants interacted with the physical environment as systems (Ricklefs 1990). • Wright and Thompson wrote: *Fauna of the National Parks of the U.S.*, observed that parks were not fully functional ecosystems "by virtue of boundary and size limitation." Wright lobbied for increasing park size but died prematurely.		
1936		• Regional Forester C.J. Buck directed national forest in Oregon and Washington to implement selective logging—a departure from clearcutting. Intense debate followed between Forest Service and academic researchers. Primarily economic reasoning. Issue resolved in 1950s in favor of clear cutting. • First North American Wildlife conference called by Roosevelt. Assembled biologists, administrators and sportsmen to discuss wildlife problems.	
1937			• Passage of **O&C Sustained Yield Act** which embodied Mason's philosophy. This required the management of the Oregon and California forests be managed for sustained yield.
1939	• Gleason published paper that first identified that Clements ideas of discrete plant communities was inappropriate and a single species may exist in several successional stages.		

TIME	ECOLOGICAL TIMELINE	MANAGEMENT TIMELINE	LEGAL TIMELINE
1942	• Lindeman published a paper on aquatic ecosystems which stated that they were energy-transforming and could be understood as thermodynamic systems (Ricklefs 1990).		
1946			• Modern BLM established by combining General Land Office, U.S. Grazing Service, and O&C Administration. • 3rd American Forest Congress.
1947			• **Forest Pest Control Act**: Policy to protect all lands regardless of ownership from destructive forest insects and diseases. First legislation to direct management across ownership and political boundaries.
1949	• Leopold wrote *Sand County Almanac and Sketches Here and There*—wrote foundation of conservation science and philosophy, and environmental ethics.		
1950		• Ecological Society's Committee for plant and animal communities proposed strategy to implement Shelford's nature sanctuary inventory. • Beginning of trend to increase planting on industrial forest lands from 2833 hectares/year in 1945 to 0.5 m hectares/year in 1980.	
1953	• Odum published text *Fundamentals of Ecology*. Text popularized idea of using energetics (i.e. energy flow diagrams) and the process-functional approach to studying ecosystems in contrast		• 4th American Forest Congress: dispelled much of the distrust in the relationship between private and public forestry.

TIME	ECOLOGICAL TIMELINE	MANAGEMENT TIMELINE	LEGAL TIMELINE
	to the population-community approach (O'Neill et al. 1986). This balanced energy input and outputs of ecosystems. • Whittaker introduced the "climax pattern hypothesis" that combined gradient analysis and successional theory (O'Neill et al. 1986). This linked space and time even though no explicit dealing with space changes with time, temporal changes restricted to successional time and the focus was on species individuality (O'Neill et al. 1986).		
1954	• Egler first introduced the concept of understanding the mechanisms controlling species replacement in succession.		
1955	• Crocker and Major published a well documented study that linked soil and vegetation development along a primary successional sequence in Alaska.		
1960s	• Active research programs on ecosystems in the U.S. mainly supported by the U.S. Atomic Energy Commission (Golley 1993). Broad range of ecosystems examined: warm and cold deserts, sandy coastal plains, deciduous forests, tropical forests, coral reefs. • The Ecosystem concept was accepted as an organizing concept in ecology. However, dominant research pursued idea that ecology should be interpreted at the individual organism level (Golley 1993). There did not exist an organized body of	• Clear-cut and terracing cut over slopes at Bitterroot National Forest protested, series of news articles followed. University of Montana study team commissioned by Senator Metcalf.	• **Multiple Use Sustained Yield Act**—Created multiple use planning, brought in new specialists like soil scientists, wildlife biologists. Some saw it as redefining old ways of operation (i.e. some rangers interpreted legislation as timber sales) while others it became an opportunity to open vistas, access to hunting, recreation, browse, increase irrigation potential, fire roads, wood for the nation's housing etc. • Clear cutting of Monongahela National Forest in West Virginia led to lawsuit on behalf of Izaak

TIME	ECOLOGICAL TIMELINE	MANAGEMENT TIMELINE	LEGAL TIMELINE
	knowledge in ecosystem studies. Ecosystem studies conducted were very descriptive—with little analysis of functional attributes (Golley 1993). • Olson, Van Dyne and Patten at Oak Ridge National Laboratory began applying systems modeling to analyzing ecosystems. This approach used models to predict and test ideas on ecosystems that were felt to be too complicated to experimentally manipulate (Golley 1993).		Walton League (turkey hunters) on premise that the 1897 Organic act did not allow clear cutting. In 1973 Federal District Court ruled against U.S. Forest Service. After fourth circuit court of appeals also ruled against the agency, Forest Service and Congress decided law had to be changed to allow clear cutting. • Tax code changes to base timber land taxes on bare land values. • U.S. Fish and Wildlife Service (USFWS) formed.
1962	• Rachel Carson's Silent Spring—DDT, role of chemical pesticides, raised consciousness about the environment. Preston first time formalized the relationship between land area and the number of species present (created a mathematical model)—suggesting a linear relationship between land area and species number.		• **McIntire-Stennis Act:** boosted forest research.
1963	• MacArthur and Wilson proposed the Island Biogeography Theory where the number of species on an island is dependent on the number of immigrants migrating in and the extinction rate of pre-existing on the island—a dynamic equilibrium hypothesis. Linking animal species to land area.		• 5th American Forest Congress: 1st congress to devote attention to products of the forest other than timber.
1964	• Global program launched (International Biological Program). First phase (1964-67) was planning and development of methodology for research, 2nd phase (1967-74) initiation of the biome studies		• **Classification and Multiple Use Act.** Like MUSY of 1960 required USFS to manage lands for multiple uses and to sustain in perpetuity outputs of various renewable

TIME	ECOLOGICAL TIMELINE	MANAGEMENT TIMELINE	LEGAL TIMELINE
	(Golley 1993). Program identified many variables (biological, physical, chemical) to quantitatively collect data collaboratively in tundra, boreal forest, temperate forest, grassland and tropical forests around the world. This standardized methods across all sites. Impact of this program on the development of ecosystem theory was minor but did produce the next generation of scientists that had a systems approach (Golley 1993).		resource commodities and other uses (sustained-yield principle). • **Wilderness Act.** Established National Wilderness Preservation System. Forest Service opposed act because thought it was unnecessary due to Multiple Use Act. Also they had been managing administrative wilderness since 1924.
1966			• **Administrative Procedures Act.** This act gave access to decisions and process for the next three decades. Spawned lawsuits on cases of national importance. Ushered in an era of citizens being able to question and litigate federal agency decisions as not complying with federal law. Was not generally used by citizens and groups until 1970s, especially when tied to NEPA. • **Endangered Species Preservation Act.** Authorized the Secretary of Interior to acquire land and funds to protect endangered species.
1969	• Van Dyne presented Ecosystem Concept in Natural Resource Management. Paine coined the word Keystone Species. • Odum published article *The Strategy of Ecosystem Development* which presented several hypotheses needing to be studied which related ecosystem concepts to successional processes in plant communities. • Bormann and Likens published first		• **National Environmental Policy Act.** Signed into law Jan. 1, 1970. Contained a citizen suit provision. Required environmental impact statements. "encourage productive and enjoyable harmony between man and his environment." • **Endangered Species Conservation Act.** Authorized Secretary of Interior to generate

TIME	ECOLOGICAL TIMELINE	MANAGEMENT TIMELINE	LEGAL TIMELINE
	article entitled *Nutrient Cycling* which introduced a new approach (i.e. watershed) for studying ecosystems. This replaced energy flow with nutrient cycling as the dominant process to monitor in ecosystems and first time specified defined boundaries of ecosystems (watersheds) (Golley 1993). Introduced new conceptual model of ecosystems with 4 components: the atmosphere, available nutrients in soil, living and dead organic matter, soil and rock minerals (Golley 1993). Concept of the terrestrial ecosystems being closely linked to the hydrologic cycle introduced (Golley 1993).		endangered species list and ban their import.
1960s			• **Federal Advisory Committee Act** (FACA). Interpreted in 1990s (legally and politically) to require more openness and formality in public/private sector interactions.
1970	• Lynton Cladwell published an article that advocated using ecosystems as the basis for public policy. • Likens, Bormann, Johnson, Fisher, Pierce first time introduced concept that forest management practice may affect other ecosystem functions (i.e. nutrient cycles).		• **Resource Recovery Act.** construction grant for innovative solid waste management.
1971–2	• Trophic model of ecosystems proposed by Wiegert and Owen. They defined three major groups (i.e. autotrophs, biophages, saprophages) and distinguished between living and dead organic material, and stressed the interactions between the groups (Golley 1993).	• Forest Service voluntarily agreed to stay within Church guidelines for clear-cut size.	• Congressional Hearings 1971–1972, Bitterroot and Monongahela spurred a series of congressional hearings over clear cutting. Senator Church offered analysis report which led to Church Guidelines for clear-cut size.

TIME	ECOLOGICAL TIMELINE	MANAGEMENT TIMELINE	LEGAL TIMELINE
1972	• One of the first reports of the impact of human generated activity (i.e. acid rain) on forest ecosystems at large spatial scales in Scandinavia (Abrahamsen 1980). Brought attention to the fact that ecosystems were susceptible to degradation based on human activities.		• **Federal Water Pollution Control Act** (amended Clean Water act of 1948). EPA forbid the use of DDT in the U.S. except with permission.
1973			• **Endangered Species Act.** Bans taking species on endangered list and allows preservation of critical habitats. • Izaak Walton League wins lawsuit versus Forest Service on grounds that 1897 Organic Act did not allow clearcutting—first in the District Court then in the 4th Circuit Court after Forest Service appealed (re: Monongahela National Forest, W. Virginia)
1974			• **Freedom of Information Act.** Required federal agencies to provide information to citizens on request and established procedures for this.
1975	• Whittaker wrote *Communities and Ecosystems*. Suggested species presence determined by how they responded to gradients in the physical environment.		• 6th American Forest Congress.
1976	• O'Neill presented grouping of ecosystem components using functional emphasis comprised of energy capture, nutrient retention and rate regulation (O'Neill 1976). These emphasized analysis of the ecosystem and not individuals.		• **National Forest Management Act**—enacted as a result of several years of hearings and negotiations. Required U.S. Forest Service to develop long-range land-use planning. Required outside scientific committee who were not officers or employees of Forest Service to

TIME	ECOLOGICAL TIMELINE	MANAGEMENT TIMELINE	LEGAL TIMELINE
			provide scientific advice and counsel. Required the preservation and enhancement of plant and animal community diversities as equivalent as possible to natural forest. • **Federal Land Policy and Management Act**. Required BLM to develop long-range land-use planning and/or general management plans for its lands.
1977			• **Clean Air Act** (amended Clean Air Act of 1955).
1978			• **National Parks and Recreation Act**. Required National Park Service to develop long-range land-use or general management plans for their lands.
1979	• Bormann and Likens published synthesis book on Hubbard Brook Experimental Forest research entitled *Pattern and Process in a Forested Ecosystem*. Discussed the structure and function of ecosystems in New Hampshire. • Kramer and Kozlowski published *Physiology of Woody Plants*. This book synthesized very effectively the literature on woody plant physiology, was written for foresters and horticulturalists.	• John and Frank Craighead Grizzly Bear Study. In order to conserve bear need to conserve ecosystem (=Yellowstone National Park + surrounding forests, Federal, private, and state lands). Emergence of Ecosystem Approach. Based on 12 years of population research, indicated that Yellowstone population required at least 2,023,500 hectares of protected habitat. Set precedent: the area must provide primary habitat necessary to sustain the largest carnivore in the region.	
1980	• Long-Term Ecological Research sites established in the United States by the National Science Foundation. 18 sites were established that varied from coastal, to grasslands,		• Alaska National Interest Lands Conservation Act. Protected 41.2 million hectares of Alaskan land as national wilderness, wildlife refuges and parks. Super fund established.

TIME	ECOLOGICAL TIMELINE	MANAGEMENT TIMELINE	LEGAL TIMELINE
	hardwood deciduous forest, to coniferous forests to tropical forests. National Atmospheric Precipitation Assessment Program (NAPAP) established to study impact of pollution on different forest and aquatic ecosystems, on humans and on building materials in the United States.		
1982	• First documented case of air pollution causing unusual mortality of spruce in Vermont, U.S.A (Siccama, Bliss, Vogelmann). • Vogt, Edmonds, Grier document how much ecosystem production was transferred to mycorrhizal associations in young and mature coniferous forest.		
1983	• Annual terrestrial inventories conducted in Europe to determine the amount of damage in forests due to air pollution (Huettl 1989).	• Greater Yellowstone Ecosystem Coalition formed = 75+ organizational members. Their goal bring ecosystem management to the region.	
1984	• Harris published *The Fragmented Forest. Island Biogeography Theory and the Preservation of Biotic Diversity.* Synthesized scientific information from western Cascades, Oregon to evaluate the utility of island biogeography theory as a guide for managing old-growth forest ecosystems.	• Society of American Foresters report *Scheduling the Harvest of Old Growth Timber* called for mapping old growth, applying ecological definitions and setting up additional reserves.	
1985	• Newmark's work compared legal and biotic boundaries of some parks and reserves in western North America—reinforced Craighead's ideas.		• **Food Security Act** (Farm Bill). Limited benefits to producers on highly erodible lands or converted wetlands.

TIME	ECOLOGICAL TIMELINE	MANAGEMENT TIMELINE	LEGAL TIMELINE
Late 1980s		• Ecosystem approach being advocated by many federal agency officials, scientists and natural resource policy analysts to better address declining ecological condition (Dufftus Testimony, 9/20/94). Proposals focused on specific region such as greater Yellowstone area.	
1986	• O'Neill, DeAngelis, Waide and Allen published *A Hierarchical Concept of Ecosystems*. Presented new theory on scaling issues in ecosystems and that there exist hierarchical structure in ecological systems.		• **Super Fund Amendment & Reauthorization.** • **Safe Drinking Water Act.**
1987		• Franklin and Forman publish article *Creating landscape patterns by forest cutting: Ecological consequences and principles* which documents the importance of forest cutting patterns in modifying landscape structural characteristics and therefore their ecological characteristics. They suggest the importance of combining the ecosystem approach with landscape level analysis in management.	• **Clean Water Act Amendments.**
1988		• Agee and Johnson write first book on ecosystem management. Suggested that essential elements include: ecologically defined boundaries, clearly stated management goals, interagency cooperation, monitoring of management results, and leadership at national policy level. Also humans must be brought into the equation.	• **Ocean Dumping Act.** Ended all ocean disposal of sewage sludge and industrial waste.

TIME	ECOLOGICAL TIMELINE	MANAGEMENT TIMELINE	LEGAL TIMELINE
1989		• Nearly all forest NFMA forest plans (USFS) and FLPMA district plans (BLM) were in final or draft form. Tried to integrate science of forestry and range management, and other resource disciplines—but had to ignore many issues. • These issues carried over in 1990s and dealt with outside NFMA and FLPMA planning process: i.e. spotted owls and other focus species, roadless areas, old growth, ancient forest, Pacific yew, clear cutting, monoculture, herbicides. Issues now being decided by courts and Congress. In Pacific Northwestern U.S., Fish and Wildlife is considering how far the act (ESA) should protect spotted owl on private lands (section 4(d) rule). • Keiter—legal aspects of ecosystem management examined.	
1980s–early 1990s		• USFS released series of EIS statements and management plans designed to protect spotted owl habitat.	
Early 1990s		• Millions spent on NFMA and FLPMA planning, public disputes led to tree-sitting, tree spiking, etc., threats and intimidation, lawsuits, appeals, court decisions. • Brown (1990) calls Franklin former USFS and Maser former BLM 'gurus' of the new forestry approach. Research in H.J. Andrews Forest led to new clues about long-term health and productivity. Push for change came from inside agencies,	• **Oil Pollution Prevention Response, Liability and Compensation Act of 1990.** • **Pollution Prevention Act.** • **Clean Air Act Amendments.**

453

TIME	ECOLOGICAL TIMELINE	MANAGEMENT TIMELINE	LEGAL TIMELINE
		not external pressure from interest groups or Congress. • New Forestry would address NFMA requirement for Biodiversity. Some felt New Forestry = Old Forestry. Changes = i.e. leaving large logs on ground after harvest to replenish soil nutrients; leaving dead trees for wildlife vs. felling for safety; creating wildlife snags by blasting commercially viable trees; creating corridors; riparian buffer zones; cool or no burning for site preparation; replanting mixed species vs. monocultures of the past.	
1990		• First interagency attempt to implement ecosystem management on federal lands, draft form, became hotly debated, intense political debate led to its demise (Lichtman and Clark in press). • Forest Service embraces "New Perspectives" as a top-down approach to complement new forestry's bottom up approach. It was an attempt to upgrade multiple use for 1990s.	
1991	• Mooney, Drake, Luxmoore, Oechel, Pitelka publish article *Predicting Ecosystem Responses to Elevated CO_2 Concentrations* which calls for new long-term ecosystem level research and a decreased emphasis on laboratory experiments. • Ecological Society of America published the Lubchenco et al. report entitled *The Sustainable Biosphere Initiative: An Ecological Research Agenda* that identified key issues	• Final EIS for spotted owl released to reflect recommendations by Interagency Scientific Committee's 1990 report *A conservation Strategy for the Northern Spotted Owl*. But debate between timber, fishing, and conservationists continued re. if whether "planned habitat conservation areas" offered adequate protection. • California was first state to address biological impoverishment through a	• Law suit brought by environmental interest groups resulted in a Federal Court injunction against further timber sales in District's National Forests, and a stay of implementation of U.S. Forest Service's management plan, until U.S. Forest Service could demonstrate compliance with all legal obligations imposed under ESA and NFMA.

TIME	ECOLOGICAL TIMELINE	MANAGEMENT TIMELINE	LEGAL TIMELINE
	to be addressed by the ecological community with ecosystem sustainability as a strong focal point.	policy largely based on ecosystem management concepts. • "Gang of Four," report suggests an integrated ecosystem management approach to managing "owl forests."	
1992	• Grumbine published *Ghost Bears: Exploring the Biodiversity Crisis*.	• USFS altered its resource-based management focus to fit the agency's particular vision of EM. (USDA FS 1992).	• USFS Chief, Dale Robertson, makes ecosystem management concept forest service policy, but equates it with "multiple-use." New Forestry and New Perspectives programs replaced by more comprehensive term "ecosystem management" which came from 90+ yrs of science and laws.
Feb. 1992		• Fish, Wildlife Service (FWS) decision to take a multi-species approach rather than species-by-species approach to protecting plants and animals.	
June 4, 1992			• USFS Chief announces new policy of multiple-use ecosystem management on national forests and grasslands.
Late 1992		• FWS establishes a working group to develop its policies and strategy for biodiversity management.	
April 2 1993		• Clinton's Forest Conference ("Timber Summit") resulted in Presidential Directive creating interagency Forest Ecosystem Management Assessment Team.	
July 1993		• FEMAT report issued, followed by EIS a precedent setting attempt to develop an "ecosystem management" plan for 7.9 million hectares west-side forests and 1.1	

TIME	ECOLOGICAL TIMELINE	MANAGEMENT TIMELINE	LEGAL TIMELINE
		million hectares BLM in region. Option 9, the recommended alternative, supposed to represent both economic needs of timber dependent communities, the legal mandate to provide sufficient habitat pursuant to ESA and NFMA, need to protect commercially important watersheds, goal of preserving old growth ecosystem integrity.	
Sept. 1993	• Sustainable Forestry symposium and book presents definitions of ecosystem management and links the concept to "sustainability."	• Report of the *National Performance Review: Creating a Government that Works Better and Costs Less* recommended that the President issue Executive Order establishing ecosystem management policies across federal agencies.	
1993	• Mills, Soule, Doak publish article *The Keystone-Species Concept in Ecology and Conservation* which suggests that the keystone species concept should be replaced by studying the strengths of interactions within and between species and their habitats. • Gosz publishes article *Ecotone hierarchies* which supports conducting research at ecotones or transition zones because many pattern-process relationships occur across several scales in ecological systems, the amount and type of information needed to understand systems increase as one traverses from larger to smaller spatial scales, and the thresholds for ecological processes are easier detected at the ecotones since the variability is higher.	• White House Office on Environmental Policy created by President which established an Interagency Ecosystem Management Task Force to implement an ecosystem approach to environmental management. • Slocombe publishes *Implementing Ecosystem-based Management* which supports that most management activities fail to use scientific knowledge of the environment to meet socio-economic goals. • Clark published *Creating and Using Knowledge for Species and Ecosystem Conservation: Science, Organizations and Policy*. • Franklin publishes article *Preserving Biodiversity: Species, Ecosystems, or Landscapes* which suggests the ecosystem approach is the only way to conserve habitats and species	

TIME	ECOLOGICAL TIMELINE	MANAGEMENT TIMELINE	LEGAL TIMELINE
		especially since so little is known about the majority of species.	
1994		• Clark and Minta published *Greater Yellowstone Future: Prospects for Ecosystem Science, Management, and Policy.* • Gordon published Ecosystem *Management: An Idiosyncratic Overview* which identified 5 descriptors of ecosystem management: manage where you are; manage with people in mind; manage across boundaries; manage based on mechanisms instead of algorithms; and manage without externalities. • Interagency Task Force issues a draft "Ecosystem Management Initiative Overview."	
April 19, 1994		• CRS report to Congress. Ecosystem Management: Federal Agencies Activities (94-339 ENR).	
Mid-1994		• National Park Service establishes a working group to develop its ecosystem management policies and strategies. Approach to pursue new partnerships, alliances, coalitions and promoting comprehensive regional ecosystem restoration and management.	
August 1994		• GAO Ecosystem Management Report entitled *Ecosystem Management Additional Actions Needed to Adequately Test a Promising Approach.*	
1995		• Fiscal Year Budget: The Interagency Ecosystem Management Task Force developed 1995 budget requesting $610 million in discretionary	

TIME	ECOLOGICAL TIMELINE	MANAGEMENT TIMELINE	LEGAL TIMELINE
		spending for ecosystem management initiatives, including acceleration of 3 ongoing interagency restoration efforts which are pilot ecosystem management projects. • Clinton administration proposed in its fiscal year 1995 budget to begin studying four ecosystem management pilot projects using ecological rather than political or administrative boundaries (old growth forests in the Pacific Northwest U.S.; natural resources damaged by Exxon Valdez oil spill in the Copper River Delta, Alaska; ecological health of south Florida, U.S. including the Everglades and Florida Bay; ecological health of Anacostia River in Maryland and Washington DC).	

Index